Introduction to the Theory of Linear Partial Differential Equations

STUDIES IN MATHEMATICS AND ITS APPLICATIONS

VOLUME 14

NORTH-HOLLAND PUBLISHING COMPANY – AMSTERDAM • NEW YORK • OXFORD

INTRODUCTION TO THE THEORY OF LINEAR PARTIAL DIFFERENTIAL EQUATIONS

JACQUES CHAZARAIN and ALAIN PIRIOU
Professors at the Université de Nice

English version edited, prepared and produced by
TRANS-INTER-SCIENTIA
P.O. Box 16, Tonbridge, TN11 8DY, England

1982

NORTH-HOLLAND PUBLISHING COMPANY – AMSTERDAM • NEW YORK • OXFORD

ISBN 0 444 864520

Translation of:
Introduction à la Théorie des Equations
aux Dérivées Partielles Linéaires
© Bordas (Dunod), Paris, 1981

Publishers:
NORTH-HOLLAND PUBLISHING COMPANY
AMSTERDAM • NEW YORK • OXFORD

Sole distributors for the U.S.A. and Canada:
ELSEVIER SCIENCE PUBLISHING COMPANY, INC.
52 VANDERBILT, NEW YORK, N.Y. 10017

Library of Congress Cataloging in Publication Data

Chazarain, Jacques.
Introduction to the theory of linear partial differential equations.

(Studies in mathematics and its applications ; 14)
Translation of: Introduction a la théorie des équations aux dérivées partielles linéaires.
Bibliography: p.
Includes index.
1. Differential equations, Partial. I. Piriou, Alain. II. Title. III. Series.
QA374.C4513 1982 515.3'53 82-8226
ISBN 0-444-86452-0 (Elsevier)

PRINTED IN THE NETHERLANDS

TABLE OF CONTENTS

FOREWORD

Since the fifties, the theory of partial differential equations has undergone considerable development, yet it is certainly one of those fields where we find both the largest number of research articles and the smallest number of works of synthesis. The specialist can come to terms with such a situation, but the student who desires an introduction to this theory finds it rather a handicap. Thus, in the present book, which is not a treatise but an introduction, we aim to present a reasonably large range of recent methods: pseudo-differential operators, oscillatory integrals, stationary phase expansion, microlocalisation.... We show how these methods permit the solution of numerous classical problems in an elegant and general manner: elliptic equations, boundary problems, evolution equations, mixed hyperbolic problems,... . We have allotted considerable room to the equations of Physics because, historically, this has been the origin of important problems in partial differential equations and remains to the present day the principal source of interesting questions.

As regards the form of the book, we have endeavoured to give very detailed proofs in such a way as to render the work accessible to a reader at post-graduate level. With this in mind, we have sometimes preferred to restrict to some degree the generality of certain results in order to avoid obscuring the ideas underlying the techniques being used. Each chapter begins, in general, with a brief description of the subject matter and the methods which will be developed therein, and ends with supplements in the form of exercises together with hints for solution (an asterisk indicates those which will be utilised in the text which follows). For the convenience of the reader, the main extensions are listed within the table of contents. The bibliography given is deliberately restricted: anyone writing on such a subject refers to such a large number of sources that an exhaustive list of these would be unwieldy.

It seems to us to be possible to use this book at two levels; on the one hand the first four Chapters (Distributions and Operators; Sobolev Spaces; Symbols and Oscillatory Integrals; Pseudo-differential Operators) can serve as the basis for an introductory post-graduate course. On the other hand, the last four Chapters (Elliptic boundary problems; Evolution equations; Mixed hyperbolic problems; Microlocalisation) are aimed more particularly at Ph.D. students and at research workers who will find contained therein numerous results formerly dispersed amongst specialised journals.

CHAPTER I

DISTRIBUTIONS AND OPERATORS

1. SPACES OF DISTRIBUTIONS

The considerable expansion of the theory of partial differential equations during the last twenty years has, in large measure, its origin in the development by L. Schwartz of the theory of distributions. This theory clarifies and unifies several concepts which were previously rather unclear, like for example the concepts of: weak derivative, generalised Fourier transform, fundamental solution. The theory of distributions is nowadays universally known; however, for the convenience of the reader, we shall here summarise the elements of this which will prove useful later. The most standard of the proofs will be omitted, these being available in SCHWARTZ [1], HORMANDER [1], TREVES [1].

Letting $\alpha = (\alpha_1, \ldots, \alpha_n) \in \mathbb{N}^n$ be a multi-index, we put

$$\partial^\alpha = \partial_1^{\alpha_1} \ldots \partial_n^{\alpha_n} \qquad \text{with} \qquad \partial_j = \frac{\partial}{\partial x_j}$$

$$D^\alpha = D_1^{\alpha_1} \ldots D_n^{\alpha_n} \qquad \text{with} \qquad D_j = \frac{1}{i}\frac{\partial}{\partial x_j}$$

$$|\alpha| = \alpha_1 + \ldots + \alpha_n \quad , \qquad \alpha! = (\alpha_1!) \times \ldots \times (\alpha_n!);$$

if $x = (x_1, \ldots, x_n) \in \mathbb{R}^n$ we put $|x| = (x_1^2 + \ldots + x_n^2)^{1/2}$. We let X be an open subset of $\mathbb{R}^n$, and we suppose $m \in \mathbb{N} \cup \{+\infty\}$. We denote by $C^m(X)$ the space of complex-valued functions whose derivatives of order less than or equal to m are continuous. The family of semi-norms

$$C^m(X) \ni \varphi \longmapsto p_{K,j}(\varphi) = \sup_{\substack{x \in K \\ |\alpha| \leq j}} |\partial^\alpha \varphi(x)| ,$$

where K is a compact subset of X and j is an integer such that $0 \leq j \leq m$, equips $C^m(X)$ with a Fréchet-space topology. A subset B of $C^m(X)$ is said to be bounded if, for any semi-norm $p_{K,j}$, we have $\sup_{\varphi \in B} p_{K,j}(\varphi) < +\infty$.

The support of a function $\varphi \in C^0(X)$ is by definition the smallest closed subset of X outside of which the restriction of φ is zero (or alternatively, the closure of the set of points of X where φ is non-zero), and we denote it by supp φ . If K is a compact subset of X, we denote by $C_K^m(X)$ or $\mathcal{D}_K(X)$ (or more concisely C_K^m or $\mathcal{D}_K$) the closed topological subspace of $C^m(X)$ formed by the functions with support in K. The vector subspace of $C^m(X)$ formed by the functions with compact support is denoted by $C_0^m(X)$ or $\mathcal{D}^m(x)$ (or more concisely $\mathcal{D}(X)$ when $m = +\infty$); we equip it with the inductive limit topology of the $C_K^m(X)$ when K varies over the family of compact subsets of X. In particular, a linear mapping from $C_0^m(X)$ into a topological vector space F is continuous if and only if its restriction to each $C^m_K(X)$ is continous, so it is sufficient to base our reasoning on sequences. A bounded subset of $C^m_0(X)$ is a bounded subset of $C^m_K(X)$ for a certain compact subset K.

DEFINITION 1.1 : *A distribution* u *on the open set* X *is a continuous linear form on* $C^\infty_0(X) = \mathcal{D}(X)$, *i.e. a linear form* u *on* $\mathcal{D}(X)$ *such that for any compact subset* $K \subset X$, *there exists* $C \geq 0$, $j \in \mathbb{N}$ *with* $|u(\varphi)| \leq C\, p_{K,j}(\varphi)$ *for all* $\varphi \in \mathcal{D}_K(X)$. *For* $\varphi \in \mathcal{D}(X)$ *and* $u \in \mathcal{D}'(X)$, *we write* $u(\varphi) = \langle u, \varphi \rangle$.

The injection of $\mathcal{D}(X)$ into $\mathcal{D}^m(X)$ is continuous, and also of dense image (consider a regularising sequence, i.e. a sequence of functions ρ_K of the form

$$(1.1.1.) \qquad \rho_k(x) = k^n \rho(kx) ,$$

where $\rho \in \mathcal{D}(\mathbb{R}^n)$ is such that $\int \rho(x)\, dx = 1$; suppose $u \in \mathcal{D}^m(X)$; we consider the regularisations $u_k = u * \rho_k$ of u defined by $u_k(x) = \int u(y)\, \rho_k(x-y)\, dy$ and we can easily show that $u_k \to u$ in $\mathcal{D}^m(X)$ if $k \to +\infty$). Thus the dual topological space $\mathcal{D}'^m(X)$ of $\mathcal{D}^m(X)$ is identified with a vector subspace of $\mathcal{D}'(X)$ (we identify a continuous linear form on $\mathcal{D}^m(X)$ with its restriction to $\mathcal{D}(X)$). $\mathcal{D}'^m(X)$ is termed the space of distributions on X of order $\leq m$. In particular the distributions of order ≤ 0 are measures. For example, for $a \in X$, we define the Dirac measure

$\delta_a \in \mathcal{D}^{0}(X)$ at the point a by $< \delta_a, \varphi > = \varphi(a)$ for $\varphi \in \mathcal{D}^{0}(X)$. Let $L^1_{loc}(X)$ be the space of classes of functions which are integrable over any compact subset of X for the Lebesgue measure $dx = dx_1 \ldots dx_n$ in X. For $f \in L^1_{loc}(X)$, the measure $f\,dx \in \mathcal{D}^{0'}(X)$ is identified with the distribution (also denoted by f) defined by

$$< f, \varphi > = \int f \varphi \, dx \qquad (\varphi \in \mathcal{D}(X)) \quad .$$

In particular we shall consider the case where f is the Heaviside function Y in $\mathbb{R}$ defined by $Y(x) = 1$ if $x \geq 0$, and by $Y(x) = 0$ if $x < 0$.

We now indicate without proof several topological properties of $\mathcal{D}(X)$, $\mathcal{D}'(X)$ which will be useful later on. There are two natural topologies on $\mathcal{D}'(X)$: the topology $\mathcal{T}_s$ of simple convergence, and the strong dual topology, $\mathcal{T}_b$, of $\mathcal{D}(X)$, i.e. the topology of uniform convergence on the bounded subsets of $\mathcal{D}(X)$ (since a bounded subset of $\mathcal{D}(X)$ is relatively compact, the topology of uniform convergence on the compact subsets of $\mathcal{D}(X)$ coincides with $\mathcal{T}_b$). When we equip $\mathcal{D}'(X)$ with topology $\mathcal{T}_b$, the strong dual of $\mathcal{D}'(X)$ is identified algebraically and topologically with $\mathcal{D}(X)$. Furthermore, $\mathcal{D}(X)$ is a barreled space, which means that the bounded sets of $\mathcal{D}'(X)$ are the same for $\mathcal{T}_s$ and $\mathcal{T}_b$, and that they coincide with the equicontinuous subsets of

$\mathcal{D}'(X)$. We thereby deduce that $\mathcal{T}_s$ and $\mathcal{T}_b$ coincide on the bounded subsets of $\mathcal{D}'(X)$. In particular, in the case of a sequence of distributions, convergence in the sense of $\mathcal{T}_s$ and convergence in the sense of $\mathcal{T}_b$ coincide, and we then speak of a convergent sequence of distributions. Thus, if A is a mapping from a metrizable space into $\mathcal{D}'(X)$, the continuity of A does not depend upon the topology ($\mathcal{T}_s$ or $\mathcal{T}_b$) with which we equip $\mathcal{D}'(X)$. The same is also true if A is a linear operator from $\mathcal{D}(Y)$ into $\mathcal{D}'(X)$, where Y is an open subset of $\mathbb{R}^p$, since the continuity of A is equivalent to that of its restriction to each $\mathcal{D}_K(Y)$. This being the case, we shall simply refer to a continuous linear operator from $\mathcal{D}(Y)$ into $\mathcal{D}'(X)$. In particular, the injection of $\mathcal{D}(X)$ into $\mathcal{D}'(X)$ is continuous, and we shall see that it is of sequentially dense image (see Remark 2.3).

We can generalise to distributions certain operations on functions. If Y is an open subset of X and if $u \in \mathcal{D}'(X)$, we define the restriction $u_{|Y} \in \mathcal{D}'(Y)$ of u to Y by
$< u_{|Y}, \varphi > = < u, \varphi >$ for all $\varphi \in \mathcal{D}(Y)$. The product of a distribution $u \in \mathcal{D}'(X)$ and a function $a \in C^\infty(X)$ is the distribution $au \in \mathcal{D}'(X)$ defined by $< au, \varphi > = < u, a\varphi >$ for all $\varphi \in \mathcal{D}(X)$. The derivative $\partial^\alpha u$ of u is the distribution on X defined by $< \partial^\alpha u, \varphi > = (-1)^{|\alpha|} < u, \partial^\alpha \varphi >$ for all $\varphi \in \mathcal{D}(X)$. If $P(x, D) = \sum_{|\alpha| \leq m} a_\alpha(x) D^\alpha$ is a differential operator with coefficients $a_\alpha \in C^\infty(X)$, we put $P(x, D)u = \sum_{|\alpha| \leq m} a_\alpha D^\alpha u$.

Leibniz' formula gives

$$D_j(au) = D_j a \,.\, u + a \,.\, D_j u$$

and reasoning by recurrence over the extent of the multi-index β, we obtain

$$(1.1.2.) \qquad D^\beta(a\,u) = \sum_{\alpha \leq \beta} \binom{\beta}{\alpha} D^{\beta-\alpha}a \,.\, D^\alpha u$$

where $\binom{\beta}{\alpha} = \frac{\beta!}{(\beta-\alpha)!\,\alpha!}$ and where $\alpha \leq \beta$ means that $\alpha_j \leq \beta_j$ for $j = 1, \ldots, n$.

We term the *complete symbol* of the differential operator $P(x, D)$ the function (polynomial in ξ) $P(x, \xi) = \sum_{|\alpha| \leq m} a_\alpha(x)\, \xi^\alpha$, where $\xi^\alpha = \xi_1^{\alpha_1} \ldots \xi_n^{\alpha_n}$. We term the *principal symbol* of P the function (homogeneous of degree m in ξ) $P_m(x, \xi) = \sum_{|\alpha|=m} a_\alpha(x)\, \xi^\alpha$. We denote by $P^{(\alpha)}(x, D)$ the differential operator with complete symbol $\partial_\xi^\alpha P(x, \xi)$, and we have the Leibniz-Hörmander formula, which can be deduced from (1.1.2) by linearity:

$$(1.1.3.) \qquad P(x, D)(au) = \sum_\alpha \frac{1}{\alpha!} (P^{(\alpha)}(x, D)a) \,.\, D^\alpha u \quad .$$

For example, if $a(x) = \exp(ix.\xi)$ where $x.\xi = x_1\xi_1 + \ldots + x_n\xi_n$, we have

$$(1.1.4.) \qquad e^{-ix.\xi} P(x, D)(e^{ix.\xi} u) = \sum_\alpha \frac{1}{\alpha!} \partial_\xi^\alpha P(x, \xi)\, D^\alpha u \;.$$

We define the conjugate $\bar{u} \in \mathcal{D}'(X)$ of a distribution $u \in \mathcal{D}'(X)$ by $< \bar{u}, \varphi > = \overline{< u, \bar{\varphi} >}$ for all $\varphi \in \mathcal{D}(X)$.

The distributions form a sheaf on X ; this means that given an arbitrary covering of X by open subsets V_j and distributions $u_j \in \mathcal{D}'(V_j)$ such that $u_j = u_k$ in $V_j \cap V_k$, then there exists a unique $u \in \mathcal{D}'(X)$ such that $u = u_j$ in V_j for all j. If we suppose in fact that φ_j is a partition of unity in X subordinate to the given covering, i.e. a family of functions $\varphi_j \in C^\infty(X)$ such that: $\operatorname{supp} \varphi_j \subset V_j$, the family $\operatorname{supp} \varphi_j$ is locally finite (i.e. for any compact subset K of X, the set of indices j such that $\operatorname{supp} \varphi_j \cap K \neq \emptyset$ is finite), $\sum_j \varphi_j = 1$, and $\varphi_j \geq 0$. We can then show that the formula $u = \sum_j \varphi_j u_j$ defines a distribution $u \in \mathcal{D}'(X)$ such that $u_{|V_j} = u_j$ for all j. The uniqueness of u is obtained in the same manner by decomposing each test function φ into $\varphi = \sum_j \varphi_j \varphi$.

This sheaf property shows in particular that if a distribution is zero (resp: C^∞) in open subsets V_j of X, then it is zero (resp: C^∞) in their union; this allows us to state:

DEFINITION 1.2 : *We term the support (resp: singular support) of a distribution* $u \in \mathcal{D}'(X)$ *the smallest closed subset of* X *outside of which* u *is zero (resp:* C^∞*). We denote this by* supp u *(resp:* sing supp u*). If* F *is a closed subset of* X*, we denote by* $\mathcal{D}'_F(X)$ *the subspace of* $\mathcal{D}'(X)$ *formed by the distributions* *with support in* F.

If we let P = P(x, D) be a differential operator with C^∞ coefficients on X, it is clear that

(1.2.1.) $\operatorname{supp} Pu \subset \operatorname{supp} u$, $\operatorname{sing\,supp} Pu \subset \operatorname{sing\,supp} u$

for $u \in \mathcal{D}'(X)$.

The injection of $C_0^\infty(X) = \mathcal{D}(X)$ into $C^\infty(X)$ is continuous, and also of dense image. This is proved by truncation: consider an exhaustive sequence of compact subsets of X, i.e. a sequence of compact subsets K_j of X such that $K_j \subset \overset{\circ}{K}_{j+1}$, $\bigcup_j K_j = X$. Note that if K is an arbitrary compact subset of X, then $K \subset K_j$ for j sufficiently large. Let there be functions $\chi_j \in C_0^\infty(X)$ such that

(1.2.2) $\chi_j = 1$ in K_j .

If $\varphi \in C^\infty(X)$ it immediately follows that $\chi_j\varphi \underset{j}{\to} \varphi$ in $C^\infty(X)$.

Thus the dual $\mathcal{E}'(X)$ of $C^\infty(X)$ is identified with a vector subspace of $\mathcal{D}'(X)$. We can show that this subspace is the subspace of distributions with compact support in X, and that the distribution u with compact support in X is identified with the continuous linear form on $C^\infty(X)$ defined by $\varphi \to \langle u, \theta\varphi \rangle$, where θ is an arbitrary function in $C_0^\infty(X)$ with value one in the neighbourhood of supp u. We shall also denote by $\langle u, \varphi \rangle$ the value at $\varphi \in C^\infty(X)$ of $u \in \mathcal{E}'(X)$. More generally, if $u \in \mathcal{D}'(X)$ and

$\varphi \in C^{\infty}(X)$ are such that

(1.2.3.) $\quad K = \operatorname{supp} u \cap \operatorname{supp} \varphi$ is compact,

we put $< u, \varphi > = < u, \theta\varphi >$ where $\theta \in C_o^{\infty}(X)$ has the value 1 in the neighbourhood of K.

For the pair $C^{\infty}(X)$, $\mathcal{E}'(X)$ we have the same topological properties as those which have been described for the pair $\mathcal{D}(X)$, $\mathcal{D}'(X)$.

Let X_j be open subsets of $\mathbb{R}^{n_j}$ $(j = 1, 2)$. We can define the tensor product $u_1 \otimes u_2$ of two distributions $u_j \in \mathcal{D}'(X_j)$ $(j = 1, 2)$. Firstly, we have

LEMMA 1.3: *The space of finite linear combinations of functions of the form* $\varphi_1 \otimes \varphi_2$ *(where* $\varphi_j \in \mathcal{D}(X_j)$*) is sequentially dense in* $\mathcal{D}(X_1 \times X_2)$, *and we denote it by* $\mathcal{D}(X_1) \otimes \mathcal{D}(X_2)$.

In fact, suppose we have $\varphi \in \mathcal{D}(X_1 \times X_2)$. Consider a closed cube K of side T such that $\operatorname{Supp} \varphi \subset \overset{o}{K}$. We extend $\varphi_{|K}$ periodically, with period T, into $\tilde{\varphi} \in C^{\infty}(\mathbb{R}^{n_1+n_2})$. Suppose we have functions $\theta_j \in C_o^{\infty}(R^{n_j})$ $(j = 1, 2)$ such that $\operatorname{supp}(\theta_1 \otimes \theta_2) \subset K$, with $\theta_1 \otimes \theta_2 = 1$ in $\operatorname{supp} \varphi$. We have $\varphi = \theta_1 \otimes \theta_2 \tilde{\varphi}$, and the Fourier series expansion of $\tilde{\varphi}$ can be written

$$\tilde{\varphi}(x_1, x_2) = \sum_{\substack{\alpha_j \in \mathbb{N}^{n_j} \\ j=1,2}} c_{\alpha_1,\alpha_2} e^{ix_1 \cdot \alpha_1 2\pi/T} e^{ix_2 \cdot \alpha_2 2\pi/T}$$

where $c_{\alpha_1,\alpha_2} =$

$$T^{-n_1-n_2} \int_{\mathbb{R}^{n_1} \times \mathbb{R}^{n_2}} e^{-ix_1 \cdot \alpha_1 2\pi/T} e^{-ix_2 \cdot \alpha_2 2\pi/T} \varphi(x_1, x_2)\, dx_1 dx_2 .$$

By integration by parts, we see that for all N there exists a constant C_N such that

$$(1.3.1.) \quad |c_{\alpha_1,\alpha_2}| \leq C_N(1 + |\alpha_1| + |\alpha_2|)^{-N}$$

$$\text{for all} \quad \alpha_j \in \mathbb{N}^{n_j} \qquad (j = 1, 2) .$$

We have $\varphi(x_1, x_2) = \sum_{\alpha_1,\alpha_2} c_{\alpha_1,\alpha_2} \varphi_{\alpha_1}(x_1) \varphi_{\alpha_2}(x_2)$

where

$$\varphi_{\alpha_j}(x_j) = \theta_j(x_j) e^{ix_j \cdot \alpha_j 2\pi/T} \qquad (j = 1, 2) ,$$

and (1.3.1) shows that this series converges in $\mathcal{D}(X_1 \times X_2)$, hence the lemma.

We consider the bilinear mapping

$$(\varphi_1, \varphi_2) \longmapsto \varphi_1 \otimes \varphi_2 \quad \text{from } \mathcal{D}(X_1) \times \mathcal{D}(X_2) \text{ into } \mathcal{D}(X_1 \times X_2) .$$

This extends in unique manner to a continuous bilinear mapping from $\mathcal{D}'(X_1) \times \mathcal{D}'(X_2)$ into $\mathcal{D}'(X_1 \times X_2)$, which we also denote by

$$(u_1, u_2) \longrightarrow u_1 \otimes u_2 \ .$$

Uniqueness in fact follows from Lemma 1.3. In order to prove existence, it may be shown that we can put, for $\varphi \in \mathcal{D}(X_1 \times X_2)$:

$$< u_1 \otimes u_2, \varphi > \ = \ < u_1, \psi_1 > \ = \ < u_2, \psi_2 > \quad , \text{ where}$$

$$\psi_1(x_1) = < u_2, \varphi(x_1, .) > \ , \quad \psi_2(x_2) = < u_1, \varphi(., x_2) > \quad .$$

We remark that

$$< u_1 \otimes u_2, \varphi_1 \otimes \varphi_2 > \ = \ < u_1, \varphi_1 > < u_2, \varphi_2 >$$

$$\text{for} \quad \varphi_j \in \mathcal{D}(X_j) \quad , \quad j = 1, 2 \quad .$$

$$\partial_{x_1}^{\alpha_1} \partial_{x_2}^{\alpha_2} (u_1 \otimes u_2) = \partial^{\alpha_1} u_1 \otimes \partial^{\alpha_2} u_2$$

$$\operatorname{supp} (u_1 \otimes u_2) = \operatorname{supp} u_1 \times \operatorname{supp} u_2 \quad .$$

1.4 JUMP FORMULA: We consider a function $u \in C^\infty(\mathbb{R}^n)$, and we denote by u^o the function which is equal to u for $x_n \geq 0$ and to 0 for $x_n < 0$. We can show by recurrence on the integer j that

$$D_n^j(u^o) = (D_n^j u)^o + \frac{1}{i} \sum_{k=0}^{j-1} \gamma_{j-k-1} u \otimes D^k \delta(x_n)$$

where

$$(\gamma_k u)(x') = (D_n^k u)(x', 0) \in C^\infty(R^{n-1}) \quad , \quad x' = (x_1, \ldots, x_{n-1}) \quad ,$$

and where δ is the Dirac measure at 0 in $\mathbb{R}$.

More generally, suppose:

$$P(x, D) = \sum_{j=0}^{m} P_j(x, D')D_n^k \qquad (\text{with} \quad D' = (D_1, \ldots, D_{n-1}))$$

is a differential operator with C^∞ coefficients in $\mathbb{R}^n$. We have the "jump formula"

$$(1.4.1.) \; P(u^0) = (Pu)^0 + \sum_{\ell+k+1 \leq m} P_{\ell+k+1}(x, D') \, \gamma_\ell u \otimes D^k \delta(x_n).$$

1.5 DISTRIBUTIONS REGULAR WITH RESPECT TO ONE VARIABLE:

Suppose we have $U \in C^0(R, \mathcal{D}'(R^{n-1}))$; we can show that with it we can associate injectively the distribution $u \in \mathcal{D}'(\mathbb{R}^n)$ by putting:

$$< u, \varphi > = \int_R < U(x_n), \varphi(., x_n) > dx_n \qquad \text{for} \quad \varphi \in \mathcal{D}(R^n).$$

Such a distribution u is said to be continuous with respect to x_n with values which are distributions with respect to x' and we define its "sectional trace on $\{x_n = 0\}$" $\gamma_0 u \in \mathcal{D}'(R^{n-1})$ by $\gamma_0 u = U(0)$. We note that $\partial_{x'}^{\alpha'} u$ is the distribution associated with the function $\partial^{\alpha'} U$. If $U \in C^k(R, \mathcal{D}'(R^{n-1}))$, we say that u is of class C^k with respect to x_n with values which are distributions with respect to x'. We note that, for $0 \leq j \leq k$, $\partial_n^j u$ is the distribution associated with the function $\dot{U}^{(j)}$, and we define the sectional trace $\gamma_j u \in \mathcal{D}'(R^{n-1})$ by $\gamma_j u = \gamma_0(D_n^j u)$. In practice we shall make no distinction between the notations U and u.

2. CONVOLUTION AND FOURIER TRANSFORMATION OF DISTRIBUTIONS

The abelian group structure on $\mathbb{R}^n$ allows the definition of additional operations on $\mathcal{D}'(\mathbb{R}^n) = \mathcal{D}'$ (we shall omit the indication of the open subset X in the spaces of distributions when $X = \mathbb{R}^n$).

2.1 CONVOLUTION: If we have $u \in L^1$, $v \in C_0^\infty$, we define the convolution product

(2.1.1.) $$(u * v)(x) = \int u(x-y)\, v(y)\, dy = \int u(y)\, v(x-y)\, dy$$

and we show that

(2.1.2.) $$\|u * v\|_{L^2} \leq \|u\|_{L^1} \cdot \|v\|_{L^2}$$; consequently

(2.1.1) remains meaningful (almost everywhere) for $v \in L^2$ and we have $u * v \in L^2$. If furthermore $\varphi \in \mathcal{D}$, we can show that

$$< u * v, \varphi > = \iint u(x)\, v(y)\, \varphi(x+y)\, dx\, dy \quad .$$

In view of (1.2.3) this formula leads us to state

DEFINITION 2.1: *Two distributions* $u, v \in \mathcal{D}'$ *are said to be convolvable if the mapping* $(x, y) \overset{s}{\longmapsto} x+y$ *from* $\operatorname{supp} u \times \operatorname{supp} v$ *into* $\mathbb{R}^n$ *is proper. Under these conditions the convolution product* $u * v$ *of* u *and* v *is the distribution on* $\mathbb{R}^n$ defined by

$$< u * v, \varphi > \; = \; < u \otimes v, \varphi \circ s > \qquad \text{for all} \quad \varphi \in \mathcal{D} \quad .$$

The principal properties of convolution are summarised in the following:

PROPOSITION 2.2: *Suppose* u, v *are two convolvable distributions. Then*

$$u * v = v * u$$

$$\partial^\alpha (u * v) = \partial^\alpha u * v = u * \partial^\alpha v$$

$\operatorname{supp}(u * v) \subset \operatorname{supp} u + \operatorname{supp} v$, *where the right-hand side is closed.*

If $u \in \mathcal{D}'$, $v \in \mathcal{D}$ *or if* $u \in \mathcal{E}'$, $v \in C^\infty$, *then*

(2.2.1.) $u * v \in C^\infty$ *and* $(u * v)(x) = < u, \tau_x \check{v} >$, *where*

$$(\tau_x \check{v})(y) = v(x-y) \quad .$$

REMARK 2.3: Let ρ_j be a regularising sequence (see (1.1.1)). If $u \in \mathcal{D}'$, we can show that the C^∞ functions $u * \rho_j$ converge towards u in $\mathcal{D}'$; suppose χ_j is a sequence of truncation functions as in (1.2.2); we can show that the functions $\rho_j * (\chi_j u) \in \mathcal{D}$ converge towards u in $\mathcal{D}'$. This proves the sequential density of $\mathcal{D}$ in $\mathcal{D}'$.

2.4 FOURIER TRANSFORMATION: If $f \in L^1 = L^1(R^n)$, its Fourier transform is denoted by $\mathcal{F}f$ where $\hat{f}$ is defined by

$$(2.4.1.) \qquad \hat{f}(\xi) = \int e^{-ix\cdot\xi} f(x)\, dx \qquad \text{for} \qquad \xi \in R^n.$$

We have $\hat{f} \in C^0(\mathbb{R}^n)$ and $\sup\limits_{\xi \in R^n} |\hat{f}(\xi)| \leq \|f\|_{L^1}$.

Let us now define a subspace of L^1 which is invariant under Fourier transformation:

DEFINITION 2.4: *We denote by* $\mathcal{S}$ *the space of functions* $\varphi \in C^\infty(\mathbb{R}^n)$ *such that, for any integer* h, *the quantity*

$$q_h(\varphi) = \sup_{\substack{x \in R^n \\ |\alpha| \leq h}} (1 + |x|^2)^h \, |\partial^\alpha \varphi(x)|$$

is finite. $\mathcal{S}$ *is called the space of functions which are rapidly decreasing, along with all their derivatives.*

We equip $\mathcal{S}$ with the topology defined by the family of semi-norms $q_h (h \in \mathbb{N})$. $\mathcal{S}$ is a Fréchet space. We recall without proof the following:

THEOREM 2.5: *The Fourier transformation* $\mathcal{F}$ *establishes a topological vector isomorphism of* $\mathcal{S}$ *onto itself, with inverse* $\mathcal{F}^{-1}$ *satisfying* $(\mathcal{F}^{-1} g)(x) = \int e^{ix.\xi} g(\xi)\, đ\xi$ *for* $g \in \mathcal{S}$ *where* $đ\xi = (2\pi)^{-n}\, d\xi$. *If* $\varphi, \psi \in \mathcal{S}$, *we have*

$$(2.5.1.) \qquad \int \hat{\varphi}(\xi)\, \psi(\xi)\, d\xi = \int \varphi(x)\, \hat{\psi}(x)\, dx \quad .$$

If we have $\varphi \in \mathcal{D}$, Definition (2.4.1) for $\hat{\varphi}$ shows that it extends into an entire function on $\mathbb{C}^n$ defined by

$$(2.5.2.) \qquad \zeta \in \mathbb{C}^n \longrightarrow \int e^{-ix\zeta}\, \varphi(x)\, dx \quad .$$

We shall now show that the behaviour of $\hat{\varphi}$ in $\mathbb{C}^n$ determines the convex envelope of the support of φ.

To do this, we define the indicator function I_K of a compact subset K of $\mathbb{R}^n$ by

$$\eta \in \mathbb{R}^n \longrightarrow I_K(\eta) = \sup_{x \in K} \langle x, \eta \rangle \in \mathbb{R} \quad .$$

For example, if K is the ball B with centre O and radius R, we have

$$I_B(\eta) = |\eta| \cdot R \quad .$$

The indicator of K determines, for each hyperplane direction, the smallest closed half-space which contains K. Consequently, the Hahn-Banach theorem shows that I_K determines the convex envelope $\widehat{K}$ of K and $I_K = I_{\widehat{K}}$.

We can state the following:

THEOREM 2.6: (Paley-Wiener). *Let* K *be a convex compact subset of* $\mathbb{R}^n$. *An entire function* U *on* $\mathbb{C}^n$ *is the Fourier transform of a function belonging to* $\mathcal{D}_K$ *if and only if, for all* N, *there exists* C_N *such that*

$$(2.6.1.) \qquad |U(\zeta)| \leq C_N(1+|\zeta|)^{-N} e^{I_K(\eta)} ,$$

$$\zeta = \xi + i\eta \in C^n .$$

Proof of necessity. For any multi-index α, the equality (2.5.1) implies that

$$\zeta^\alpha \cdot \hat{\varphi}(\zeta) = \int e^{-ix\zeta} D^\alpha \varphi(x)\, dx$$

which gives the upper bound

$$|\zeta|^\alpha |\hat{\varphi}(\zeta)| \leq C_\alpha e^{I_K(\eta)} \sup_{x \in K} |D^\alpha \varphi(x)|$$

and (2.6.1) follows from this immediately.

Proof of sufficiency. The upper bounds (2.6.1) imply that $U(\xi)$ is of rapid decrease; it is thus the Fourier transform of a C^∞ function φ, given by

$$(2.6.2.) \qquad \varphi(x) = \int e^{ix\xi} U(\xi)\, d\!\!\!/\xi .$$

In order to show that supp $\varphi \subset K$, it is sufficient to verify that if $x^0 \notin K$ we have $\varphi(x^0) = 0$. By Hahn-Banach, there exists a hyperplane which separates x^o and the compact convex set K. In other words, there exists $\eta_o \in \mathbb{R}^n \setminus 0$ and $c \in \mathbb{R}$ such that

$$(2.6.3.) \qquad K \subset \{ x \mid x.\eta_o < c \} \quad \text{and} \quad x^o.\eta_o > c \quad .$$

After a linear change of coordinates, we can assume $\eta_o = (0,\ldots, 0, 1)$; we put $\xi = (\xi', \xi_n)$.

For $t > 0$, Cauchy's theorem allows the integral (2.6.2) to be replaced by

$$(2.6.4.) \quad \varphi(x) = \int_{R^{n-1}} e^{ix'\xi'} \left(\int_R e^{ix_n(\xi_n+it)} U(\xi', \xi_n+it) d\!\!/\xi_n \right) d\!\!/\xi'$$

since (2.6.1) implies $\lim\limits_{\substack{|\xi| \to \infty \\ |\eta_n| \leq 1}} |U(\xi', \xi_n + i\eta_n)| = 0$.

Taking $N = n+1$, we deduce from (2.6.1) and (2.6.4) the upper bound

$$(2.6.5.) \quad |\varphi(x^o)| \leq C \int e^{-t(x^o_n - I_K(\eta_o))} (1 + |\xi|)^{-n-1} \, d\xi \quad .$$

Since (2.6.3) implies $x^o_n > I_K(\eta_o)$, the right-hand side of (2.6.5) tends to 0 when $t \to +\infty$; hence $\varphi(x^o) = 0$.

It can easily be seen that the injections $\mathcal{D} \hookrightarrow \mathcal{S} \hookrightarrow C^\infty$ are continuous and of dense image (to prove that $\mathcal{D}$ is dense in $\mathcal{S}$, we proceed by truncation using functions

$\chi_j(x) = \chi(x/j)$, where $\chi \in \mathcal{D}(\mathbb{R}^n)$ is such that $\chi(x) = 1$ for $|x| \le 1$). Thus the dual $\mathcal{S}'$ of $\mathcal{S}$ is naturally identified with a vector subspace of $\mathcal{D}'$ containing $\mathcal{E}'$. $\mathcal{S}'$ is termed the space of tempered distributions.

For $u \in \mathcal{S}'$ and $\varphi \in \mathcal{S}$, we again denote by $\langle u, \varphi \rangle$ the value of u at φ, and for the pair $\mathcal{S}, \mathcal{S}'$ we have topological properties analogous to those of the pair $\mathcal{D}, \mathcal{D}'$. Formula (2.5.1) permits the Fourier transformation to be extended to $\mathcal{S}'$:

DEFINITION 2.7: *For* $u \in \mathcal{S}'$, *we define the Fourier transform* $\mathcal{F}u = \hat{u} \in \mathcal{S}'$ *of* u *by* $\langle \mathcal{F}u, \psi \rangle = \langle u, \mathcal{F}\psi \rangle$ *for all* $\psi \in \mathcal{S}$. *The Fourier transformation* $\mathcal{F}$ *establishes a topological vector isomorphism of* $\mathcal{S}'$ *(equipped with one of the two topologies* $\mathcal{T}_s$ *or* $\mathcal{T}_b$*) onto itself.*

In the case of distributions with compact support, we have:

THEOREM 2.8: *If* $u \in \mathcal{E}'$, *then* $\mathcal{F}u$ *is the* C^∞ *function defined by*

$$(\mathcal{F}u)(\xi) = \langle u, e^{-ix\xi} \rangle \quad \text{for} \quad \xi \in R^n .$$

$\mathcal{F}u$ *extends the entire function on* $\mathbb{C}^n$ *defined by*

$$(\mathcal{F}u)(\zeta) = \langle u, e^{-ix.\zeta} \rangle \quad \text{for} \quad \zeta \in R^n .$$

$\mathcal{F}u$ *is a slowly increasing function in* $\mathbb{R}^n$, *i.e. there exist* C *and* μ *such that*

$$(2.8.1.)\qquad |(\mathcal{F}u)(\xi)| \leq C\,(1+|\xi|)^{\mu} \quad \text{for all} \quad \xi \in \mathbb{R}^n \ .$$

More generally, if B *is a bounded subset of* $\mathcal{E}'$ *we can find* C and μ *such that* (2.8.1) *is valid for all* $u \in B$.

We point out merely that the upper bound (2.8.1) derives from the upper bound of the linear form u on C^{∞} by a semi-norm $p_{K,\mu}$; furthermore, we can employ an identical semi-norm for any $u \in B$ because a bounded subset is an equicontinuous part of $\mathcal{E}'$.

We have the analogue of Theorem 2.6 as follows:

THEOREM 2.9: (Paley-Wiener-Schwartz). *Let* K *be a convex compact subset of* $\mathbb{R}^n$. *An entire function* U *is the Fourier transform of a distribution of* $\mathcal{E}'_K$ *if and only if there exist* C *and* N *such that*

$$(2.9.1.)\qquad |U(\zeta)| \leq C(1+|\zeta|)^{N}\, e^{I_K(\eta)} \quad , \quad \zeta = \xi + i\eta \in \mathbb{C}^n .$$

We shall not have to make use of this theorem and we postpone a proof until Exercise (10.1).

The principal properties of Fourier transformation in $\mathcal{S}'$ are summarised in the following theorem:

THEOREM 2.10: *Suppose we have* $u \in \mathcal{S}'$ *and* $v \in \mathcal{E}'$. *Then*

$$(2.10.1.) \qquad \mathcal{F}(u * v) = \mathcal{F}u . \mathcal{F}v$$

$$\mathcal{F}(D^\alpha u) = \xi^\alpha \mathcal{F}u$$

$$(2.10.2.) \qquad \mathcal{F}(P(D)u) = P(\xi)\, \mathcal{F}u \quad ;$$

$\mathcal{F}$ *establishes a linear homeomorphism of* L^2 *and we have Parseval's formula*:

$$\int u \bar{v}\, dx = \int \hat{u} \bar{\hat{v}}\, d\!\!/\xi \qquad (u, v \in L^2)$$

$$\int u v\, dx = \int \hat{u}(\xi)\, \hat{v}(-\xi)\, d\!\!/\xi .$$

Formula (2.10.2) shows that, by Fourier transformation, the differential operator $P(D)$ is transformed into the operator of multiplication by $P(\xi)$. This property makes Fourier transformation a basic tool for the study of differential operators.

REMARK 2.11: Sometimes convolution products can be defined without a support condition. Thus, for $u \in \mathcal{S}'$ and $v \in \mathcal{S}$, we put, just as in (2.2.1), $(u * v)(x) = \langle u, \tau_x \check{v} \rangle$. We have $u * v \in C^\infty(\mathbb{R}^n)$, and we can show that $u * v \in \mathcal{S}'$ and that formula (2.10.1) remains valid.

3. SINGULAR SPECTRUM OF A DISTRIBUTION

From the definition of the singular support of $u \in \mathcal{D}'(X)$, we have the equivalence

$$x^\circ \notin \text{sing supp } u \Leftrightarrow \left\{ \begin{array}{l} \text{there exists an open neighbourhood} \\ V \text{ of } x^\circ \text{ such that for all } \alpha \in \mathcal{D}(V) \\ \text{we have } \alpha u \in \mathcal{D}(V) \end{array} \right\}$$

and by Fourier transformation, the condition $\alpha u \in \mathcal{D}(V)$ is equivalent to the existence for all N of a constant C such that

(3.1.1.) $$|\widehat{\alpha u}(\xi)| \leq C\,(1 + |\xi|)^{-N} \quad \text{for } \xi \in \mathbb{R}^n.$$

In the latter form, we can microlocalise the concept of singular support by distinguishing the directions ξ in which we have an upper bound of the type (3.1.1.).

DEFINITION 3.1. *A point* $(x^\circ, \xi^\circ) \in X \times (\mathbb{R}^n - \{0\})$ *is not within the singular spectrum of the distribution* $u \in \mathcal{D}'(X)$ *if there exists a neighbourhood* V *of* x° *and a neighbourhood* W *of* ξ° *such that for all* $\alpha \in \mathcal{D}(V)$ *and all* N *there exists* C *such that*

(3.1.2.) $$|\widehat{\alpha u}(\tau\xi)| \leq C\,\tau^{-N} \quad \text{for } \tau \geq 1 \text{ and } \xi \in W.$$

The singular spectrum of u *is a closed subset of* $X \times (\mathbb{R}^n - \{0\})$ *denoted by* WFu.

This concept was introduced and investigated by HORMANDER [5] (under the name "wave-front set") following the introduction by SATO of the concept of the "singular spectrum" of a hyperfunction (see for example SATO [1]). See also DUISTERMAAT [1] and GUILLEMIN-STERNBERG [1] for the concept of a singular spectrum.

Knowledge of the singular spectrum determines the singular support because

(3.1.3.) $\pi_X\,(WFu) = \text{sing supp } u$ where

$$\pi_X : X \times (\mathbb{R}^n \setminus 0) \ni (x, \xi) \longmapsto x \in X.$$

In fact, if $x^o \notin$ sing supp u, it is clear that $(x^o, \xi^o) \notin WFu$ for all $\xi \in \mathbb{R}^n - \{ 0 \}$. Conversely, if $x^o \notin \pi_X(WFu)$, we associate with all $\xi \in \mathbb{R}^n - \{ 0 \}$ such that $|\xi| = 1$ neighbourhoods V_ξ and W_ξ of x^o and ξ in which we have an inequality of the type (3.1.2). If we extract a finite covering (W_{ξ_j}) of the sphere $|\xi| = 1$ and we put $V = \cap_j V_{\xi_j}$, it is clear that with this choice of V we have the condition (3.1.1) and consequently $x^o \notin$ sing supp u.

EXAMPLE 3.2 : The immediate equalities

$$|< \delta_a, \alpha\, e^{-ix\xi} >| = |\alpha(a)|, \quad < Y, \alpha\, e^{-ix\xi} > = \frac{i\alpha(0)}{\xi} + O(\frac{1}{\xi^2})$$

show that $WF(\delta_a) = \{a\} \times (\mathbb{R}^n \setminus 0)$ and $WF(Y) = \{0\} \times (\mathbb{R} \setminus 0)$

4. OPERATORS AND KERNELS.

Suppose X, Y are open subsets of $\mathbb{R}^n$, $\mathbb{R}^p$; we investigate continuous linear operators from $\mathcal{D}(Y)$ into $\mathcal{D}'(X)$.

EXAMPLE 4.1 : Let K be a function in $C^\infty(X \times Y)$. Then the integral operator A which associates with each $\psi \in \mathcal{D}(Y)$ the function $A\psi$ defined by

$$(A\psi)(x) = \int K(x, y)\, \psi(y)\, dy$$

is continuous from $\mathcal{D}(Y)$ into $C^\infty(X)$. Moreover, A extends into a linear operator $\tilde{A}$ from $\mathcal{E}'(Y)$ into $C^\infty(X)$ when we put $(\tilde{A}v)(x) = < v, K(x, .) >$ for $v \in \mathcal{E}'(Y)$. More precisely, if $p_{H,j}$ is a semi-norm on $C^\infty(X)$, we have

$$p_{H,j}(\tilde{A}v) = \sup_{\substack{x \in H \\ |\alpha| \leq j}} |< v, \partial_x^\alpha K(x, .) >|$$

and since the functions $\partial_x^\alpha K(x, .)$ describe a bounded subset of $C^\infty(Y)$ when x describes H, we see that $\tilde{A}$ is a continuous linear operator from $\mathcal{E}'(Y)$ (equipped with $\mathcal{T}_b$) into $C^\infty(X)$.

More generally, let us consider a distribution $K \in \mathcal{D}'(X \times Y)$. With it we associate an operator $A = Op(K)$ which is continuous and linear from $\mathcal{D}(Y)$ into $\mathcal{D}'(X)$, by putting

(4.1.1.) $< A\psi, \varphi > = < K, \varphi \otimes \psi >$ for all $\psi \in \mathcal{D}(Y)$, $\varphi \in \mathcal{D}(X)$.

EXAMPLE 4.2 : Let $D = \text{diag}\,(X \times X) = \{ (x, x) \mid x \in X \}$ be the diagonal of X x X. We denote by δ_D the distribution on X x X defined by $< \delta_D, \Phi > = \int \Phi(x, x)\, dx$ for all $\Phi \in \mathcal{D}(X \times X)$. Then it is clear that $Op(\delta_D)$ = Identity.

EXAMPLE 4.3 : Suppose we have $T \in \mathcal{D}'(\mathbb{R}^n)$. With it we associate the distribution on $\mathbb{R}^n \times \mathbb{R}^n$ denoted, by an abuse of notation, by $T(x - y)$ and defined by

$$< T(x - y), \Phi > = < T(x), \Phi(x + y, y) > dy , \quad \Phi \in \mathcal{D}(R^n \times R^n)$$

(thus $\delta_D = \delta(x - y)$). Then, $Op\,(T(x - y))$ is the operator of convolution with T: $\psi \to T * \psi$.

We leave to the reader the task of verifying that the singular support of a distribution of the form $T(x - y)$ is equal to the set

(4.3.1.) $\{ (x, y) \in R^n \times R^n \mid (x - y) \in \text{sing supp } T \}$.

We now return to the linear mapping $K \to A = \mathrm{Op}(K)$ from $\mathcal{D}'(X \times Y)$ into $\mathcal{L}(\mathcal{D}(Y), \mathcal{D}'(X))$ defined by (4.1.1). This mapping is injective in view of Lemma 1.3. We shall now see that it is surjective:

THEOREM 4.4 (*Schwartz's kernels theorem*) : *For any continuous linear operator* A *from* $\mathcal{D}(Y)$ *into* $\mathcal{D}'(X)$, *there exists a unique distribution* $K \in \mathcal{D}'(X \times Y)$ *such that* $A = \mathrm{Op}(K)$; K *is called the kernel of the operator* A.

PROOF : The bilinear form

$$\mathcal{D}(X) \times \mathcal{D}(Y) \ni (\varphi, \psi) \longrightarrow \langle A\psi, \varphi \rangle \in \mathbb{C}$$

induces a linear form K on the algebraic tensor product $\mathcal{D}(X) \otimes \mathcal{D}(Y)$. Explicitly, for $\Phi \in \mathcal{D}(X \times Y)$ we have

$$(4.4.1.) \qquad \langle K, \Phi \rangle = \sum_{j=1}^{N} \langle A\,\psi_j, \varphi_j \rangle$$

where

$$(4.4.2.) \qquad \Phi = \sum_{j=1}^{N} \varphi_j \otimes \psi_j \ , \quad \varphi_j \in \mathcal{D}(X) \ , \quad \psi_j \in \mathcal{D}(Y)$$

is a decomposition of Φ, and (4.4.1) is independent of the choice of the decomposition of Φ.

To conclude, it follows from Lemma 1.3 that it suffices to show that K is a continuous linear form on $\mathcal{D}(X) \otimes \mathcal{D}(Y)$ for the topology induced by $\mathcal{D}(X \times Y)$. Letting H, L be compact subsets of X, Y, we show that there exists a semi-norm $p_{H \times L, m}$ such that that

$$(4.4.3.) \qquad |\langle K, \Phi \rangle| \le C\, p_{H \times L, m}(\Phi) \ , \quad \text{for } \Phi \in \mathcal{D}(Y) \text{ with supp } \Phi \subset H \times L.$$

Let H_1, L_1 be compact neighbourhoods of H, L; by truncation, we can always write (4.4.2) with $\varphi_j \in \mathcal{D}_{H_1}(X)$, $\psi_j \in \mathcal{D}_{L_1}(Y)$ if $\operatorname{supp} \Phi \subset H \times L$.

Suppose we have $f \in C_o^\infty(X)$ equal to 1 on H_1 and we put $H_2 = \operatorname{supp} f$; likewise suppose we have $g \in C_o^\infty(Y)$ equal to 1 on L_1 and we put $L_2 = \operatorname{supp} g$. We have $\psi_j(y) = \int g(y)\, e^{iy.\eta}\, \hat{\psi}_j(\eta)\, d\!\!/\eta$; by considering this integral as an integral of a function with values in $\mathcal{D}_{L_2}(Y)$, the continuity of A permits us to write

$A\,\psi_j = \int A(g(y)\, e^{iy.\eta})\, d\!\!/\eta$. Likewise, by writing

$\varphi_j(x) = \int f(x)\, e^{ix.\xi}\, \hat{\varphi}_j(\xi)\, d\!\!/\xi$, we obtain

$$(4.4.4.) \qquad \langle A\,\psi_j, \varphi_j \rangle = \iint F(\xi, \eta)\, \hat{\varphi}_j(\xi)\, \hat{\psi}_j(\eta)\, d\!\!/\xi\, d\!\!/\eta$$

where

$$F(\xi, \eta) = \langle A(g(y)\, e^{iy.\eta}) , f(x)\, e^{ix\xi} \rangle .$$

The continuity of A implies that the bilinear mapping $(\varphi, \psi) \longrightarrow \langle A\psi, \varphi \rangle$ is separately continuous from $\mathcal{D}_{H_2}(X) \times \mathcal{D}_{L_2}(Y)$ into $\mathbb{C}$; thus, using a result from functional analysis, it is continuous. Also, there exists C, h, ℓ such that

$$|\langle A\psi, \varphi \rangle| \leq C\, p_{H_2,h}(\varphi).p_{L_2,\ell}(\psi) \quad \text{for} \quad \varphi \in \mathcal{D}_{H_2} , \ \psi \in \mathcal{D}_{L_2}$$

In particular, it follows from the definition of F that there exists C such that

$$(4.4.5.) \qquad |F(\xi, \eta)| \le C(1 + |\xi|)^{\ell}(1 + |\eta|)^{h} \quad \text{for } (\xi, \eta) \in R^n \times R^p .$$

By linearity, we deduce from (4.4.1) and (4.4.4) that

$$(4.4.6.) \qquad \langle K, \Phi \rangle = \int F(\xi, \eta)\, \hat{\Phi}(\xi, \eta)\, d\!\!/\xi\, d\!\!/\eta \quad .$$

We choose an integer m such that $h+\ell-m < -(n+p)$; integrating by parts in the integral defining $\hat{\Phi}$, we obtain:

$$|\hat{\Phi}(\xi, \eta)| \le C'(1 + |\xi| + |\eta|)^{-m} P_{H \times L, m}(\Phi) \quad ,$$

which allows us to majorise (4.4.6) and obtain (4.4.3) by virtue of (4.4.5).

4.5 TRANSPOSE AND ADJOINT OF A LINEAR OPERATOR: The transpose (resp. the adjoint) of a continuous linear operator $A : \mathcal{D}(Y) \longrightarrow \mathcal{D}'(X)$ is the linear operator ${}^tA$ (resp. A^*) from $\mathcal{D}(X)$ into $\mathcal{D}'(Y)$ defined by

$$(4.5.1.) \quad \langle {}^tA\, \varphi, \psi \rangle = \langle \varphi, A\psi \rangle \quad (\langle A^* \varphi, \overline{\psi} \rangle = \langle \varphi, \overline{A\psi} \rangle) \quad \text{for } \varphi \in \mathcal{D}(X), \ \psi \in \mathcal{D}(Y) .$$

We immediately have:

PROPOSITION 4.5: *The operators* ${}^tA$ *and* A^* *are continuous linear operators from* $\mathcal{D}(X)$ *into* $\mathcal{D}'(Y)$. *The kernel of* ${}^tA$

(resp. of A^**) is deduced from the kernel* K *of* A *(resp. of* $\overline{K}$*) by interchanging the roles of the variables* x, y. *We have*
$^t(^tA) = A$ *and* $(A^*)^* = A$.

The operator A *is continuous from* $\mathcal{D}(Y)$ *into* $C^\infty(X)$ *if and only if* ^tA *extends into a continuous operator from* $\mathcal{E}'(X)$ *into* $\mathcal{D}'(Y)$ *(both equipped with the topology* $\mathcal{T}_b$*).*

For example, if A is a differential operator
$\sum_{|\alpha|\leq m} a_\alpha(x) D^\alpha$ with C^∞ coefficients in X, then ^tA is the differential operator $\sum_{|\alpha|\leq m} (-1)^{|\alpha|} D^\alpha(a_\alpha(x).)$ and A^* is the differential operator $\sum_{|\alpha|\leq m} D^\alpha(\overline{a_\alpha(x)}.\ )$.

4.6 REGULARISING OPERATORS:

DEFINITION 4.6: *A continuous linear operator* $A : \mathcal{D}(Y) \longrightarrow \mathcal{D}'(X)$ *is said to be regularising if it extends into a continuous operator from* $\mathcal{E}'(X)$ *into* $C^\infty(X)$, *where* $\mathcal{E}'(Y)$ *is equipped with* $\mathcal{T}_b$).

We have the following characterisation:

THEOREM 4.7: *An operator* A *is regularising if and only if its kernel* K_A *is in* $C^\infty(X \times Y)$.

PROOF: We saw in connection with Example 4.1 that the condition is sufficient. To show that it is necessary, we consider the mapping

$$X \times Y \ni (x, y) \longrightarrow a(x, y) = (A\,\delta_y)(x) \in C \,.$$

By hypothesis, the function a is C^∞ separately with respect to x and y, and this being so the theorem arises out of

LEMMA 4.8: *The function* a *is in* $C^\infty(X \times Y)$ *and* $K_A = a$.

PROOF: We begin by showing that a is continuous.

Supposing $(x_0, y_0) \in X \times Y$, we have:

$$|a(x, y) - a(x_0, y_0)| \leq$$
$$|(A\,\delta_y)(x) - (A\,\delta_{y_0})(x)| + |(A\,\delta_{y_0})(x) - (A\,\delta_{y_0})(x_0)| .$$

If we have $\varepsilon > 0$, then since $(A\,\delta_{y_0}) \in C^\infty(X)$, there exists $\eta' > 0$ such that

$$\text{(4.8.1.)} \quad |(A\,\delta_{y_0})(x) - (A\,\delta_{y_0})(x_0)| \leq \frac{\varepsilon}{2} \quad \text{when} \quad |x - x_0| \leq \eta' .$$

Furthermore, the continuity of A implies the existence of $\eta'' > 0$ such that

$$\text{(4.8.2.)} \quad \sup_{|x - x_0| \leq \eta'} |(A\,\delta_y)(x) - (A\,\delta_{y_0})(x)| \leq \frac{\varepsilon}{2} \quad \text{when} \quad |y - y_0| \leq \eta'' .$$

The continuity of a follows from (4.8.1) and (4.8.2).

In order to prove that a is C^∞, it is sufficient to reason in like manner for the functions

$$(\partial_x^\alpha \partial_y^\beta)\, a(x, y) = \partial_x^\alpha (A\, \delta_y^{(\beta)})(x) .$$

If we denote by A_1 the operator with kernel a, it is thus regularising; by construction, we have for any $y \in Y$ and any multi-index β

$$A_1(\delta_y^{(\beta)}) = A\,(\delta_y^{(\beta)})$$

which proves that $A = A_1$ since linear combinations of distributions of the type $\delta_y^{(\beta)}$ are dense in $\mathcal{E}'(Y)$.

REMARK 4.9: Theorems 4.4 and 4.7 are in fact applications of the theory of nuclear spaces; for this approach, see for example TREVES [1].

5. OPERATORS AND SUPPORT PROPERTIES

PROPOSITION 5.1: *If we let* A *be an operator with kernel* $K_A \in \mathcal{D}'(X \times Y)$, *we have for* $v \in C_0^\infty(Y)$

(5.1.1) $$\operatorname{supp} Av \subset (\operatorname{supp} K_A) \circ \operatorname{supp} v = \left\{ x \;\middle|\; \begin{array}{l} \textit{there exists } (x,y) \in \operatorname{supp} K_A \\ \textit{such that } y \in \operatorname{supp} v \end{array} \right\}$$

PROOF: Since supp v is compact, the set supp $K_A \circ$ supp v is closed in X. We now have to show that $\langle Av, u \rangle = 0$ for those $u \in C_0^\infty(X)$ which are such that

$\operatorname{supp} u \cap (\operatorname{supp} K_A \circ \operatorname{supp} v) = \emptyset$; now, this condition means that $(\operatorname{supp} u \times \operatorname{supp} v) \cap \operatorname{supp} K_A = \emptyset$, hence the result.

We denote by p_X (resp. p_Y) the projection of $X \times Y$ on X (resp. Y).

DEFINITION 5.2: *We say that the operator* A *with kernel* $K_A \in \mathcal{D}'(X \times Y)$ *is proper if the projections*

$$p_X|_{\operatorname{supp} K_A} \quad \text{and} \quad p_Y|_{\operatorname{supp} K_A} \quad \text{are proper.}$$

We have

PROPOSITION 5.3: *Suppose* A *is a proper operator; then*

i) *For any compact subset* $K \subset Y$, *the set* $\operatorname{supp} K_A \circ K = K'$ *is a compact subset of* X *and consequently* A *sends* $\mathcal{D}_K(Y)$ *to* $\mathcal{D}'_{K'}(X)$.

ii) *For any compact subset* $K' \subset X$, *there exists a compact subset* K *of* Y *such that:*

$$(\operatorname{supp} v \cap K = \emptyset) \Rightarrow (\operatorname{supp} Av \cap K' = \emptyset).$$

Consequently, the operator A *transforms a locally finite sum of functions in* $\mathcal{D}(Y)$ *into a locally finite sum of distributions in*

$\mathcal{E}'(X)$. *Furthermore, it extends uniquely into a continuous linear mapping from* $C^\infty(Y)$ *into* $\mathcal{D}'(X)$ *which again satisfies* (5.1.1).

PROOF OF i): We have by definition

$$K' = p_X\left(p_Y^{-1}(K) \cap \operatorname{supp} K_A\right)$$

so this is a compact subset of X because $p_Y|_{\operatorname{supp} K_A}$ is proper.

PROOF OF ii): If we put $K = p_Y\left(p_X^{-1}(K') \cap \operatorname{supp} K_A\right)$, this is a compact subset of Y and, by construction, it is clear that if supp v does not intersect K then supp Av does not intersect K'. Let (φ_α) be a partition of unity on Y by means of functions $\varphi_\alpha \in C_0^\infty(Y)$. For $v \in C^\infty(Y)$, we have a locally finite sum $v = \sum_\alpha \varphi_\alpha v$; thus the sum $\sum_\alpha A(\varphi_\alpha v)$ is also locally finite and we denote it by Av. We have thus defined an extension of A into a linear mapping, again denoted by A, from $C^\infty(Y)$ into $\mathcal{D}'(X)$. The continuity of this extension follows immediately from the definition of the strong topologies. The uniqueness comes from the density of $C_0^\infty(Y)$ in $C^\infty(Y)$.

Finally, by using (5.1.1), we have for $v \in C^\infty(Y)$:

$$\operatorname{supp} Av \subset \bigcup_\alpha \operatorname{supp} A(\varphi_\alpha v) \subset \bigcup_\alpha \left(\operatorname{supp} K_A \circ \operatorname{supp} \varphi_\alpha v\right) = \operatorname{supp} K_A \circ \operatorname{supp} v$$

For the singular support we have

PROPOSITION 5.4: *Let* A *be an operator with kernel* $K_A \in \mathcal{D}'(X \times Y)$. *We assume that* A *extends into a continuous operator from* $\mathcal{E}'(Y)$ *into* $\mathcal{D}'(X)$ *such that*

$A(C_0^\infty(Y)) \subset C^\infty(X)$. *We then have the inclusion*

$$\text{(5.4.1.)} \quad \text{sing supp } Av \subset \text{sing supp } K_A \circ \text{sing supp } v \ , \quad v \in \mathcal{E}'(Y) \ .$$

PROOF: Suppose we have x_0 which does not belong to the right-hand side of (5.4.1); in other words,

$$(\{x_0\} \times \text{sing supp } v) \cap \text{sing supp } K_A = \emptyset \ .$$

Since sing supp v is compact, there exist open neighbourhoods U, V of x_0 and sing supp v such that

$$(U \times V) \cap \text{sing supp } K_A = \emptyset \ .$$

Suppose we have $\alpha \in C_0^\infty(V)$, identical to 1 in the neighbourhood of sing supp v. We write $v = \alpha v + (1-\alpha)v$, and we have

$$\text{sing supp } Av = \text{sing supp } A(\alpha v) \text{ since } (1-\alpha)v \in C_0^\infty(Y) \ .$$

Since $K_A|_{U \times V}$ is C^∞, we thereby deduce that $A(\alpha v)_{|U}$ is in $C^\infty(U)$ which proves that x_0 does not belong to the left-hand side of (5.4.1).

6. DIFFERENTIAL OPERATORS WITH CONSTANT COEFFICIENTS

6.1 FUNDAMENTAL SOLUTION: Let $P(D) = \sum_{\text{finite}} a_\alpha D^\alpha$ be a differential operator with constant coefficients $a_\alpha \in \mathbf{C}$. We say that a distribution $E \in \mathcal{D}'(\mathbb{R}^n)$ is a fundamental solution of P(D) if

(6.1.1.) $$P(D).E = \delta_0 \quad .$$

Fundamental solutions have long been known for the more classical operators, and we tabulate these below:

	Operator P	A fundamental solution E
(6.1.2.)	$\frac{\partial}{\partial x_n}$	$1(x') \otimes Y(x_n)$
(6.1.3.)	$\frac{\partial}{\partial \bar{z}} = \frac{1}{2}(\frac{\partial}{\partial x} + i\,\frac{\partial}{\partial y})$	$\frac{1}{\pi(x+iy)}$
(6.1.4.)	$\Delta = \partial_1^2 + \ldots + \partial_n^2$	$\frac{1}{(n-2)\sigma_{n-1}} \cdot \lvert x\rvert^{2-n}$, $n \geq 3$ σ_{n-1} = area of the unit sphere in $\mathbb{R}^n$
(6.1.5.)	$\partial_t - \Delta$ (heat)	$\frac{Y(t)}{(4\pi t)^{n/2}} \exp\,(-\frac{\lvert x\rvert^2}{4t})$
(6.1.6.)	$\frac{1}{i}\,\partial t + \Delta$ (Schrödinger)	$\frac{Y(t)}{(4\pi t)^{n/2}} \exp\,(-\frac{\lvert x\rvert^2}{4it} - in\,\frac{\pi}{4})$
(6.1.7.)	$\partial_{tt}^2 - \Delta$ (waves)	$< E, \varphi > = \int \frac{\hat{\varphi}(\xi,\, \tau - i\gamma)}{\lvert\xi\rvert^2 - (\tau - i\gamma)^2}\, d\xi\, d\tau$ for $\varphi \in C_0^\infty(R^{n+1})$, γ constant > 0.

We postpone the proofs of these results until the exercises.

In the general case, we have

THEOREM 6.1: *Any non-zero differential operator* P(D) *admits a fundamental solution.*

This general theorem has been proved by EHRENPREISS and by MALGRANGE, and we give here a constructive proof which is due to HORMANDER [1].

If $P(\xi) \neq 0$ for $\xi \in \mathbb{R}^n$, it is natural to try to define E as the inverse Fourier transform of $\frac{1}{P(\xi)}$, i.e.

$$< E, \varphi > \;=\; \int_{\mathbb{R}^n} \frac{\tilde{\varphi}(\xi)}{P(\xi)}\, d\xi \;, \qquad \varphi \in C_o^\infty \quad \text{and} \quad \tilde{\varphi} = \mathcal{F}^{-1} \varphi \;.$$

In the general case, the idea is to integrate in the complex domain in order to avoid the zeros of $P(\zeta)$, $\zeta \in \mathbb{C}^n$. We utilise the following:

LEMMA 6.2: *If we have* $P(D) \neq 0$, *there exists a sequence* $\theta_k \in \mathbb{R}^n$ *with* $|\theta_k| \leq 1$, *a partition of unity* (ρ_k) *with* $\rho_k \in C^\infty(\mathbb{R}^n)$, $k \in \mathbb{N}$ *and a constant* $c > 0$ *such that*

$$(6.2.1.) \qquad |P(\xi + z\,\theta_k)| \geq c$$

$$\text{for} \quad \xi \in \operatorname{supp} \rho_k \,, \; z \in \mathbb{C} \;\text{ and }\; |z| = 1 \,.$$

PROOF: Letting $\mathcal{P}$ be the finite-dimensional vector space on $\mathbb{C}$ of the polynomials Q in n variables of degree $\leq m$, we equip it with the norm $N(Q) = (\sum_\alpha |Q^{(\alpha)}(0)|^2)^{1/2}$.

We can show without too much difficulty that the function

$$r(Q) = \sup_{\substack{|\theta| \leq 1 \\ \theta \in \mathbb{R}^n}} \left(\inf_{\substack{|z| = 1 \\ z \in \mathbb{C}}} Q(z\theta) \right)$$

is also a norm on $\mathcal{P}$; consequently, there exists a constant $C_1 > 0$ such that

$$r(Q) \geq C_1 N(Q) \quad , \quad Q \in \mathcal{P} \quad .$$

For Q fixed, the definition of r(Q) shows that there exists some $\theta \in \mathbb{R}^n$, $|\theta| \leq 1$, such that

$$(6.2.2.) \qquad Q(z\theta) \geq C_1 N(Q) \quad , \quad z \in \mathbb{C} \quad , \quad |z| = 1 \quad .$$

We thus deduce that there exists some neighbourhood of Q in which (6.2.2) remains true with the constant $C_2 = \frac{1}{2} C_1$.

For all $\xi \in \mathbb{R}^n$, we define the translated polynomial $P_\xi(\zeta) = P(\xi + \zeta)$, and the preceding discussion shows that there exists a θ_ξ and a neighbourhood V_ξ of ξ such that we have

$$(6.2.3.) \qquad |P(\eta + z\,\theta_\xi)| \geq C_2 N(P_\eta) \quad , \quad |z| = 1 \text{ and } \eta \in V_\xi .$$

From the covering of $\mathbb{R}^n$ by the V_ξ, we can, by making use of the compactness of the spheres $|\xi| = \text{const}$, extract a countable subcovering V_{ξ_k} which is locally finite. We let (ρ_k) be a partition of unity subordinate to this covering and we put $\theta_k = \theta_{\xi_k}$. Since the terms of Q which are of maximal degree are invariant with respect to translation, there exists some $C > 0$ such that

$$C_2 \, N(P_\xi) \geq C \qquad \text{for all} \qquad \xi \in \mathbb{R}^n ,$$

which when combined with (6.2.3) gives (6.2.1).

We define E by putting, for $\varphi \in C_0^\infty$:

$$\text{(6.2.4.)} \quad \langle E, \varphi \rangle = \sum_{k \in \mathbb{N}} \int_{\mathbb{R}^n} \rho_k(\xi) \Big(\frac{1}{2i\pi} \int_{|z|=1} \frac{\tilde{\varphi}(\xi + z\,\theta_k)}{P(\xi + z\,\theta_k)} \frac{dz}{z} \Big) \, d\xi .$$

It can easily be shown that E is a distribution by using (6.2.1) and the immediate inequality

$$\sup_{\substack{|z| = 1 \\ k \in \mathbb{N}}} |\tilde{\varphi}(\xi + z\,\theta_k)| \leq C \, p_{n+1}(\varphi) \, (1 + |\xi|)^{-n-1} , \quad \xi \in \mathbb{R}^n, \ \varphi \in \mathcal{D}_K .$$

Finally, Cauchy's formula shows

$$\langle P(D)E, \varphi \rangle =$$

$$\sum_k \int \rho_k(\xi) \left(\frac{1}{2\pi i} \int_{|z|=1} \frac{(\widetilde{P(-D)\varphi})(\xi + z\,\theta_k) dz}{P(\xi + z\,\theta_k)} \right) d\xi$$

$$= \sum \int \rho_k(\xi) \left(\frac{1}{2\pi i} \int_{|z|=1} \tilde{\varphi}(\xi + z\,\theta_k) \frac{dz}{z} \right) d\xi$$

$$= \sum_k \int \theta_k(\xi) \, \tilde{\varphi}(\xi) \, d\xi = \varphi(0) .$$

The existence of a fundamental solution E shows that any operator $P(D) \neq 0$ with constant coefficients is locally solvable in distribution space i.e. for any $f \in \mathcal{E}'$,the equation $P(D)u = f$ admits at least one solution $u = E * f \in \mathcal{D}'$. We can also investigate the global solvability of P(D), i.e. the surjectivity of the linear mappings

$$C^\infty(X) \ni u \longrightarrow P(D)u \in C^\infty(X)$$

$$\mathcal{D}'(X) \ni u \longrightarrow P(D)u \in \mathcal{D}'(X)$$

depending on the open subset $X \subset \mathbb{R}^n$. For these questions, the treatises of HORMANDER [1] and TREVES [2] can be referred to.

6.3 HYPOELLIPTICITY OF OPERATORS WITH CONSTANT COEFFICIENTS: After the questions of existence, a further important question concerns the investigation of the regularity of the solutions. For example, in which cases do the solutions of $P(x, D)u = f$ have no singularities other than those of f? We once again have the inclusion

sing supp $P(x, D)u \subset$ sing supp u , which leads us to:

DEFINITION 6.3: *A differential operator* P(x, D) *with coefficients in* $C^\infty(X)$ *is said to be hypoelliptic in the open subset* X *if*

$$\text{sing supp } P(x, D)\, u = \text{sing supp } u \quad , \qquad u \in \mathcal{D}'(X) \quad . \tag{6.3.1.}$$

For operators with constant coefficients, we have:

PROPOSITION 6.4: *The following assertions are equivalent:*

i) *The operator* P(D) *is hypoelliptic in* $\mathbb{R}^n$

ii) *The operator* P(D) *admits a fundamental solution which is* C^∞ *in* $\mathbb{R}^n - \{0\}$.

iii) *The operator* P(D) *admits a parametrix* (*i.e. a distribution* $F \in \mathcal{D}'$ *such that* $P(D)F - \delta_0 \in C^\infty(\mathbb{R}^n)$) *, which is* C^∞ *in* $\mathbb{R}^n - \{0\}$.

PROOF: The implications **i ⇒ ii ⇒ iii** are trivial. We now prove that **iii ⇒ i** . Letting F be a parametrix of P(D), to within the multiplication of F by a function from $\mathcal{D}$ identical to 1 in the neighbourhood of 0, we may assume that $F \in \mathcal{E}'$; this allows us to write

$$(6.4.1.) \qquad P(D).F = \delta_0 + \theta \quad , \quad \text{where} \quad \theta \in C_0^\infty .$$

For any open subset $X \subset \mathbb{R}^n$, we need to show that if a distribution $u \in \mathcal{D}'(X)$ satisfies $P(D)u \in C^\infty(X)$ then $u \in C^\infty(X)$. For this, it is sufficient to show that for any open subset X_1 such that $\overline{X}_1 \subset\subset X$, we have $u \in C^\infty(X_1)$. Suppose we have $\alpha \in C_0^\infty(X)$ equal to 1 in the neighbourhood of $\overline{X}_1$; we put

$$(6.4.2.) \qquad g = P(\alpha u).$$

We have $g \in C^\infty(X_1) \cap \mathcal{E}'(\mathbb{R}^n)$ since $\alpha u = u$ on the open subset X_1. By convolution of equation (6.4.2) with F, we obtain

$$(6.4.3.) \qquad \alpha u = F * g - \theta * (\alpha u) .$$

The kernel distribution F(x - y) of the operator of convolution by F satisfies, from (4.3.1) and iii),

$$\text{sing supp } F(x-y) \subset \text{Diag}(\mathbb{R}^n \times \mathbb{R}^n) .$$

By applying Proposition 5.4, we can thus deduce that

$$\text{sing supp } (F * g) \subset \text{sing supp } g .$$

The right-hand side of (6.4.3) is therefore C^∞ in X_1 and consequently u is also.

EXAMPLE 6.5: The formulas (6.1.3), (6.1.4), (6.1.5) show that the operators $\frac{\partial}{\partial \bar{z}}$, Δ, $\partial_t - \Delta$ are hypoelliptic.

We now give an important class of hypoelliptic operators.

DEFINITION 6.6: *A differential operator* $P(x, D)$ *with* C^∞ *coefficients in an open subset* $X \subset \mathbb{R}^n$ *is said to be elliptic in* X *if its principal symbol does not vanish on* $X \times \mathbb{R}^n - \{0\}$.

We shall prove in Chapter IV that elliptic operators are hypoelliptic; for the moment we prove in particular the following:

PROPOSITION 6.7: *If* $P(D)$ *is an elliptic operator with constant coefficients, then* $P(D)$ *is hypoelliptic in* $\mathbb{R}^n$.

PROOF: We shall now construct a parametrix F of P(D) which will be C^∞ in $\mathbb{R}^n - \{0\}$. If m is the degree of P(D), the assumption of ellipticity is equivalent to

$$\inf_{\substack{\xi \in R^n \\ |\xi| = 1}} |P_m(\xi)| = c > 0 .$$

By homogeneity, we deduce from this that $|P_m(\xi)| \geq c\,|\xi|^m$, and by choosing R sufficiently large, we have the lower bound

(6.7.1.) $$|P(\xi)| \geq \frac{c}{2}\,|\xi|^m$$

for $\xi \in R^n$, $|\xi| \geq R$.

If we have $\theta \in C^\infty$, identical to 1 for $|\xi| \geq R+1$ and zero for $|\xi| \leq R$, the lower bound (6.7.1) allows us to define

$$F = \mathcal{F}^{-1}\left(\frac{\theta(\xi)}{P(\xi)}\right) \in \mathcal{S}' .$$

This is a parametrix of P(D), since we have

$$P(D)F - \delta_0 = \mathcal{F}^{-1}(\theta(\xi) - 1) \in \mathcal{S} .$$

The regularity of F away from the origin will follow from

LEMMA 6.8: *For any multi-index* α *we have*

$$x^\alpha F \in C^q(R^n) \quad \text{with} \quad q = m + |\alpha| - n - 1 .$$

PROOF OF THE LEMMA: The equality

(6.8.1.) $$x^\alpha F = \mathcal{F}^{-1}\left(D^\alpha_\xi \cdot \left(\frac{\theta(\xi)}{P(\xi)}\right)\right)$$

shows that in the calculation of $D^\alpha_\xi(\frac{\theta}{P})$, the terms which involve the derivatives of the function θ provide a C^∞ contribution in $x^\alpha F$. It is thus sufficient to consider only the differentiations which relate to $\frac{1}{p}$. We can show by recurrence on $|\alpha|$ that for $|\xi| \geq R$ we have

$D^{\alpha}_{\xi}\left(\frac{1}{P(\xi)}\right)$ = a linear combination of terms of the form

$$(6.8.2.) \qquad \frac{(\partial^{\beta_1} P).(\partial^{\beta_2} P) \ldots (\partial^{\beta_j} P)}{P^{j+1}}$$

with $|\beta_1| + \ldots + |\beta_j| = |\alpha|$.

Furthermore, for any multi-index β we have

$$|\partial^{\beta} P(\xi)| = O(|\xi|^{m-|\beta|}) , \quad \text{for} \quad |\xi| \geq R \quad ;$$

consequently, taking into account (6.7.1) each term on the right-hand side of (6.8.2) is bounded above by $O(|\xi|^{-m-|\alpha|})$ for $|\xi| \geq R$. These upper bounds imply that the inverse Fourier transforms are certainly in $C^q(\mathbb{R}^n)$ because $|\xi|^{q-m-|\alpha|}$ is integrable for $|\xi| \geq R$.

This lemma shows that, for any integer p ,

$|x|^{2p} F \in C^q(R^n)$ with $q = m + 2p - n - 1$; hence $F \in C^{\infty}(R^n - \{ 0 \})$.

For operators with constant coefficients, HORMANDER has given a characterisation of the hypoellipticity of $P(D)$ in terms of the polynomial $P(\xi)$ which is expressed in the following theorem:

THEOREM 6.9: *The following assertions are equivalent:*

i) *The operator* P(D) *is hypoelliptic*

ii) *If we have* $N = \{ \zeta \in C^n \mid P(\zeta) = 0 \}$ *, then for*

(6.9.1.) $\zeta \in N$ we have $|\zeta| \to +\infty$ when $|\mathrm{Re}\,\zeta| \to +\infty$.

iii) *There exists* $\delta > 0$ *and* $C > 0$ *such that for any multi-index* α

(6.9.2.) $$\left| \frac{P^{(\alpha)}(\xi)}{P(\xi)} \right| \leq C\,|\xi|^{-\delta|\alpha|}$$

for $\xi \in R^n$, $|\xi| \geq C$.

PROOF OF i) $\Rightarrow$ ii): If $F \in \mathcal{E}'$ is a parametrix of P(D), and we put

$$P(D)\,F = \delta_0 + \theta \quad , \quad \text{where} \quad \theta \in \mathcal{D} \quad ,$$

then by Fourier transformation this equality becomes

(6.9.3.) $$P(\zeta).\hat{F}(\zeta) = 1 + \hat{\theta}(\zeta) \qquad \text{for} \qquad \zeta \in C^n .$$

From the Paley-Wiener theorem, there exists C such that

(6.9.4.) $$|\hat{\theta}(\zeta)| \leq \frac{C\,e^{A|\mathrm{Im}\,\zeta|}}{(1 + |\zeta|^2)} \qquad \text{for} \qquad \zeta \in C^n$$

where

$$A = \sup_{x \in \mathrm{supp}\,\theta} |x| \quad .$$

When $\zeta \in N$, (6.9.3) can be written

$$1 = -\hat{\theta}(\zeta)$$

which, having regard to (6.9.4), implies

$$(1 + |\zeta|^2) \leq C\, e^{A|\mathrm{Im}\ \zeta|} \qquad \text{for} \qquad \zeta \in N$$

and proves (6.9.1).

The proof of ii) $\Rightarrow$ iii) is based on purely algebraic properties for which we refer the reader to HORMANDER [1] (Chapter IV).

Finally, the implication iii) $\Rightarrow$ i) is proved by constructing a parametrix F of P(D) in the form $F = \mathcal{F}^{-1}\left(\frac{\theta(\xi)}{P(\xi)}\right)$ and by showing that this is C^∞ in $\mathbb{R}^n - \{0\}$, proceeding in an entirely analogous manner to that of the proof of Proposition 6.7; this is possible due to the upper bounds (6.9.2).

In the case of operators P(D) which are not hypoelliptic, we make use of quite general theorems to describe the singular support of u from that of Pu. We refer to the monograph of HORMANDER [6] for these questions.

7. OPERATORS AND DISTRIBUTIONS ON A MANIFOLD

7.1 CHANGE OF VARIABLES IN DISTRIBUTIONS: Let $\chi : X \to Y$ be a diffeomorphism between two open subsets of $\mathbb{R}^n$. For $v \in L^1_{loc}(Y)$, the function $u = v \circ \chi = \chi^* v \in L^1_{loc}(X)$, considered as a distribution, satisfies for $\varphi \in C_0^\infty(X)$:

$$(7.1.1) \quad \langle u,\varphi \rangle = \int u(\chi^{-1}(y))\, \varphi(\chi^{-1}(y))\, |J(\chi^{-1})(y)|\,dy$$

$$= \langle v, \quad \varphi \circ \chi^{-1} . |J(\chi^{-1})| \rangle$$

where $J(\chi^{-1}) = (\frac{Dx}{Dy})$ = the Jacobian determinant of the change of variables $y \to x = \chi^{-1}(y)$.

If now $v \in \mathcal{D}'(Y)$, the formula (7.1.1) defines a distribution $u \in \mathcal{D}'(X)$ denoted by $\chi^* v$ and called the inverse image of v under the diffeomorphism χ. If χ_1 is a diffeomorphism of Y onto an open subset Z of $\mathbb{R}^n$ and if $w \in \mathcal{D}'(Z)$, we can immediately show that $(\chi_1 \circ \chi)^* w = \chi^*(\chi_1^* w)$. We define the operation of direct image χ_* under χ as being the inverse image under χ^{-1}.

REMARK 7.1: The presence of the factor $|J|$ alongside $\varphi \circ \chi^{-1}$ in (7.1.1) shows that it is necessary to consider the test functions φ not as functions but as C^∞ densities of measure $\varphi(x)\,dx$. This remark will be fundamental to the definition of distributions on a manifold.

7.2 DIRAC MEASURE ON A HYPERSURFACE: If S is a C^∞ hypersurface of an open subset X of $\mathbb{R}^n$, the Euclidian structure of $\mathbb{R}^n$ induces on S a Riemannian structure, and we denote by $d\sigma_S$ the associated volume element on S.

We describe as the Dirac delta function on S the distribution $\delta_S \in \mathcal{D}'(X)$ defined by

$$< \delta_S, \varphi > = \int_S \varphi(x)\, d\sigma_S(x) \quad , \quad \varphi \in \mathcal{D}(X) \quad .$$

More generally, let $f : X \to \mathbb{R}$ be a C^∞ mapping such that $\operatorname{grad} f(x) = (\frac{\partial f}{\partial x_1}(x), \ldots, \frac{\partial f}{\partial x_n}(x)) \neq 0$ on the set $S = \{ x \in X \mid f(x) = 0 \}$. Then S is a C^∞ hypersurface of X and we define the distribution $\delta(f) \in \mathcal{D}'(X)$ on S by putting

$$< \delta(f), \varphi > = \int_S \frac{\varphi(x)}{|\operatorname{grad} f(x)|}\, d\sigma_S(x) \quad , \quad \varphi \in \mathcal{D}(X).$$

For example, if $f(x) = x_n$, we once again have the definition of the distribution $\delta(x_n = 0)$.

If χ is a diffeomorphism $Y \to X$ such that $(\chi^* f)(y) = y_n$, then it can easily be shown that a distribution $\chi^*(\delta(f))$ is precisely the distribution $\delta(y_n = 0)$. It is to obtain this transformation law that we need to introduce the factor $|\operatorname{grad} f|$ into the definition of $\delta(f)$. For a thorough study of this type of distribution, we refer to GELFAND and SHILOV [1].

7.3 CHANGE OF VARIABLES IN DIFFERENTIAL OPERATORS: Let X be an open subset of $\mathbb{R}^n$ and $P(x, D) = \sum_{|\alpha| \leq m} a_\alpha(x) D^\alpha$ be a differential operator of degree m with coefficients in $C^\infty(X)$. The following is a characterisation of these operators:

PROPOSITION 7.3 *Let* P *be a continuous linear mapping from* $C_0^\infty(X)$ *into* $C^\infty(X)$. *Then* P *is a differential operator of degree* m *if and only if for all real-valued functions* $\varphi \in C^\infty(X)$ *and any function* $a \in C_0^\infty(X)$ *the expression*

$$\mathbb{R} \ni \tau \longrightarrow e^{-i\tau\varphi} P(a\, e^{i\tau\varphi})(x) \tag{7.3.1.}$$

is a polynomial of degree $\leq m$ *with respect to* τ *for every* x in X.

PROOF: The necessity is obvious. Conversely, let us take $\varphi(x) = x.\xi$ with $\xi \in \mathbb{R}^n$ and put

$$e^{-i\tau\, x.\xi} P(a\, e^{i\tau x\xi}) = \sum_{k=0}^{m} \sigma_k(a, x, \xi)\, \tau^k \quad . \tag{7.3.2.}$$

The coefficient $\sigma_k(a, x, \xi)$ of τ^k is a C^∞ function of x and ξ since the same is also true of the left-hand side. Moreover, we can show (by replacing ξ by $\frac{\xi}{t}$ and τ by $t\tau$ with $t \in \mathbb{R} - \{ 0 \}$) that σ_k is a homogeneous function of degree k with respect to ξ; consequently, it is a homogeneous polynomial of degree k with respect to ξ since it is C^∞ for $\xi \in \mathbb{R}^n$. If we let K be a compact subset of X and we have $a \in \mathcal{D}(X)$ identical to 1 in the neighbourhood of K, for $u \in \mathcal{D}_K(X)$, we obtain by inverse Fourier transformation

$$u(x) = \int e^{ix\xi}\, \hat{u}(\xi)\, đ\xi = \int a(x)\, e^{ix\xi}\, \hat{u}(\xi)\, đ\xi \quad .$$

By considering this latter integral as an integral with respect to ξ with values in $C_0^\infty(X)$, the continuity of P implies that

$$(Pu)(x) = \int e^{ix\xi} . e^{-ix\xi} P(a\, e^{ix\xi})\, \hat{u}(\xi)\, đ\xi$$

and by virtue of (7.3.2), we obtain

$$(Pu)(x) = \sum_{k=0}^{n} \int e^{ix\xi} \sigma_k(a, x, \xi)\, \hat{u}(\xi)\, đ\xi ,$$

so that

$$(7.3.3.) \qquad (Pu)(x) = \sum_{k=0}^{n} \sigma_k(a, x, D_x)u \quad , \text{ for } u \in \mathcal{D}_K .$$

Therefore, P is given by a differential operator of degree m when it acts on functions from $\mathcal{D}_K$. By considering an exhaustive sequence of compact subsets K_j and associated functions a_j, the equality (7.3.3) shows that the differential operators

$$\sum_{k=0}^{m} \sigma_k(a_j, x, D_x)$$ piece together into a differential operator of degree m with coefficients in $C^\infty(X)$; this concludes the proof of the proposition.

The expression (7.3.1) provides a new means of calculating the principal symbol of P. In fact, Leibniz' formula shows that the coefficient of τ_m in (7.3.1) is given by $P_m(x, d\varphi(x))\, a(x)$. In particular, if we have $x_o \in X$, and if we put $d\varphi(x_o) = \xi_o$ and assume that $a(x_o) = 1$, then the coefficient of τ^m is given by $P_m(x_o, \xi_o)$.

If $\chi : X \to Y$ is a diffeomorphism between two open subsets of $\mathbb{R}^n$, and P is a differential operator of degree m on X, we define the transported operator $\chi_* P$ of P by χ:

$$v \in C^\infty(Y) \longrightarrow (\chi_* P)\, v(y) = P(v \circ \chi)(\chi^{-1}(y)) ,$$

i.e.

$$\chi_* P = \chi_* \circ P \circ \chi^* \qquad (7.3.4.)$$

Proposition 7.3 shows immediately that $\chi_* P$ is a differential operator of degree m on Y. In order to calculate its principal symbol at the point $(y_o, \eta_o) \in Y \times \mathbb{R}^n$, we calculate the coefficient of τ^m in the polynomial

$$e^{-i\tau \psi(y_o)}(\chi_* P)(e^{i\tau\psi})(y_o) = e^{-i\tau \psi(y_o)} P(e^{i\tau \psi \circ \chi})(\chi^{-1}(y_o))$$

where ψ is such that $d\psi(y_o) = \eta_o$. We obtain

$$(\chi_* P)_m (y_o, \eta_o) = P_m(x_o, {}^t\chi'(x_o)\, \eta_o) \qquad (7.3.5.)$$

where $x_o = \chi^{-1}(y_o)$.

We now give a brief outline of the generalisation of the preceding discussion to the case of systems of differential operators. A differential system (q,r) of degree m consists of an operator of the form

$$P(x, D_x) = \sum_{|\alpha| \leq m} a_\alpha(x) D^\alpha$$

where the coefficients are matrices $a_\alpha(x) \in \mathcal{L}(\mathbb{C}^q, \mathbb{C}^r)$ with C^∞ dependence on x in X. It is clear that such an operator defines a continuous linear mapping from $C_o^\infty(X; \mathbb{C}^q)$ into $C^\infty(X; \mathbb{C}^r)$. Proposition 7.3 generalises immediately to this case, it being sufficient to take the function a in $C_o^\infty(X; \mathbb{C}^q)$ and to demand that the expression (7.3.1) be a polynomial in τ with coefficients in $\mathbb{C}^r$.

We describe as the principal symbol of P at a point

$(x, \xi) \in X \times \mathbb{R}^n$ the linear mapping

$$P_m(x, \xi) = \sum_{|\alpha| = m} a_\alpha(x)\, \xi^\alpha \in \mathcal{L}(C^q, C^r) .$$

The coefficient of τ^m in (7.3.1) is equal to $P_m(x, d\varphi(x)).a(x)$, which provides a new definition of $P_m(x, \xi)$ and allows us to prove that (7.3.5) remains true.

7.4 DISTRIBUTION ON A MANIFOLD: Let X be a manifold of dimension n (we shall assume in this book that a manifold is C^∞ and countable at infinity).

We equip $C^\infty(X)$ with the topology defined by the family of semi-norms $\varphi \to p(\chi_* \varphi)$, where χ describes the family of local charts $\chi : X \supset U \to \tilde{U} \subset \mathbb{R}^n$ of X, and p describes the family of semi-norms defined on $C^\infty(\tilde{U})$. By using a partition of unity, we can show that it actually suffices to consider only the semi-norms associated with an atlas of X and we can show that $C^\infty(X)$ is a Fréchet space.

With an obvious notation, we define as in section 1 the topological vector spaces $C^\infty_K(X) = \mathcal{D}_K(X)$, $C^\infty_o(X) = \mathcal{D}(X)$. If $\mathcal{F}$ is a complex vector bundle of rank N on X, we denote by $C^\infty(X; \mathcal{F})$ the space of C^∞ sections of $\mathcal{F}$ and we equip it with a topology as above by reducing to $C^\infty(\tilde{U}; C^N) = (C^\infty(\tilde{U}))^N$ by means of local trivialisations of $\mathcal{F}$.

We define similarly the spaces $\mathcal{D}_K(X; \mathcal{F})$ and $\mathcal{D}(X; \mathcal{F}) = C_0^\infty(X; \mathcal{F})$.

In order to define distributions on a manifold X, we shall need first to define the bundle of densities on X.

A density ω on a vector space E of dimension n is a mapping from $\Lambda^n E \setminus 0$ into $\mathbb{C}$ such that

(7.4.1.) $\omega(\lambda w) = |\lambda|\, \omega(w)$, for $\lambda \in \mathbb{R} \setminus 0$ and $w \in \Lambda^n E \setminus 0$.

In particular, if $v_1, \ldots, v_n$ is a basis for E and if $A \in GL(E)$, we have

(7.4.2.) $\omega(Av_1, \wedge \ldots \wedge Av_n) = |\det A|\, \omega(v_1 \wedge \ldots \wedge v_n)$.

More generally, if we have $\alpha \in \mathbb{R}$, we say that ω is a density of order α on E if (7.4.1) is satisfied when we replace $|\lambda|$ by $|\lambda|^\alpha$; we then need to use $|\det A|^\alpha$ in (7.4.2.).

The set of densities of order α on E constitutes a complex vector space of dimension 1 which we denote by $|E|_\alpha$; for example $|R^n|_1 = \{ a\, dx_1 \ldots dx_n \mid a \in \mathbb{C} \}$. By multiplication, we define the bilinear mapping

(7.4.3.) $|E|_\alpha \times |E|_\beta \ni (\omega_1, \omega_2) \longmapsto \omega_1 . \omega_2 \in |E|_{\alpha+\beta}$,

and this shows that $|E|_\alpha \otimes |E|_\beta \simeq |E|_{\alpha+\beta}$ which allows us to identify the dual of $|E|_\alpha$ with $|E|_{-\alpha}$ since $|E|_0 \simeq \mathbb{C}$.

If X is a manifold of dimension n, the family of vector spaces $|T_x X|_\alpha$ above X arranges itself naturally into a vector bundle of rank 1, termed bundle of densities of order α on X and written $\Omega_\alpha(X)$. In fact, if $\chi_1 : X \supset U_1 \longrightarrow \tilde{U}_1 \subset \mathbb{R}^n$ is a chart of X, it defines a chart $\Lambda\chi_1 : \Lambda^n TU_1 \longrightarrow \Lambda^n T\tilde{U}_1$ and hence a bijection $|\chi_1|_\alpha$ of $\Omega_\alpha(U_1)$ onto $\Omega_\alpha(\tilde{U}_1) \simeq \tilde{U}_1 \times \mathbb{C}$. If $\chi_2 : X \supset U_2 \longrightarrow \tilde{U}_2 \subset \mathbb{R}^n$ is another chart of X, formula (7.4.2) (with $|\det A|^\alpha$) shows that, in the common open subset, the coordinate transition function $|\chi_2|_\alpha \circ (|\chi_1|_\alpha)^{-1}$ constitutes multiplication by $|\det(\chi_1 \circ \chi_2^{-1})'|^\alpha$.

Let ρ be a density of order α on X, i.e. a C^∞ section of $\Omega_\alpha(X)$; its restriction $\rho_{|U_j}$ is transported by $|\chi_j|_\alpha$ into a density of order α on $\tilde{U}_j$; we denote this by ρ_j and we put

$$\rho_j(\tilde{x}_j) = a_j(\tilde{x}_j)\ |d\tilde{x}_j|^\alpha \qquad (j = 1, 2)$$

with $a_j(.) \in C^\infty(\tilde{U}_j)$. The transformation law of the bundle $\Omega_\alpha(\chi)$ implies that in the open subset $U_1 \cap U_2$, we have

$$a_2(\tilde{x}_2) = a_1(\tilde{x}_1)\ .\ |\det(\chi_1 \circ \chi_2^{-1})'|^\alpha \quad . \tag{7.4.4.}$$

Formula (7.4.4) shows that we can speak of real-valued densities and of strictly positive densities. To lessen the notational burden, we put

$$C^\infty(X;\Omega_\alpha(X)) = C^\infty(|X|_\alpha) \quad , \quad C^\infty(|X|_1) = C^\infty(|X|) \quad ,$$
$$C^\infty(|X|_0) = C^\infty(X) \quad .$$

Since the densities of order 1 are measures on X, we can integrate them; having regard to (7.4.3.), this allows us to define the bilinear mapping

$$(7.4.5.)\quad C^\infty(|X|_\alpha) \times C_0^\infty(|X|_\beta) \ni (\omega_1, \omega_2) \longmapsto \int_X \omega_1 . \omega_2 \in \mathbb{C},$$

when $\alpha + \beta = 1$.

The particular case $\alpha = 0$, $\beta = 1$ shows that $C^\infty(X)$ functions define continuous linear forms on $C_0^\infty(|X|)$ which leads us, in accordance with Remark 7.1, to state:

DEFINITION 7.5: *A distribution on the manifold* X *is a continuous linear form on the space* $C_0^\infty(|X|)$ *of densities of order* 1 *on* X. *The space of distributions is denoted by* $\mathcal{D}'(X)$. *More generally, if* $\mathcal{F}$ *is a complex vector bundle over* X, *we describe as the space of distributional sections of* $\mathcal{F}$ *the space of continuous linear forms on* $C_0^\infty(X;\, \Omega_1(X) \otimes \mathcal{F}^*)$ *and we denote it by* $\mathcal{D}'(X;\, \mathcal{F})$. ($\mathcal{F}^*$ *denotes the dual bundle of* $\mathcal{F}$).

We have the injection of $C^\infty(X;\, \mathcal{F})$ in $\mathcal{D}'(X;\, \mathcal{F})$ defined by the pairing

$$< f, \varphi > \;=\; \int_X < f(x), \varphi(x) >_{\mathcal{F},\, \mathcal{F}^*} .$$

Taking into account (7.4.3) and (7.4.5), we can show that the dual of the space $C_0^\infty(X)$ is identified with the space of distributional sections of $\Omega_1(X)$. When X is equipped with a strictly positive density μ (we can always construct one of these by using a partition of unity for X) we are able to identify these spaces, since we can use the topological vector isomorphism

$$(7.5.1.) \qquad C^\infty(X) \ni f \longrightarrow f.\mu \in C^\infty(|X|) \quad .$$

We now give a brief outline of the principal properties of distributions on a manifold, limiting ourselves, for simplicity, to the case $\mathcal{F} = \mathbf{C}$. We have for the pair $\mathcal{D}(X)$, $\mathcal{D}'(X)$ the same topological properties as those mentioned in Section 1, for the case where X is an open subset of $\mathbb{R}^n$.

We have the continuous injection of $L^1_{loc}(X)$ into $\mathcal{D}'(X)$. If we let V be an open subset of X, a distribution $u \in \mathcal{D}'(X)$ defines by restriction to $C_0^\infty(|V|)$ a distribution $u_{|V} \in \mathcal{D}'(V)$. The same proof as was used in the case of an open subset of $\mathbb{R}^n$ shows that the spaces $\mathcal{D}'(V)$ form a sheaf on X. The concepts of support and singular support generalise in the same way to the case of a manifold: the space $\mathcal{E}'(X)$ of distributions with compact support is identified with the dual of $C^\infty(|X|)$.

We can also give a definition of distributions with the aid of local charts. Letting $\chi_\alpha : X \supset U_\alpha \longrightarrow \tilde{U}_\alpha \subset \mathbf{R}^n$ be a chart of X, then if $u \in \mathcal{D}'(X)$ the diffeomorphism χ_α transports it into a continuous linear form $\chi_\alpha^* \quad (u_{|U_\alpha})$ on $C_0^\infty(|\tilde{U}_\alpha|) \simeq C_0^\infty(\tilde{U}_\alpha)$, i.e. into a distribution $\tilde{u}_\alpha$ on $\tilde{U}_\alpha$. If χ_β is another chart, we have the chart transition formula for the restrictions of $\tilde{u}_\alpha$ and $\tilde{u}_\beta$ to the open subsets $\chi_\alpha(U_\alpha \cap U_\beta)$ and $\chi_\beta(U_\alpha \cap U_\beta)$

(7.5.2.) $$\tilde{u}_\beta = (\chi_\beta \circ \chi_\alpha^{-1})^* \tilde{u}_\alpha .$$

Conversely, if (χ_α) is an atlas of X, in each open subset $\tilde{U}_\alpha$ we take a distribution $\tilde{u}_\alpha \in \mathcal{D}'(\tilde{U}_\alpha)$ such that (7.5.2) is satisfied for all α,β. Then, there exists a unique distribution $u \in \mathcal{D}'(X)$ such that $\chi_\alpha^*(\tilde{u}_\alpha) = u_{\alpha|U_\alpha}$. In fact, the distributions $\tilde{u}_\alpha$ lift into distributions $u_\alpha \in \mathcal{D}'(U_\alpha)$ which piece together into a unique distribution $u \in \mathcal{D}'(X)$ by virtue of the sheaf property.

7.6 DIFFERENTIAL OPERATORS ON A MANIFOLD: If X is a manifold of dimension n, Proposition 7.3 leads us to state:

DEFINITION 7.6: *Let* P *be a continuous linear mapping from* $C_o^\infty(X)$ *into* $C^\infty(X)$; *we say that this is a differential operator of degree* m *if for all real-valued* $\varphi \in C^\infty(X)$ *and all* $a \in C_o^\infty(X)$, *the function*

(7.6.1.) $$\mathbb{R} \ni \tau \longrightarrow e^{-i\tau\varphi(x)} P(a\, e^{i\tau\varphi})(x) \in \mathbb{C}$$

is a polynomial of degree m *with respect to* τ *for all fixed* x *in* X.

This definition shows that the restriction $P_{|U}$ of P to an open subset U of X is a differential operator on U.

Let $\chi : X \supset U \longrightarrow \tilde{U} \subset \mathbb{R}^n$ be a chart of X, and P a linear mapping from $C_0^\infty(X)$ into $C^\infty(X)$; we define the linear operator P_χ from $C_0^\infty(\tilde{U})$ into $C^\infty(\tilde{U})$ by putting

$$P_\chi = \chi_* \circ (P_{|U}) \circ \chi^* .$$

With this notation we have:

PROPOSITION 7.7: *Let* P *be a continuous linear mapping from* $C_0^\infty(X)$ *into* $C^\infty(X)$. *Then,* P *is a differential operator of degree* m *if and only if for any chart* χ *from an atlas of* X *the operator* P_χ *is a differential operator of degree* m.

PROOF: The necessity follows from Proposition 7.3. The sufficiency is immediate by using a partition of unity subordinate to the covering of X by charts from the given atlas.

We can define the principal symbol of a differential operator P of degree m on X. If $\chi : U \to \tilde{U}$ is a chart of X, the diffeomorphism χ allows us to define on the cotangent space T^*U a homogeneous function P_m^U by lifting the principal symbol of $\chi_* P$. The equality (7.3.5) shows that the functions P_m^U combine together into a C^∞ function on T^*X. This function is denoted by $P_m(x, \xi)$ and is called the principal symbol of P; it is homogeneous of degree m with respect to $\xi \in T_x^* X$.

By calculating with a chart, we can show that the coefficient of τ^m in the polynomial (7.6.1) is given by

$$P_m(x,\ d\varphi(x)).\ a(x) \tag{7.7.1.}$$

The concept of a differential system generalises into the concept of a differential operator acting upon the sections of vector bundles. Suppose $\mathcal{F}$ and $\mathcal{G}$ are complex vector bundles of rank q and r above X.

DEFINITION 7.8: *A continuous linear mapping* $P : C_0^\infty(X,\ \mathcal{F}) \longrightarrow C^\infty(X,\ \mathcal{G})$ *is a* $(\mathcal{F},\ \mathcal{G})$ *differential operator of degree* m *if for all real-valued* $\varphi \in C^\infty(X)$ *and all* $a \in C_0^\infty(X;\ \mathcal{F})$ *the function*

$$\tau \longrightarrow e^{-i\tau\varphi(x)}\ P(a\ e^{i\tau\varphi})(x) \tag{7.8.1.}$$

is, for all $x \in X$, *a polynomial of degree* m *with respect to* τ *with coefficients in the fibre* $\mathcal{G}_x$ *of* $\mathcal{G}$ *above* x.

Let $\chi : U \longrightarrow \tilde{U} \subset \mathbb{R}^n$ be a chart above which the bundles $\mathcal{F}$ and $\mathcal{G}$ admit trivialisations f and g ;

$$\begin{array}{ccc} \tilde{U} \times C^q & \xleftarrow{\ f\ } & \mathcal{F}_{|U} \\ \downarrow & & \downarrow \\ \tilde{U} & \xleftarrow{\ \chi\ } & U \end{array} \qquad\qquad \begin{array}{ccc} \mathcal{G}_{|U} & \xrightarrow{\ g\ } & \tilde{U} \times C^r \\ \downarrow & & \downarrow \\ U & \xrightarrow{\ \chi\ } & \tilde{U} \end{array} .$$

Taking P to be a continuous linear mapping from $C_0^\infty(X;\ \mathcal{F})$ into $C^\infty(X,\ \mathcal{G})$, we define by transport a continuous linear mapping $P_{f,g}$ from $C_0^\infty(\tilde{U},\ C^q)$ into $C^\infty(\tilde{U};\ C^r)$ by putting:

(7.8.2.) $$P_{f,g} = g_* \circ (P_{|U}) \circ f^* \quad .$$

With this notation, Proposition 7.7 generalises immediately by replacing the operators of type P_X therein by operators of the type $P_{f,g}$ associated with a covering of X by open coordinate patches above which the bundles $\mathcal{F}$ and $\mathcal{G}$ are trivialisable.

If P is a $(\mathcal{F}, \mathcal{G})$ differential operator of degree m , then by utilising local trivialisations of $\mathcal{F}$ and $\mathcal{G}$ we can show that for x fixed, the coefficient of τ^m in (7.8.1) is a function which depends only on $d\varphi(x)$ and $a(x)$. Furthermore, this function is homogeneous of degree m with respect to the variable $d\varphi(x) = \xi$ and is a linear mapping from $\mathcal{F}_x$ into $\mathcal{G}_x$ with respect to the variable $v = a(x)$, and we denote this by $P_m(x, \xi)$; this is by definition the value of the principal symbol of P at the point $(x, \xi) \in T^* X$.

EXAMPLE 7.9: Let k be an integer contained between 0 and n-1; we consider the bundles

$\mathcal{F} = (\Lambda^k T^*X) \otimes C$, $\mathcal{G} = (\Lambda^{k+1} T^*X) \otimes C$ over X. The exterior differential d defines a continuous linear mapping from $C_0^\infty(X, \mathcal{F})$ into $C^\infty(X, \mathcal{G})$. In this case, the function (7.8.1) is given by

$$e^{-i\tau\varphi(x)} d(e^{i\tau\varphi}.a)(x) = i\tau\, d\varphi(x) \wedge a(x) + da(x) \quad ,$$

consequently d is a $(\mathcal{F}, \mathcal{G})$ differential operator of degree

1. The coefficient of τ shows that the principal symbol of d at the point $(x, \xi) \in T^* X$ is the linear mapping from $(\Lambda^k T_x^* X) \otimes C$ into $(\Lambda^{k+1} T_x^* X) \otimes C$ defined by left exterior multiplication by the one-form $i\xi$.

8. OPERATORS AND KERNEL DISTRIBUTIONS ON A MANIFOLD

Let X, Y be manifolds equipped respectively with strictly positive densities μ and ν. We now show how the definitions and the results of Sections 4 and 5 generalise to this situation.

With any distribution $K \in \mathcal{D}'(X \times Y)$, we can associate a continuous linear operator $A : C_0^\infty(Y) \longrightarrow \mathcal{D}'(X)$ by putting

$$(8.1.1.) \quad \langle A\psi, \varphi.\mu \rangle = \langle K, \varphi\mu \otimes \psi.\nu \rangle, \text{ for all } \varphi \in \mathcal{D}(X), \ \psi \in \mathcal{D}(Y).$$

Conversely, we have the kernels theorem:

THEOREM 8.1: *Any continuous linear operator* $A : C_0^\infty(Y) \longrightarrow \mathcal{D}'(X)$ *is defined by a unique kernel distribution* $K_A \in \mathcal{D}'(X \times Y)$ *such that we have* (8.1.1).

PROOF: Let (U_α) and (V_β) be atlases of X and Y. By transport with these charts, Theorem 4.4 shows that for any α, β there exists a unique distribution $K_{\alpha,\beta} \in \mathcal{D}'(U_\alpha \times V_\beta)$ such that

$$< A\psi, \varphi \circ \mu > \;=\; < K_{\alpha,\beta}, \varphi.\mu \otimes \psi\nu > , \varphi \in \mathcal{D}(U_\alpha), \quad \psi \in \mathcal{D}(V_\beta).$$

Making use of the bundle property, we deduce from this that there exists a unique distribution $K \in \mathcal{D}'(X \times Y)$ such that $K|_{U_\alpha \times V_\beta} = K_{\alpha,\beta}$ for any α,β. By partition of unity, we thereby deduce that K satisfies (8.1.1).

REMARK 8.2: It is clear from (8.1.1) that the kernel K_A of an operator $A : C_0^\infty(Y) \longrightarrow \mathcal{D}'(X)$ does not depend on the density μ chosen on X. Furthermore, if we have diffeomorphisms $\chi : X \to \tilde{X}$, $\theta : Y \to \tilde{Y}$ we define the transported operator $\tilde{A} : C_0^\infty(\tilde{Y}) \longrightarrow \mathcal{D}'(\tilde{X})$ obtained from A by $\tilde{A} = \chi_*.A.\theta^*$. We specify a positive density $\tilde{\nu}$ on $\tilde{Y}$; we can immediately show that the kernel $K_{\tilde{A}}$ of $\tilde{A}$ satisfies

$$(8.2.1.) \qquad K_{\tilde{A}} = (\tilde{K}_A) . f$$

where $\tilde{K}_A$ is obtained by transporting K_A under the diffeomorphism $(\theta, \chi) : X \times Y \longrightarrow \tilde{X} \times \tilde{Y}$ and where f is the C^∞ function on Y defined by $\theta_* \nu = f . \tilde{\nu}$.

In particular, a change of density on Y manifests itself on the kernel K through multiplication by a strictly positive C^∞ function of the second variable.

Finally we note that if the manifold Y is not endowed with a density, we still have a kernels theorem as long as we seek K in the space $\mathcal{D}'(X \times Y ; \Omega_0(X) \otimes \Omega_1(Y))$. The method of proof used for Theorem 8.1 allows us to prove the following:

THEOREM 8.3: *A continuous linear operator* $A : C_0^\infty(Y) \to \mathcal{D}'(X)$ *is regularising if and only if its kernel* K_A *is in* $C^\infty(X \times Y)$.

If A is a continuous linear operator from $C_0^\infty(Y)$ into $\mathcal{D}'(X)$, the densities μ and ν allow us to define linear operators ${}^tA$ and A^* from $C_0^\infty(X)$ into $\mathcal{D}'(Y)$, by putting

$$(8.3.1.) \quad \langle A\psi, \varphi.\mu \rangle = \langle \nu.\psi, {}^tA\varphi \rangle = \langle \nu\psi, \overline{A^* \bar{\varphi}} \rangle$$

for all $\varphi \in \mathcal{D}(X)$, $\psi \in \mathcal{D}(Y)$. It is clear that ${}^t({}^tA) = A$ and $(A^*)^* = A$.

REMARK 8.4: Proposition 4.5 generalises word for word to this case.

THEOREM 8.5: *We assume that the manifold* Y *is compact. Then in this case any regularising operator* $A : C^\infty(Y) \longrightarrow C^\infty(X)$ *is a compact operator.*

PROOF: Since A extends continuously to $\mathcal{D}'(Y)$, it extends a fortiori into a continuous operator from $C^0(Y)$ into $C^\infty(X)$, giving

the commutative diagram

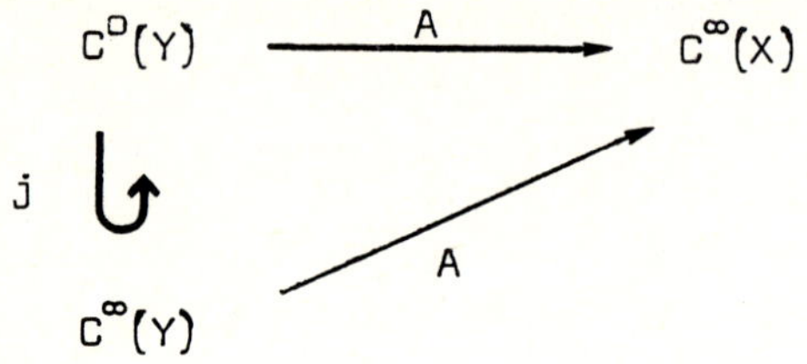

It is thus sufficient to prove the following:

LEMMA 8.6: *If* Y *is a compact manifold, then the injection* $j : C^\infty(Y) \longrightarrow C^0(Y)$ *is compact.*

PROOF: We need to show that there exists a neighbourhood $\mathcal{V}$ of 0 in $C^\infty(Y)$ which is relatively compact for the topology induced by $C^0(Y)$. Let (V_j) be a finite covering of Y by open coordinate patches. We define $\mathcal{V}$ to be the set of functions $v \in C^\infty(Y)$ which transport under each chart to functions having their first derivatives bounded by 1. By construction, $\mathcal{V}$ is a neighbourhood of 0 in $C^\infty(Y)$ composed of equicontinuous functions; thus it is relatively compact for the topology of $C^0(Y)$.

9. REGULAR OPEN SUBSETS OF $\mathbb{R}^n$ AND MANIFOLDS WITH BOUNDARY

We introduce in this section some concepts which will be useful in the study of boundary problems.

DEFINITION 9.1 : *Let* X *be an open subset of* $\mathbb{R}^n$ *with closure* $\bar{X}$ *and with boundary* ∂X. *We say that* X *is a regular open subset of* $\mathbb{R}^n$ *if* ∂X *is a hypersurface of* $\mathbb{R}^n$ *and if* X *is locally situated on only one side of* ∂X. *This means that for any* $x \in \partial X$, *there exists an open neighbourhood* U *of* x *in* $\mathbb{R}^n$ *and a diffeomorphism* $\chi : U \to \tilde{U}$ *of* U *onto an open subset* $\tilde{U}$ *of* $\mathbb{R}^n$ *such that:*

$$\chi(U \cap X) = \{ \tilde{x} \in \tilde{U} \mid \tilde{x}_n > 0 \}$$

$$\chi(U \cap \partial X) = \{ \tilde{x} \in \tilde{U} \mid \tilde{x}_n = 0 \} .$$

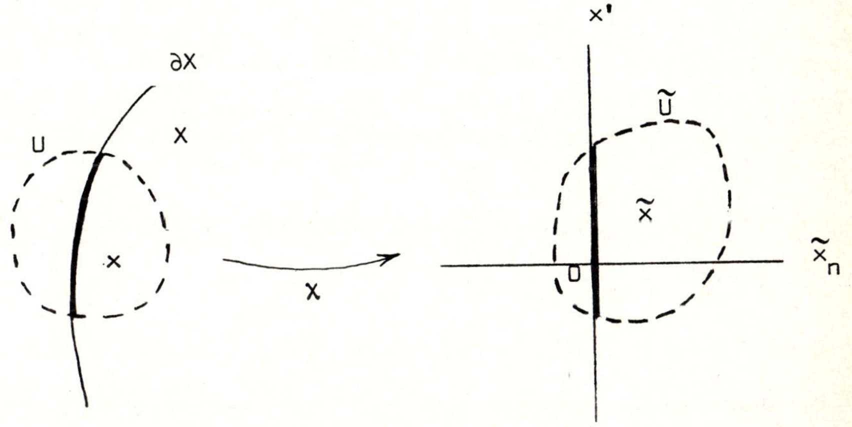

In the preceding situation we say that χ is a local chart of $\bar{X}$ at the boundary. We note that $\chi_{|U \cap \partial X}$ is a local chart of the hypersurface ∂X at x.

If X is an arbitrary open subset of $\mathbb{R}^n$, we denote by

$C^\infty(\bar{X})$ the space of functions $\varphi \in C^\infty(X)$ such that $\partial^\alpha\varphi \in C^0(\bar{X})$ for any multi-index α. Equipped with the family of semi-norms $\varphi \longmapsto p_{K,j}(\varphi) = \sup_{\substack{x \in K \\ |\alpha| \leq j}} |\partial^\alpha \varphi(x)|$, where K is an arbitrary compact subset of $\bar{X}$ and j is an arbitrary integer ≥ 0, $C^\infty(\bar{X})$ is a Fréchet space. If K is a compact subset of $\bar{X}$, we denote by $\mathcal{D}_K(\bar{X}) = C^\infty_K(\bar{X})$ the (Fréchet) subspace of $C^\infty(\bar{X})$ composed of the functions which are zero in $X \setminus K$, and by $C^\infty_0(\bar{X}) = \mathcal{D}(\bar{X})$ the inductive limit of $\mathcal{D}_K(\bar{X})$ when K varies over the family of compact subsets of $\bar{X}$. We henceforth assume that the open subset X is regular; we shall see that we can define $C^\infty(\bar{X})$ as the space of the restrictions to X of C^∞ functions in $\mathbb{R}^n$. More precisely, we have:

PROPOSITION 9.2 (SEELEY) : *There exists a continuous linear extension operator* $\rho : C^\infty(\bar{X}) \longrightarrow C^\infty(\mathbb{R}^n)$ *whose restriction to* $\mathcal{D}(\bar{X})$ *is a continuous linear extension operator (again denoted by* ρ*)* $\rho : \mathcal{D}(\bar{X}) \longrightarrow \mathcal{D}(\mathbb{R}^n)$.

PROOF: We begin with the special case $X = \mathbb{R}^n_+ = \{ x \in \mathbb{R}^n \mid x_n > 0 \}$, in which we use a SEELEY extension. More precisely, for $\varphi \in C^\infty(\overline{\mathbb{R}^n_+})$, we define $\rho(\varphi)$ by

$$\begin{cases} \rho(\varphi)(x) = \varphi(x) & \text{if } x_n \geq 0 \\ \rho(\varphi)(x) = \sum_{k=1}^{\infty} \lambda_k \, (\chi\varphi)\, (x', -2^k x_n) & \text{if } x_n < 0 , \end{cases}$$

where $\chi \in C_0^\infty(\mathbb{R})$ is equal to 1 in the neighbourhood of 0 and where (λ_k) is a sequence of real numbers such that

(9.2.1.) for all $j \in \mathbb{N}$, $\sum_{k \geq 1} \lambda_k 2^{kj}$ is an absolutely convergent series with sum $(-1)^j$.

It is immediately evident that the operator ρ so constructed matches the requirement.

In the case where X is an arbitrary regular open subset of $\mathbb{R}^n$, we consider a locally finite covering of a neighbourhood of ∂X in $\mathbb{R}^n$ by (relatively compact) open subsets U_j of local charts χ_j of $\bar{X}$ at the boundary. We put $U_0 = \bar{X} \setminus (\bigcup_j U_j)$.
Suppose we have $\varphi_0 \in C^\infty(U_0)$ and $\varphi_j \in C_0^\infty(U_j)$ such that

$\varphi_0 + \sum_{j \geq 1} \varphi_j = 1$ in the neighbourhood of $\bar{X}$.

Suppose $\psi_j \in C_0^\infty(U_j)$ is such that $\psi_j = 1$ in supp φ_j. For $\varphi \in C^\infty(\bar{X})$, we have $\varphi = \varphi_0 \varphi + \sum_j \varphi_j \varphi$.
We extend φ into $\Phi = \rho(\varphi) \in C^\infty(R^n)$ by putting

$\Phi = \Phi_0 + \sum_j \Phi_j$, where Φ_0 is obtained by extending $\varphi_0 \varphi$ by 0 outside of U_0 and where

$$\Phi_j = \psi_j \chi_j^* \left[\rho(\widetilde{\varphi_j \varphi})|_{\widetilde{U}_j} \right], \quad \text{with} \quad \widetilde{\varphi_j \varphi} = \chi_{j*} (\varphi_j \varphi)$$

It is immediately obvious that the operator ρ so constructed meets the requirement. It remains to prove the existence of a sequence λ_k satisfying (9.2.1). With $N \in \mathbb{N}$, we consider the Cramer system

$$(9.2.2.) \qquad \sum_{k=1}^{N} \lambda_{k,N}\, 2^{kj} = (-1)^j \qquad j = 0, \ldots, N-1 .$$

A calculation of Vandermonde determinants shows that

$$\lambda_{k,N} = \prod_{\substack{j=1 \\ j \neq k}}^{N} \left(\frac{1 + 2^{-j}}{1 - 2^{k-j}} \right) .$$

Since the series $\sum_{j \geq 1} 2^{-j}$ converges, there exist constants C and $c > 0$ such that

$$\prod_{j=1}^{+\infty} (1 + 2^{-j}) \leq C \quad , \qquad c \leq \prod_{j=1}^{\infty} (1 - 2^{-j})$$

and

$$\lambda_{k,N} \xrightarrow[N \to \infty]{} \lambda_k = \prod_{\substack{j=1 \\ j\neq k}}^{\infty} \left(\frac{1 + 2^{-j}}{1 - 2^{k-j}} \right) .$$

We have the upper bounds

$$|\lambda_{k,N}| \leq C \prod_{\substack{j=1 \\ j\neq k}}^{N} |1 - 2^{k-j}|^{-1} \leq \frac{C}{c} \prod_{j=1}^{k-1} (2^{k-j} - 1)^{-1}$$

and since $\dfrac{1}{2^{\ell} - 1} \leq \dfrac{1}{2^{\ell-1}}$ for $\ell \geq 1$, we thereby deduce

$$(9.2.3.) \qquad |\lambda_{k,N}| \leq \frac{C}{c}\, 2^{-\frac{1}{2}(k-1)(k-2)} .$$

Letting $N \to \infty$, we obtain

$$(9.2.4.) \qquad |\lambda_k| \leq \frac{C}{c}\, 2^{-\frac{1}{2}(k-1)(k-2)} .$$

Consequently (9.2.1) follows from (9.2.2), (9.2.3) and (9.2.4).

We now consider the restriction operator $\mathcal{D}(R^n) \longrightarrow \mathcal{D}(\overline{X})$. It is clearly continuous, and Proposition 9.2 shows that it is surjective. By transposition, we obtain an injection of the dual $\mathcal{D}'(\overline{X})$ of $\mathcal{D}(\overline{X})$ into $\mathcal{D}'(\mathbb{R}^n)$. We identify a continuous linear form ℓ on $\mathcal{D}(\overline{X})$ with its image under this injection, i.e. with the distribution u on $\mathbb{R}^n$ defined by

(9.2.5.) $$\langle u, \Phi \rangle = \ell(\Phi_{|X}) \quad \text{for all} \quad \Phi \in \mathcal{D}(\mathbb{R}^n).$$

We then have:

PROPOSITION 9.3 : *By way of the identification* (9.2.5), *the dual* $\mathcal{D}'(\overline{X})$ *of* $\mathcal{D}(\overline{X})$ *coincides with* $\mathcal{D}'_{\overline{X}}(\mathbb{R}^n)$. *This equality is algebraic and topological, both of these two spaces being equipped either with* $\mathcal{T}_s$ *or with* $\mathcal{T}_b$. *In the same manner, the dual* $\mathcal{E}'(\overline{X})$ *of* $C^\infty(\overline{X})$ *coincides with* $\mathcal{E}'_{\overline{X}}(\mathbb{R}^n)$.

PROOF: If u is defined by (9.2.5), we have $\langle u, \Phi \rangle = 0$ if $\operatorname{supp} \Phi \subset \mathbb{R}^n \setminus \overline{X}$ and thus $u \in \mathcal{D}'_{\overline{X}}(\mathbb{R}^n)$. Conversely, we have

(9.3.1.) LEMMA: *Suppose we have* $u \in \mathcal{D}'_{\overline{X}}(\mathbb{R}^n)$ *and* $\Phi \in \mathcal{D}(\mathbb{R}^n)$ *such that* $\Phi = 0$ *in* X. *Then* $\langle u, \Phi \rangle = 0$.

This lemma shows that if $u \in \mathcal{D}'_{\overline{X}}(\mathbb{R}^n)$, we define a linear form ℓ on $\mathcal{D}(\overline{X})$ by putting $\ell(\varphi) = \langle u, \Phi \rangle$ where $\Phi \in \mathcal{D}(\mathbb{R}^n)$ is an arbitrary extension of $\varphi \in \mathcal{D}(\overline{X})$.

Making use of Proposition 9.2, we have $\ell(\varphi) = \langle u, \rho(\varphi) \rangle$, and thus $\ell \in \mathcal{D}'(\overline{X})$. This proves the algebraic equality $\mathcal{D}'(\overline{X}) = \mathcal{D}'_{\overline{X}}(\mathbb{R}^n)$. The topological equalities follow from the continuity of ρ and of the restriction operator. It now remains for us to prove Lemma (9.3.1): through a partition of unity, we can as before reduce to the case $X = \mathbb{R}^n_+$. For $\epsilon > 0$, we put $\Phi_\epsilon(x) = \Phi(x', x_n + \epsilon)$; since Φ_ε is zero in the neighbourhood of $\bar{X}$, we have $< u, \Phi_\epsilon > = 0$; we thereby deduce that $< u, \Phi > = 0$ since $\Phi_\epsilon \longrightarrow \Phi$ in $\mathcal{D}(\mathbb{R}^n)$ when $\epsilon \to 0$.

REMARK 9.4 : We have the continuous injection

$$L^1_{loc}(\bar{X}) \longrightarrow \mathcal{D}'(\bar{X}) \quad \text{defined by} \quad < f, \varphi > = \int_X f(x)\,\varphi(x)\,dx$$

for all $\varphi \in \mathcal{D}(\bar{X})$. By means of (9.2.5), f is identified with the function $f^o \in L^1_{loc}(\mathbb{R}^n)$ defined by $f^o = f$ in $\bar{X}$, $f^o = 0$ in $\mathbb{R}^n \setminus \bar{X}$.

REMARK 9.5 : $\mathcal{D}(X)$ is sequentially dense in $\mathcal{D}'(\bar{X})$: by a partition of unity we reduce to the sequential density of $\mathcal{D}(X)$ in $\mathcal{E}'_{\bar{X}}(\mathbb{R}^n)$ in the case $X = \mathbb{R}^n_+$. In order to establish this density, we proceed by regularisation (see 1.1.1) by choosing ρ with support in $\mathbb{R}^n_+$.

REMARK 9.6 : When in Definition 9.1 we replace $\mathbb{R}^n$ by a manifold M, $\bar{X}$ is called a manifold with boundary imbedded in M. All the preceding definitions and properties extend to this case without difficulty.

We now generalise to this situation the relationships between kernels and operators. Let Y be a manifold and $\bar{X}$ be a manifold with boundary imbedded in M: we assume that positive densities μ and ν are specified on M and Y; we note that $\bar{X} \times Y$ is a manifold with boundary imbedded in $M \times Y$ and that, from Lemma 1.3 and Proposition 9.2, $\mathcal{D}(\bar{X}) \otimes \mathcal{D}(Y)$ is sequentially dense in $\mathcal{D}(\bar{X} \times Y)$. We have $\mathcal{D}'(\bar{X} \times Y) = \mathcal{D}'_{\bar{X} \times Y}(M \times Y)$, $\mathcal{D}'(\bar{X}) = \mathcal{D}'_{\bar{X}}(M)$ and the space $\mathcal{L}(\mathcal{D}(Y), \mathcal{D}'(\bar{X}))$ of continuous linear operators from $\mathcal{D}(Y)$ into $\mathcal{D}'(\bar{X})$ is a subspace of $\mathcal{L}(\mathcal{D}(Y), \mathcal{D}'(M))$. By virtue of Lemma (9.3.1), it can immediately be shown that the operation of passage to the kernel (see Theorem 8.1) induces a bijection of $\mathcal{L}(\mathcal{D}(Y), \mathcal{D}'(\bar{X}))$ onto $\mathcal{D}'(\bar{X} \times Y)$ and that formula (8.1.1) is true for all $\varphi \in \mathcal{D}(\bar{X})$, $\psi \in \mathcal{D}(Y)$. Now suppose $A \in \mathcal{L}(\mathcal{D}(\bar{X}), \mathcal{D}'(Y))$; if $r : \mathcal{D}(M) \longrightarrow \mathcal{D}(\bar{X})$ is the restriction operator, we have $A \circ r \in \mathcal{L}(\mathcal{D}(M), \mathcal{D}'(Y))$, and we call the kernel $K \in \mathcal{D}'(Y \times M)$ of $A \circ r$ the kernel of A; we can show as above that the operation of passage to the kernel is a bijection of $\mathcal{L}(\mathcal{D}(\bar{X}), \mathcal{D}'(Y))$ onto $\mathcal{D}'(Y \times \bar{X})$ and that it is again characterised by: $< A\varphi, \psi\nu > = < K, \psi\nu \otimes \varphi\mu >$ for all $\varphi \in \mathcal{D}(\bar{X})$, $\psi \in \mathcal{D}(Y)$.

Finally, let $\bar{X}$, $\bar{Y}$ be two manifolds with boundary imbedded in manifolds M, N equipped with positive densities μ, ν.

Suppose we have $A \in \mathcal{L}(\mathcal{D}(\overline{Y}), \mathcal{D}'(\overline{X}))$; we define the kernel K of A by considering A as a continuous linear operator from $\mathcal{D}(\overline{Y})$ into $\mathcal{D}'(M)$, or, which amounts to the same thing, by considering the kernel of the operator $A \circ r \in \mathcal{L}(\mathcal{D}(N), \mathcal{D}'(\overline{X}))$. The above shows that $K \in \mathcal{D}'_{X \times \overline{Y}}(M \times N) \cap \mathcal{D}'_{\overline{X} \times N}(M \times N) = \mathcal{D}'_{\overline{X} \times \overline{Y}}(M \times N)$. By utilising Lemma (9.3.1), we can again show that the operation of passage to the kernel is a bijection of $\mathcal{L}(\mathcal{D}(\overline{Y}), \mathcal{D}'(\overline{X}))$ onto $\mathcal{D}'_{\overline{X} \times \overline{Y}}(M \times N)$ and that it is characterised by

$$< A\psi, \varphi\mu > \; = \; < K, \tilde{\varphi}\mu \otimes \psi\nu > \quad \text{for all} \quad \varphi \in \mathcal{D}(\overline{X}), \quad \psi \in \mathcal{D}(\overline{Y}) \; ,$$

where $\tilde{\varphi} \in \mathcal{D}(M)$ is an extension of φ , or, further, by

$$< A\psi, \varphi\mu > \; = \; < K, \varphi\mu \otimes \tilde{\psi}\nu > \quad \text{for all} \quad \varphi \in \mathcal{D}(\overline{X}), \quad \psi \in \mathcal{D}(\overline{Y})$$

where $\tilde{\psi} \in \mathcal{D}(N)$ is an extension of ψ.

We do not use the notation $\mathcal{D}'(\overline{X} \times \overline{Y})$ since we shall refrain from systematically studying here the "manifold with corners" $\overline{X} \times \overline{Y}$ embedded in $M \times N$. If the kernel of the operator $A \in \mathcal{L}(\mathcal{D}(\overline{Y}), \mathcal{D}'(\overline{X}))$ is the extension by 0 outside $\overline{X} \times \overline{Y}$ of a C^∞ function in $\overline{X} \times \overline{Y}$, we say that A is "an operator with C^∞ kernel in $\overline{X} \times \overline{Y}$". In conclusion, the kernels theorem is still valid for manifolds with boundary.

Let $\mathcal{V}$, $\mathcal{W}$ be two manifolds with or without boundary; we assume as before that positive densities μ and ν are specified. A continuous linear operator $A : \mathcal{D}(\mathcal{W}) \longrightarrow C^\infty(\mathcal{V})$ is said to be regularising if it extends into a continuous linear operator from $\mathcal{E}'(\mathcal{W})$ - equipped with its strong dual topology $\mathcal{T}_b$ -into $C^\infty(\mathcal{V})$. As in Theorem 4.7, we can prove:

THEOREM 9.7: *An operator* $A : \mathcal{D}(\mathcal{W}) \longrightarrow \mathcal{D}'(\mathcal{V})$ *is regularising if and only if its kernel is* C^∞ *in* $\mathcal{V} \times \mathcal{W}$.

As in Theorem 8.5 , we can prove:

THEOREM 9.8: *Suppose that* $\mathcal{W}$ *is compact. Then any regularising operator* $A : \mathcal{D}(\mathcal{W}) \longrightarrow C^\infty(\mathcal{V})$ *is a compact operator.*

Suppose $A \in \mathcal{L}(\mathcal{D}(\mathcal{W}), \mathcal{D}'(\mathcal{V}))$; as in (8.3.1), we define the transpose ${}^tA$ and the adjoint A^* of A by

$$< A\psi, \varphi\mu > = < \nu\psi, {}^tA\varphi > = < \nu\psi, \overline{A^* \bar{\varphi}} > .$$

We can show that the strong dual of $\mathcal{D}'(\mathcal{V})$ is identified canonically with $\mathcal{D}(\mathcal{V})$; hence we have a proposition analogous to Proposition 4.5.

10. ADDITIONAL NOTES AND EXERCISES

10.1 PALEY-WIENER-SCHWARTZ THEOREM: We use the notation of Theorem 2.9.

a) Let U be the Fourier transform of a distribution $T \in \mathcal{E}'_K$. Suppose $\theta \in C^\infty(\mathbb{R})$ is such that $\theta(t) = 1$ for $t \le 1/2$, and such that $\theta(t) = 0$ for $t \ge 1$. We put $\varphi_\zeta(x) = \theta(x.\eta - I_K(\eta))\, e^{-ix.\zeta}$. Show that $U(\zeta) = \langle T(x), \varphi_\zeta(x) \rangle$, and deduce from this the upper bound (2.9.1).

b) Conversely, let U be an entire function in $\mathbb{C}^n$ satisfying (2.9.1). We put $T = \mathcal{F}^{-1}(U_{|\mathbb{R}^n})$. Let ρ_j be a regularising sequence (see (1.1.1)), with $\text{supp}\, \rho \subset \{ |x| \le 1 \}$. We put $T_j = \rho_j * T$. Using Theorem 2.6, show that $T_j \subset \{ x \mid \text{dist.}(x, K) \le \frac{1}{j} \}$. Deduce from this that supp $T \subset K$.

10.2 HOMOGENEOUS DISTRIBUTIONS: We denote by Ω one of the two open sets $\mathbb{R}^n$ or $\mathbb{R}^n \setminus \{ 0 \}$. For $T \in \mathcal{D}'(\Omega)$ and $\lambda > 0$, let $T_\lambda \in \mathcal{D}'(\Omega)$ be given by transporting T under the diffeomorphism $\Omega \ni x \longmapsto \frac{1}{\lambda} x \in \Omega$. We say that T is homogeneous of degree m (with $m \in \mathbb{C}$) in Ω if $T_\lambda = \lambda^m T$ for all $\lambda > 0$.

a) Show that if T is homogeneous of degree m, then $D^\alpha T$ is homogeneous of degree $m - |\alpha|$.

Show that $D^\alpha \delta$ is homogeneous of degree $-n - |\alpha|$ in $\mathbb{R}^n$.

b) Suppose $T \in \mathcal{S}'(\mathbb{R}^n)$ is homogeous of degree m. Show that $\hat{T}$ is homogeneous of degree $- n - m$ in $\mathbb{R}^n$.

c) Suppose we have $T \in \mathcal{D}'(\mathbb{R}^n)$ such that $T|_{\mathbb{R}^n \setminus \{0\}}$ is C^∞ and homogeneous. Show that $T \in \mathcal{S}'(\mathbb{R}^n)$ and that $\hat{T}|_{\mathbb{R}^n \setminus \{0\}}$ is C^∞ [for $\alpha \in C_0^\infty(\mathbb{R}^n)$ identical to 1 in the neighbourhood of 0, write $T = \alpha T + (1 - \alpha)T$].

d) Suppose $\mathrm{pv}\,\frac{1}{x} \in \mathcal{D}'(\mathbb{R})$ is the derivative of the function $\mathrm{Log}\,|x|$. Show that $\mathrm{pv}\,\frac{1}{x}$ is odd and homogeneous of degree $- 1$, and that $x\,\mathrm{pv}\,\frac{1}{x} = 1$. Deduce from this that the Fourier transform of $\mathrm{pv}\,\frac{1}{x}$ is the function equal to $- i\pi$ for $\xi > 0$, and to $i\pi$ for $\xi < 0$, and that

$$\mathcal{F}^{-1} Y = - \frac{1}{2i\pi} \left(\mathrm{pv}\,\frac{1}{x} - i\pi\,\delta\right). \tag{10.2.1.}$$

(*) 10.3 FOURIER TRANSFORM OF $e^{i\lambda x^2/2}$. We recall that

$$\int_{-\infty}^{+\infty} e^{-x^2/2}\,dx = \sqrt{2\pi}\,.$$

Show that the mapping $\alpha \longmapsto T_\alpha = e^{-\alpha\,x^2/2} \in \mathcal{S}'(\mathbb{R})$ is holomorphic for $\mathrm{Re}\,\alpha > 0$ and continuous for $\mathrm{Re}\,\alpha \geq 0$. Show that for $\alpha > 0$ we have $\hat{T}_\alpha = \sqrt{2\pi}\,\alpha^{-\frac{1}{2}}\,e^{-\xi^2/2\alpha}$. Show that this equality remains true for $\mathrm{Re}\,\alpha \geq 0$, $\alpha \neq 0$, by taking the main branch of $\alpha^{-\frac{1}{2}}$. Deduce from this that for $\lambda \in \mathbb{R} \setminus \{0\}$, the Fourier transform of $e^{i\lambda x^2/2}$ is

$$\sqrt{2\pi}\,|\lambda|^{\frac{1}{2}}\,e^{i\frac{\pi}{4}\,\mathrm{sgn}\,\lambda}\,e^{\xi^2/2i\lambda}$$

10.4 FUNDAMENTAL SOLUTION OF THE CAUCHY-RIEMANN OPERATOR: We consider in $\mathbb{R}^2$ the Cauchy-Riemann operator

$$\frac{\partial}{\partial \bar{z}} = \frac{1}{2}\left(\frac{\partial}{\partial x} + i \frac{\partial}{\partial y}\right)$$

a) Let Ω be a regular bounded open subset of $\mathbb{R}^2$. By applying Stokes' formula, show that

$$(10.4.1.) \qquad \int_{\partial\Omega} u\,\varphi\,dz = 2i \iint_{\Omega} \left(\frac{\partial u}{\partial \bar{z}}\varphi + u \frac{\partial \varphi}{\partial \bar{z}} \right) dx\,dy$$

for $u, \varphi \in C^1(\bar{\Omega})$, $z = x + iy$, $\partial\Omega$ being given its canonical orientation as the boundary of a compact set.

b) Suppose we have $z_0 \in \Omega$. By applying (10.4.1) for $u(z) = \frac{1}{\pi(z - z_0)}$ after having removed from Ω a disk with centre z_0 and with radius ε, then by letting ε tend to 0, show that

$$(10.4.2.) \qquad \varphi(z_0) = \frac{1}{2i\pi} \int_{\partial\Omega} \frac{\varphi(z)}{z - z_0}\,dz - \frac{1}{\pi} \iint_{\Omega} \frac{\partial \varphi}{\partial \bar{z}} \frac{1}{z - z_0}\,dx\,dy$$

c) Deduce from this that $E = \frac{1}{\pi z}$ is a fundamental solution of the operator $\frac{\partial}{\partial \bar{z}}$.

d) By observing that $\hat{E}$ is homogeneous of degree -1 (see 10.2., b), show that $\hat{E} = \frac{2}{i(\xi + i\eta)}$.

10.5 FUNDAMENTAL SOLUTION OF THE LAPLACE OPERATOR: We consider in $\mathbb{R}^n$ the Laplace operator $\Delta = \sum_{j=1}^{n} \frac{\partial^2}{\partial x_j^2}$.

a) Let Ω be a regular open subset of $\mathbb{R}^n$. Suppose we have $u, v \in C^1(\overline{\Omega})$, at least one of the two being assumed to have compact support.

By applying Stokes' formula to the (n - 1) - form

$$\omega = \sum_{j=1}^{m} \frac{\partial u}{\partial x_j} v(-1)^{j-1} dx_1 \wedge \ldots \wedge dx_{j-1} \wedge dx_{j+1} \wedge \ldots \wedge dx_n$$

show that

$$\iint_\Omega \Delta u \cdot v \, dx + \iint_\Omega \sum_{j=1}^{n} \frac{\partial u}{\partial x_j} \frac{\partial v}{\partial x_j} dx = \int_{\partial\Omega} \omega .$$

Thus, deduce that

$$\iint_\Omega \Delta u . v \, dx + \iint_\Omega \text{grad } u . \text{ grad } v \, dx = - \int_{\partial\Omega} \frac{\partial u}{\partial N} v \, d\sigma$$

where N is the interior unit normal to $\partial\Omega$ and $d\sigma$ is the Riemannian density on $\partial\Omega$ ($\mathbb{R}^n$ being equipped with its canonical Euclidian structure).

Deduce from this that

$$\text{(10.5.1.)} \quad \iint_\Omega \Delta u . v \, dx = \iint_\Omega u . \Delta v \, dx - \int_{\partial\Omega} \left(\frac{\partial u}{\partial N} v - u \frac{\partial v}{\partial N} \right) d\sigma .$$

b) Suppose that $n \geq 3$ and $\varphi \in \mathcal{D}(\mathbb{R}^n)$. By applying (10.5.1) for $\Omega = \{ x \in \mathbb{R}^n \mid |x| > \epsilon \}$, and by putting $r = |x|$, show that

$$< \Delta (|x|^{-n+2}), \varphi > = \lim_{\epsilon \to 0} \int_{r=\epsilon} \left((2-n) r^{-n+1} \varphi - r^{2-n} \frac{\partial \varphi}{\partial r} \right) d\sigma$$

$$= (2-n) \, \sigma_{n-1} \varphi(0) ,$$

where σ_{n-1} is the area of the unit sphere of $\mathbb{R}^n$.

Deduce from this that $E = \dfrac{1}{(2-n)\sigma_{n-1}} |x|^{-n+2}$ is a fundamental solution of Δ when $n \geq 3$.

c) In the case $n = 2$, proceed in a similar manner by replacing $|x|^{-n+2}$ by Log $|x|$, and show that $E = \frac{1}{2\pi} \operatorname{Log} |x|$ is a fundamental solution of Δ .

d) Show that, for $n \geq 3$, we have $\hat{E} = \dfrac{-1}{|\xi|^2}$ (use 10.2, b)

10.6 FUNDAMENTAL SOLUTION OF THE HEAT OPERATOR: We consider in $\mathbb{R}^{n+1}$ the heat operator

$$P = P(D) = \frac{\partial}{\partial t} - \sum_{j=1}^{n} \frac{\partial^2}{\partial x_j^2}$$

We put

$$x = (x_1, \ldots, x_n) \quad , \quad \xi = (\xi_1, \ldots, \xi_n) \quad ,$$

$$P(\tau, \xi) = i\tau + |\xi|^2 .$$

a) Show that the formula

$$< E, \varphi > = \int_{\mathbb{R}^{n+1}} \frac{\hat{\varphi}(-\tau, -\xi)}{P(\tau, \xi)} \, d\!\!\!/\sigma \, d\!\!\!/\xi$$

(with $\tau = \sigma + i\gamma, \ \gamma < 0$ fixed; $\varphi \in \mathbb{R}^{n+1}$)

defines a distribution $E \in \mathcal{D}'(\mathbb{R}^{n+1})$, and that this distribution does not depend on the number $\gamma < 0$ chosen.

Show that E is a fundamental solution of P, and that

$$\operatorname{supp} E \subset \{ (t, x) \in \mathbb{R}^{n+1} \mid t \geq 0 \} .$$

b) Let $a \in \mathbb{C}$ be such that $\operatorname{Re} a > 0$. Show that the Fourier transform of $t \longmapsto Y(t)\, e^{-at}$ is

$$\sigma \longmapsto \frac{1}{a + i\sigma} .$$

c) Deduce from this that

$$< E, \varphi > \;=\; \int_0^{+\infty} dt \int_{\mathbb{R}^n} e^{-t|\xi|^2} \, \varphi(t, -\hat{\xi}) \, d\!\!\!/\xi$$

and, by using 10.3, that

$$< E, \varphi > \;=\; \int_0^{+\infty} \int_{\mathbb{R}^n} \frac{1}{(4\pi\, t)^{n/2}} \, e^{-|x|^2/4t} \, \varphi(t, x) \, dt \, dx$$

i.e. that E is the (locally integrable) function

$$\frac{Y(t)}{(4\pi\, t)^{n/2}} \, e^{-|x|^2/4t} .$$

10.7 FUNDAMENTAL SOLUTION OF THE SCHRÖDINGER OPERATOR.

We consider in $\mathbb{R}^{n+1}$ the Schrödinger operator

$$P = P(D) = \frac{\partial}{\partial t} - i \sum_{j=1}^{n} \frac{\partial^2}{\partial x_j^2} .$$

We put $P(\tau, \xi) = i(\tau + |\xi|^2)$.

a) Show that the formula

$$a \quad < E, \varphi > \;=\; \int_{\mathbb{R}^{n+1}} \frac{\hat{\varphi}(-\tau, -\xi)}{P(\tau, \xi)} \, d\!\!\!/\sigma \, d\!\!\!/\xi$$

(where $\tau = \sigma - i\gamma$, $\gamma < 0$ fixed; $\varphi \in \mathcal{D}(\mathbb{R}^{n+1})$) defines a distribution $E \in \mathcal{D}'(\mathbb{R}^{n+1})$, and that this distribution does not depend on the number $\gamma > 0$

chosen. Show that E is a fundamental solution of P, and that

$$\operatorname{supp} E \subset \{(t, x) \in \mathbb{R}^{n+1} \mid t \geq 0\} .$$

b) By proceeding as in the preceding exercise, show that

$$< E, \varphi > = \int_0^{+\infty} dt \int_{\mathbb{R}^n} e^{-it|\xi|^2} \varphi(t, -\hat{\xi})\, d\!\!\!/\xi$$

and that

$$< E, \varphi > = \int_0^{+\infty} dt \int_{\mathbb{R}^n} \frac{1}{(4\pi t)^{n/2}} e^{-\left(\frac{|x|^2}{4it} + in\frac{\pi}{4}\right)} \varphi(t, x)\, dx$$

(it will be noted that the function

$$\frac{Y(t)}{(4\pi t)^{n/2}} e^{-\left(\frac{|x|^2}{4it} + in\frac{\pi}{4}\right)}$$ is only locally integrable for $n = 1$).

10.8 FUNDAMENTAL SOLUTION OF THE WAVE OPERATOR: We consider in $\mathbb{R}^{n+1}$ the wave operator

$$P = \frac{\partial^2}{\partial t^2} - \sum_{j=1}^{n} \frac{\partial^2}{\partial x_j^2} .$$

We put

$$P(\tau, \xi) = -\tau^2 + |\xi|^2 .$$

a) Show that the formula

$$< E, \varphi > = \int_{\mathbb{R}^{n+1}} \frac{\hat{\varphi}(-\tau, -\xi)}{P(\tau, \xi)}\, d\!\!\!/\sigma\, d\!\!\!/\xi$$ (with $\tau = \sigma - i\gamma$, $\gamma < 0$ fixed; $\varphi \in \mathcal{D}(\mathbb{R}^{n+1})$) defines a distribution $E \in \mathcal{D}'(\mathbb{R}^{n+1})$, and that this distribution does not depend on the number $\gamma > 0$ chosen.

Show that E is a fundamental solution of P and that

$$\operatorname{supp} E \subset \{ (t, x) \in \mathbb{R}^n \mid t \geq 0 \} .$$

b) By observing that $P(\tau, \xi) = \frac{1}{2|\xi|} \left(\frac{1}{\tau + |\xi|} - \frac{1}{\tau - |\xi|} \right)$

and by using 10.6, b, show that

$$< E, \varphi > = \int_0^{+\infty} dt \int_{\mathbb{R}^n} \frac{\sin(t|\xi|)}{|\xi|} \varphi(t, -\hat{\xi}) \, d\!\!\!/\xi$$

$$= \int_0^{+\infty} < E_t, \varphi(t, .) > dt ,$$

where $E_t = \mathcal{F}^{-1} \frac{\sin t|\xi|}{|\xi|}$.

Show that $E_t(x) = \frac{1}{t^{n-1}} E_1(\frac{x}{t})$ for $t > 0$, and therefore that

$$(10.8.1.) \quad < E, \varphi > = \int_0^{+\infty} \frac{1}{t^{n-1}} < E_1(\tfrac{\cdot}{t}), \varphi(t, .) > dt ,$$

with $E_1 = \mathcal{F}^{-1} \frac{\sin |\xi|}{|\xi|}$.

c) Show that $\hat{E}_1$ extends into the entire function

$$U(\zeta) = \sum_{p \geq 0} \frac{(-1)^p}{(2p+1)!} (\zeta_1^2 + \dots + \zeta_n^2)^p ,$$

where $\zeta = (\zeta_1, \dots, \zeta_n) \in \mathbb{C}^n$.

We put $\zeta_1^2 + \dots + \zeta_n^2 = z^2$, $z = \alpha + i\beta$ with $\alpha, \beta \in \mathbb{R}$,

$\zeta = \xi + i\eta$ with $\xi, \eta \in \mathbb{R}^n$.

Show that $|z| = |\zeta|$ and that $|\beta| \leq |\eta|$. Thus deduce that

$|U(\zeta)| \leq \frac{1}{|\zeta|} e^{|\eta|}$ and then by using 10.1, that

$\operatorname{supp} E_1 \subset \{|x| \leq 1\}$.

By using (10.8.1), infer that

(10.8.2.) $$\operatorname{Supp} E \subset \{ (t, x) \in \mathbb{R}^{n+1} \mid |x| \leq t \} .$$

Further properties of E will be seen in Exercise 10.3 of Chapter III.

10.9 INVERSE IMAGE OF A DISTRIBUTION: Let $f : \mathbb{R}^n \longrightarrow \mathbb{R}$ be a C^∞ mapping such that $f'(x) \neq 0$ when $f(x) = 0$.

Let $v \in \mathcal{D}'(\mathbb{R})$ be such that $v \in C^\infty(\mathbb{R} \setminus 0)$. We shall now define $u = f^* v \in \mathcal{D}'(\mathbb{R}^n)$, the inverse image of v under f . We define $u = v \circ f$ on the open subset $f^{-1}(\mathbb{R}^n \setminus 0)$. In the neighbourhood of a point $x_0 \in \mathbb{R}^n$ such that $f(x_0) = 0$, we consider a local diffeomorphism $x \longmapsto \tilde{x} = \chi(x)$ of $\mathbb{R}^n$ transporting $f(x)$ into $\tilde{x}_n$, and we put

$$u_0 = \chi^* (1(\tilde{x}_1, \dots, \tilde{x}_{n-1}) \otimes v(\tilde{x}_n)) .$$

Show that the distributions u, u_0 piece together into a distribution on $\mathbb{R}^n$, which we denote by $f^* v$. Show that if a sequence $v_j \in \mathcal{D}'(\mathbb{R}) \cap C^\infty(\mathbb{R} \setminus 0)$ is such that : $v_j \longrightarrow 0^\infty$ in $\mathcal{D}'(\mathbb{R})$, $v_j \longrightarrow 0$ in $C^\infty(\mathbb{R} \setminus 0)$, then $f^* v_j \longrightarrow 0$ in $\mathcal{D}'(\mathbb{R}^n)$.

Show that

(10.9.1.) $$\operatorname{supp}(f^* v) = f^{-1}(\operatorname{supp} v) .$$

In the case where v is the Dirac distribution in $\mathbb{R}$, show that $f^* v$ coincides with the distribution $\delta(f)$ defined in Section 7.2.

We shall give in Chapter VIII a more general definition of the inverse image.

CHAPTER 2.

SOBOLEV SPACES AND APPLICATIONS

1. DIRICHLET'S PRINCIPLE

Sobolev spaces are spaces of distributions corresponding to a certain degree of regularity and they have, in addition, a Hilbert space structure. They arise in a natural manner in the study of boundary-value problems for differential operators, as the simple example which follows will show. We let X be a regular bounded open subset of $\mathbb{R}^n$ and we consider the following Dirichlet problem:

Find u such that $\Delta u - u = 0$ in X,

$$u_{|\partial X} = g \text{ in } \partial X,$$

where g is given. We can for example assume $g \in C^2(\partial X)$ and begin by seeking $u \in C^2(\bar{X})$. Physical considerations lead us to consider a quantity which we call the energy of u, defined by

$$E(u) = \int_X |\text{grad } u(x)|^2 \, dx + \int_X |u(x)|^2 \, dx \quad .$$

Then, we have

DIRICHLET'S PRINCIPLE : *The above Dirichlet problem admits at most one solution. If* u *is a solution, it minimises the energy* E *over all the functions equal to* g *on the boundary.*

In fact, suppose we have $v \in C^2(\bar{X})$ such that $v_{|\partial X} = g$. Let $h = v - u$. We have, if u is a solution:

$$E(v) - E(u) = 2 \operatorname{Re} \int_X (\operatorname{grad} u(x) . \overline{\operatorname{grad} h(x)} + u(x)\, \overline{h(x)})\, dx + E(h) .$$

From Green's formula

$$\int_X \operatorname{grad} u . \overline{\operatorname{grad} h}\, dx = - \int_X \Delta u . \bar{h}\, dx + \int_{\partial X} \frac{\partial u}{\partial n} . \bar{h}\, dx$$

(where $\frac{\partial u}{\partial n}$ denotes the derivative of u along the exterior unit normal to ∂X), we see that $E(v) - E(u) = E(h)$ since $\Delta u - u = 0$ and since $h_{|\partial X} = 0$. Now $E(h)$ is always positive or zero, and is zero if and only if h is constant, i.e. zero since $h_{|\partial X} = 0$. Thus, $E(u) \leq E(v)$ and the equality is equivalent to $v = u$.

Of course, the preceding argument does not prove the existence of a solution and in fact there does not always exist a solution u in $C^2(\bar{X})$. It is therefore necessary to seek u in a space larger than $C^2(\bar{X})$. We observe that $E(u)$ is the norm associated with the prehilbert structure on $C^2(\bar{X})$ defined by the inner product

$$(u, v)_1 = \int_X (\operatorname{grad} u . \overline{\operatorname{grad} v} + u.\bar{v})\, dx .$$

The Dirichlet principle leads us to consider the completed Hilbert space, which we denote by $H^1(\bar{X})$, and which, as we shall see, coincides with the Sobolev space $H^1(\bar{X})$

$$H^1(\bar{X}) = \left\{ u \in L^2(X) \mid \partial_j u \in L^2(X) \quad \text{for} \quad 1 \le j \le n \right\}.$$

We shall also see that we can define in a natural manner the trace $v_{|\partial X}$ of v on ∂X when $v \in H^1(\bar{X})$, and that the affine variety

$$\Sigma = \left\{ v \in H^1(\bar{X}) \mid v_{|\partial X} = g \right\}$$ is closed in $H^1(\bar{X})$. Thus, by analogy with the Dirichlet principle, we can consider the orthogonal projection u of 0 onto Σ; we shall show in section 4 that u is in fact the unique element of $H^1(\bar{X})$ which is such that

$$\begin{cases} \Delta u - u = 0 & \text{in } X \\ u_{|\partial X} = g & \text{in } \partial X . \end{cases}$$

2. THE SPACES $H^s(\mathbb{R}^n)$ AND $H^s_{loc}(X)$.

DEFINITION 2.1 : *For* $s \in \mathbb{R}$, *we denote by* $H^s(\mathbb{R}^n)$ *the space of distributions* $u \in \mathcal{S}'(\mathbb{R}^n)$ *such that* $(1 + |\xi|^2)^{s/2} \hat{u} \in L^2(\mathbb{R}^n)$. *We equip* $H^s(\mathbb{R}^n)$ *with the inner product*

$$(u, v)_s = \int (1 + |\xi|^2)^s \hat{u}(\xi) \overline{\hat{v}(\xi)} \, d\!\!{}^-\xi$$

and with the associated norm $\| \ \|_s$ *defined by*

$$\|u\|_s^2 = \int (1 + |\xi|^2)^s |\hat{u}(\xi)|^2 \, d\!\!{}^-\xi .$$

We shall frequently put $H^s(\mathbb{R}^n) = H^s$. We consider the topological vector isomorphism $\Lambda^s : \mathcal{S}' \longrightarrow \mathcal{S}'$ defined by

$$\Lambda^s u = \mathcal{F}^{-1}((1 + |\xi|^2)^{s/2} \hat{u}) \quad ;$$

we have

$$(u, v)_s = \int_{\mathbb{R}^n} \Lambda^s u \, . \, \overline{\Lambda^s v} \, dx$$

from Theorem 2.5 of Chapter I.

H^s is thus the Hilbert space obtained by transporting the Hilbert space $L^2(\mathbb{R}^n)$ by the isomorphism Λ^{-s}; we have in particular $H^0 = L^2(\mathbb{R}^n)$. Let $s_1, s_2 \in \mathbb{R}$ be such that $s_1 \leq s_2$; it is immediately clear that $H^{s_2} \subset H^{s_1}$, and that the injection of H^{s_2} into H^{s_1} is continuous. We put $H^{+\infty} = \bigcap_{s \in \mathbb{R}} H^s$, $H^{-\infty} = \bigcup_{s \in \mathbb{R}} H^s$; we have $\mathcal{S} \subset H^{+\infty}$, and, from (2.8.1) of Chapter I, $\mathcal{E}' \subset H^{-\infty}$. From Theorem 2.10 of Chapter I and the upper bound $|\xi^\alpha| \leq |\xi|^{|\alpha|}$, a differentiation operator D^α is continuous from H^s into $H^{s-|\alpha|}$. We have seen that $H^0 = L^2$; more generally we have

PROPOSITION 2.2 : *For* $m \in \mathbb{N}$, H^m *is the space of distributions* $u \in \mathcal{D}'(\mathbb{R}^n)$ *such that* $D^\alpha u \in L^2(\mathbb{R}^n)$ *for any multi-index* α *satisfying* $|\alpha| \leq m$. *Moreover, the inner product in* H^m *is equivalent to the inner product*

$$(2.2.1.) \quad ((u, v))_m = \sum_{|\alpha| \leq m} \int_{\mathbb{R}^n} D^\alpha u \, . \, \overline{D^\alpha v} \, dx \quad .$$

PROOF: Suppose $u \in \mathcal{D}'$; from Theorem 2.10 of Chapter I, we have $D^\alpha u \in L^2$ if and only if $\xi^\alpha \hat{u} \in L^2$, and in this case $\int |D^\alpha u|^2\, dx = \int |\xi^\alpha|^2\, |\hat{u}(\xi)|^2\, d\!\!/\xi$. Now, it can easily be shown that there exist positive constants C_1, C_2 such that

$$C_1(1 + |\xi|^2)^m \leq \sum_{|\alpha|\leq m} |\xi^\alpha|^2 \leq C_2(1 + |\xi|^2)^m \quad ;$$

hence Proposition 2.2.

PROPOSITION 2.3 : *The injection* $\mathcal{S} \longmapsto H^s$ *is continuous;* $\mathcal{D}$ *is dense in* H^s*; the injection of* H^s *into* $\mathcal{S}'$ *is continuous. Suppose* $u \in H^s$*; for* $h \in \mathbb{R}^n$*, we denote by* $\tau_h u$ *the image of* u *under the translation* $x \longmapsto x - h$ *Then* $\tau_h u \in H^s$, $\|\tau_h u\|_s = \|u\|_s$

and

$$\tau_h u \longrightarrow u \quad \text{in} \quad H^s \quad \text{when} \quad h \longrightarrow 0 \quad .$$

In the case where $h = (0, \ldots, 0, h_j, 0, \ldots, 0)$, *the differential quotient* $\dfrac{\tau_h u - u}{h_j}$ *tends to* $\partial_j u$ *in* H^{s-1} *when* $h_j \longrightarrow 0$.

PROOF: We now show that the injection of $\mathcal{S}$ into H^s is continuous; since Λ^s is a topological vector isomorphism of $\mathcal{S}$ onto $\mathcal{S}$ and of H^s onto H^0, it is sufficient to consider the case $s = 0$.

For $\varphi \in \mathcal{S}$ we have for example

$$\|\varphi\|_0^2 = \int (1 + |x|^2)^{-n} (1 + |x|^2)^n\, |\varphi(x)|^2\, dx$$

so that

$$\|\varphi\|_0 \le C \sup_{x \in \mathbb{R}^n} ((1 + |x|^2)^{n/2} |\varphi(x)|) \quad , \quad \text{with}$$

$$C^2 = \int (1 + |x|^2)^{-n} dx \quad ,$$

which proves the continuity which we seek.

The injection of H^s into $\mathcal{S}'$ is continuous; in fact, as previously we reduce to considering the case s = 0 and we observe that for $u \in H^0$ and $\varphi \in \mathcal{S}$, we have

$$|< u, \varphi >| \le \|u\|_0 \|\varphi\|_0 \ .$$

We now prove that $\mathcal{D}$ is dense in H^s. Since $\mathcal{S}$ is dense in L^2, we see that $\Lambda^{-s}(\mathcal{S}) = \mathcal{S}$ is dense in $\Lambda^{-s}(L^2) = H^s$; the density of $\mathcal{D}$ in $\mathcal{S}$ and the continuity of the injection $\mathcal{S} \longrightarrow H^s$ then imply the density of $\mathcal{D}$ in H^s.

For $u \in H^s$ and $h \in \mathbb{R}^n$, we have $\widehat{\tau_h u}(\xi) = e^{ih\cdot\xi}\,\hat{u}(\xi)$. Thus $\tau_h u \in H^s$, $\|\tau_h u\|_s = \|u\|_s$ and

$$\|\tau_h u - u\|_s^2 = \int (1 + |\xi|^2)^s |\hat{u}(\xi)| \, |e^{ih\xi} - 1|^2 d\xi \qquad \text{tends}$$

to 0 when $h \longrightarrow 0$ from the dominated convergence theorem.

Similarly, we obtain

$$\left\| \frac{\tau_h u - u}{h_j} - \partial_j u \right\|_{s-1}^2$$

$$= \int (1 + |\xi|^2)^{s-1} |\hat{u}(\xi)|^2 \left| \frac{e^{ih_j \xi_j} - 1}{h_j} - i\xi_j \right|^2 d\xi .$$

Now $\dfrac{e^{ih_j \xi_j} - 1}{h_j} - i\xi_j \longrightarrow 0$ when $h_j \longrightarrow 0$, with

$\left|\dfrac{e^{ih_j \xi_j} - 1}{h_j} - i\xi_j\right| \leq 2\,|\xi_j|$ from the Mean Value Theorem.

Therefore the conclusion again follows from the dominated convergence theorem.

DEFINITION 2.4 : *Let* F *be a closed subset of* $\mathbb{R}^n$; *we denote by* $H^s_F(\mathbb{R}^n)$ *the subspace of* H^s *formed by the elements with support in* F.

Since the injection $H^s \longrightarrow \mathcal{D}'$ is continuous, $H^s_F(\mathbb{R}^n)$ is a closed subspace of H^s and we endow it with the induced Hilbert structure.

THEOREM 2.5 (Sobolev) : *Suppose we have* $k \in \mathbb{N}$ *and* $s \in \mathbb{R}$ *such that* $s > n/2 + k$. *Then* $H^s \subset C^k(\mathbb{R}^n)$, *with continuous injection.*

PROOF: By differentiation, we reduce to the case where $k = 0$. If $u \in H^s$, we have $\hat{u} = f\,g$, where $f(\xi) = (1 + |\xi|^2)^{s/2}\,\hat{u}(\xi)$, $g(\xi) = (1 + |\xi|^2)^{-s/2}$. We have $f \in L^2$ and $g \in L^2$ when $s > n/2$.

Schwartz's inequality then proves that $\displaystyle\int |\hat{u}(\xi)|\, d\xi \leq C_s\, \|u\|_s$,

with $C_s^2 = \displaystyle\int (1 + |\xi|^2)^{-s}\, d\xi$. Hence $u \in C^0(\mathbb{R}^n)$, with

$$\sup_{x \in \mathbb{R}^n} |u(x)| \leq \int |\hat{u}(\xi)| \, d\!\!\!/\xi \leq C_s \|u\|_s \quad .$$

THEOREM 2.6 : *The multiplication* $(\varphi, u) \longmapsto \varphi u$ *is bilinear and continuous from* $\mathcal{S} \times H^s$ *into* H^s. *We have*

$$\|\varphi u\|_s \leq \|u\|_s \, 2^{|s|/2} \int (1 + |\xi|^2)^{|s|/2} |\hat{\varphi}(\xi)| \, d\!\!\!/\xi \quad .$$

PROOF: For $\varphi \in \mathcal{S}$ and $u \in H^s$, we know that $\varphi u \in \mathcal{S}'$ and that $\widehat{\varphi u}$ is the C^∞ function $(2\pi)^{-n} \hat{\varphi} * \hat{u}$. We thus have

$$\text{(2.6.1.)} \quad (1 + |\xi|^2)^{s/2} \widehat{\varphi u}(\xi)$$
$$= \int (1 + |\xi|^2)^{s/2} (1 + |\eta|^2)^{-s/2} \hat{\varphi}(\xi - \eta)(1 + |\eta|^2)^{s/2} \hat{u}(\eta) \, d\!\!\!/\eta \, .$$

Now, we have:

(2.6.2.) LEMMA (Peetre's Inequality) :

$$(1 + |\xi|^2)^s (1 + |\eta|^2)^{-s} \leq 2^{|s|} (1 + |\xi - \eta|^2)^{|s|} \quad \text{for } s \in \mathbb{R} \, ,$$

$\xi, \eta \in \mathbb{R}^n$.

In fact, we have

$$1 + |\xi + \eta|^2 \leq 1 + (|\xi| + |\eta|)^2$$
$$\leq 1 + 2|\xi|^2 + 2|\eta|^2 \leq 2(1 + |\xi|^2)(1 + |\eta|^2) \, ,$$

so that for

$$s \geq 0 : (1 + |\xi + \eta|^2)^s \leq 2^s (1 + |\xi|^2)^s (1 + |\eta|^2)^s \, ,$$

this proves Peetre's inequality for $s \geq 0$, (ξ being replaced by $\xi + \eta$). The case $s \leq 0$ can be deduced from this by permuting the roles of ξ, η. (2.6.2.) and (2.6.1.) show that

$$(2\pi)^{-n/2}(1+|\xi|^2)^{s/2}\,|\widehat{\varphi u}(\xi)| \leq (f*g)(\xi), \text{ with}$$

$$f(\xi) = (2\pi)^{-n/2}(1+|\xi|^2)^{s/2}\,|\hat{u}(\xi)|$$

$$g(\xi) = (2\pi)^{-n}\,2^{|s|/2}(1+|\xi|^2)^{|s|/2}\,|\hat{\varphi}(\xi)| .$$

The conclusion then follows from (2.1.2) of Chapter I.

We have seen that $\mathcal{D}$ is dense in H^s; we shall now make this statement more precise by giving a constructive procedure of approximation by truncation and regularisation.

THEOREM 2.7: *Let* $\rho \in C_0^\infty(\mathbb{R}^n)$ *be such that* $\int \rho(x)\,dx = 1$. *Let* $\chi \in C_0^\infty(\mathbb{R})$ *be such that* $\chi = 1$ *in the neighbourhood of* 0. *For* k *an integer* > 0, *we put:*

$$\rho_k(x) = k^n \rho(kx)$$

$$\chi_k(x) = \chi(x/k)$$

Suppose $u \in H^s$ $(s \in \mathbb{R})$.

Then $\rho_k * u$ *and* $\chi_k u$ *tend to* u *in* H^s *when* $k \to +\infty$.

PROOF: We have $\hat{\rho}_k(\xi) = \hat{\rho}(\xi/k)$ and $\widehat{\rho_k * u}(\xi) = \hat{\rho}(\xi/k)\,\hat{u}(\xi)$, so that $\|\rho_k * u - u\|_s^2 = \int (1+|\xi|^2)^s\,|\hat{u}(\xi)|^2\,|\hat{\rho}(\xi/k)-1|^2\,d\xi$.

Now $\hat{\rho}(\xi/k) \longrightarrow \hat{\rho}(0) = 1$ when $k \to +\infty$, and $|\hat{\rho}(\xi/k) - 1| \leq |\hat{\rho}(\xi/k)| + 1 \leq 2$, thus $\rho_k * u \to u$ in H^s

from the dominated convergence theorem.

From Theorem 2.6, we have

$$\|\chi_k u\|_s \leq C_k \|u\|_s \quad \text{, where}$$

$$C_k = 2^{|s|/2} \int (1 + |\xi|^2)^{|s|/2} k^m |\hat{\chi}(k\xi)| \, d\!\!/\xi$$

$$= 2^{|s|/2} \int (1 + |\xi/k|^2)^{|s|/2} |\hat{\chi}(\xi)| \, d\!\!/\xi \leq C_1 \quad ,$$

and thus

(2.7.1.) $\|\chi_k u\|_s \leq C_1 \|u\|_s$ for all k and all $u \in H^s$.

For $\varepsilon > 0$, the density of $\mathcal{D}$ in H^s shows that there exists $\varphi \in \mathcal{D}$ such that $\|u - \varphi\|_s < \varepsilon$. For k sufficiently large, we have $\chi_k \varphi = \varphi$, and thus

$\|\chi_k u - u\|_s = \|\chi_k(u - \varphi) - (u - \varphi)\|_s \leq C_1 \varepsilon$ from (1.7.1).

We now proceed to the study of the dual of H^s. Since H^s is a Hilbert space, we can identify it with its dual by means of the bilinear form on $H^s \times H^s$ defined by $u, v \longmapsto (u, \bar{v})_s$. But, for $s \neq 0$, this bilinear form does not extend the bilinear form on $H^s \times \mathcal{S}$ defined by $u, v \longmapsto \langle u, v \rangle$; therefore, we shall not use this identification. We have seen that the injection $\mathcal{S} \longrightarrow H^s$ is of dense image and is continuous; thus a continuous linear form ℓ on H^s is characterised by its restriction $\ell_{|\mathcal{S}}$ to $\mathcal{S}$, and we have $\ell_{|\mathcal{S}} \in \mathcal{S}'$. The injection

$\ell \longmapsto \ell_{|\mathcal{S}}$ of $(H^s)'$ into $\mathcal{S}'$ allows us to identify $(H^s)'$ with a subspace of $\mathcal{S}'$; we shall now see that this subspace is exactly H^{-s}. More precisely, we have:

THEOREM 2.8: *The bilinear form on* $\mathcal{S} \times \mathcal{S}$ *defined by* $u, v \longmapsto \langle u, v \rangle$ *extends into the continuous bilinear form on* $H^s \times H^{-s}$ *defined by* $u, v \longmapsto \int \hat{u}(\xi)\, \hat{v}(-\xi)\, d\!\!\!/\xi$, *which we again denote by* $\langle u, v \rangle$.

This bilinear form satisfies $|\langle u, v \rangle| \le \|u\|_s \|v\|_{-s}$ *for* $u \in H^s$, $v \in H^{-s}$, *and allows us to identify the dual of* H^s *with the normed space* H^{-s}.

PROOF: If $u \in H^s$ and $v \in H^{-s}$, we have

$$\hat{u}(\xi)\, \hat{v}(-\xi) = (1 + |\xi|^2)^{s/2}\, \hat{u}(\xi) . (1 + |\xi|^2)^{-s/2}\, \hat{v}(-\xi) ,$$

so that $|\langle u, v \rangle| \le \|u\|_s \|v\|_{-s}$ from Schwarz's inequality. Thus if $v \in H^{-s}$, we define the continuous linear form ℓ on H^s by $\ell(u) = \langle u, v \rangle$ for all $u \in H^s$. Conversely, suppose $\ell \in (H^s)'$; from Riesz's theorem, there exists a unique $f \in H^s$ such that $\ell(u) = (f, \bar{u})_s$ for all $u \in \mathcal{S}$, and we have $\|\ell\| = \|f\|_s$.

For $u \in \mathcal{S}$, we obtain

$$\ell(u) = \int (1 + |\xi|^2)^s\, \hat{f}(\xi)\, \hat{u}(-\xi)\, d\!\!\!/\xi = \langle u, v \rangle ,$$

where $v \in \mathcal{S}'$ is defined by $\hat{v} = (1 + |\xi|^2)^s \hat{f}$; since $f \in H^s$, we have $v \in H^{-s}$ with

$$\|v\|_{-s}^2 = \int (1 + |\xi|^2)^{-s} (1 + |\xi|^2)^{2s} |\hat{f}(\xi)|^2 d\xi = \|f\|_s^2 = \|\ell\|^2$$

COROLLARY 2.9: *For* $m \in \mathbb{N}$, H^{-m} *is the space of distributions* u *of the form*

(2.9.1.) $$u = \sum_{|\alpha| \leq m} \partial^\alpha f_\alpha \, , \quad \text{with} \quad f_\alpha \in L^2(\mathbb{R}^n) \, .$$

The norm in H^{-m} *is equivalent to the norm* $u \longmapsto |u|_{-m}$ *defined by the formula* $|u|_{-m}^2 = \operatorname{Inf} \sum_{|\alpha| \leq m} \|f_\alpha\|_0^2$, *the lower bound being taken for all possible decompositions* (2.9.1) *of* u . *This lower bound is attained.*

PROOF: We equip H^m with the norm $u \longmapsto |u|_m$ associated with the inner product (2.2.1). The dual of H^m is then identified with H^{-m}, and defines on H^{-m} a norm denoted by $u \longmapsto |u|_{-m}$, equivalent to the usual norm of H^{-m}. Riesz's theorem shows therefore that $u \in H^{-m}$ if and only if there exists $v \in H^m$ such that

(2.9.2.) $$< u, \varphi > = ((\varphi, v))_m \quad \text{for all} \quad \varphi \in \mathcal{S} \, ,$$

and it also shows that we then have

(2.9.3.) $$|u|_{-m} = \|v\|_m \, .$$

Since $((\varphi, v))_m = \sum_{|\alpha| \leq n} < (-1)^{|\alpha|} \partial^\alpha(\partial^\alpha \bar{v}), \varphi >$, we see that (2.9.2) is equivalent to $u = \sum_{|\alpha| \leq m} \partial^\alpha f_\alpha$, with $f_\alpha = (-1)^{|\alpha|} \partial^\alpha \bar{v} \in L^2(R^n)$, and that (2.9.3) is equivalent to

$$|u|^2_{-m} = \sum_{|\alpha| \leq m} \|f_\alpha\|^2_0 \quad .$$

Conversely, if (2.9.1) is satisfied, we have, for $\varphi \in \mathcal{S}$:

$$|\langle u, \varphi \rangle| = \Big| \sum_{|\alpha| \leq m} (-1)^{|\alpha|} \langle f_\alpha, \partial^\alpha \varphi \rangle \Big| \leq \sum_{|\alpha| \leq m} \|f_\alpha\|_0 \, \|\partial^\alpha \varphi\|_0$$

$$\leq \Big(\sum_{|\alpha| \leq m} \|f_\alpha\|^2_0 \Big)^{1/2} |\varphi|_m$$, hence $u \in H^{-m}$ with

$$|u|^2_{-m} \leq \sum_{|\alpha| \leq m} \|f_\alpha\|^2_0 \quad .$$

THEOREM 2.10: (Reillich). *Suppose we have* $s, t \in \mathbb{R}$ *with* $s > t$, *and that* K *is a compact subset of* $\mathbb{R}^n$. *Then the injection* $H^s_K(\mathbb{R}^n) \longrightarrow H^t_K(\mathbb{R}^n)$ *is compact.*

PROOF: Let (u_k) be a sequence in H^s_K such that $\|u_k\|_s \leq 1$. We now show that we can extract from this a subsequence which is convergent in H^t. Suppose we have $\varphi \in \mathcal{D}(\mathbb{R}^n)$ equal to 1 in the neighbourhood of K. For $u \in H^s_K$, we have

$u = \varphi u \quad , \quad \hat{u} = (2\pi)^{-n} \hat{\varphi} * \hat{u} \quad , \quad D^\alpha \hat{u} = (2\pi)^{-n} D^\alpha \hat{\varphi} * \hat{u}$. As in the proof of Theorem 2.6, we obtain

$(1 + |\xi|^2)^{s/2} \, |D^\alpha \hat{u}(\xi)| \leq (f * g_\alpha)(\xi)$, where

$$f(\xi) = (2\pi)^{-n/2} (1 + |\xi|^2)^{s/2} \, |\hat{u}(\xi)|$$

$$g_\alpha(\xi) = (2\pi)^{-n/2} \, 2^{|s|/2} (1 + |\xi|^2)^{|s|/2} \, |D^\alpha \hat{\varphi}(\xi)|$$

together with the existence of a constant C_α such that

$$(1 + |\xi|^2)^{s/2} \; |D^\alpha \hat{u}(\xi)| \leq C_\alpha \|u\|_s \quad \text{for all} \quad u \in H^s_K \,,$$
$$\text{for all} \quad \xi \in \mathbb{R}^n \,.$$

We thus have:

$$(1 + |\xi|^2)^{s/2} \; |D^\alpha u_k(\xi)| \leq C_\alpha \quad \text{for all } k \text{ and all } \xi \in \mathbb{R}^n \,,$$

which shows that the sequence $(D^\alpha u_k)_k$ is uniformly bounded on any compact subset of $\mathbb{R}^n$. By applying this result for multi-indices α such that $|\alpha| \leq 1$, and by using Ascoli's theorem, we see that we can extract from the sequence u_k a subsequence (also denoted, with an abuse of notation, by u_k) such that

(2.10.1) u_k converges uniformly on any compact subset of $\mathbb{R}^n$.

We shall now deduce from this that the subsequence u_k converges in H^t. Suppose $R > 0$. We have $\|u_k - u_\ell\|_t^2 = I_1 + I_2$, where

$$I_1 = \int_{|\xi| \leq R} (1 + |\xi|^2)^t \, |u_k(\xi) - u_\ell(\xi)|^2 \, d\xi$$

$$I_2 = \int_{|\xi| \geq R} (1 + |\xi|^2)^t \, |u_k(\xi) - u_\ell(\xi)|^2 \, d\xi \,.$$

Writing

$$I_2 = \int_{|\xi| \geq R} (1 + |\xi|^2)^{t-s} \, (1 + |\xi|^2)^s \, |u_k(\xi) - u_\ell(\xi)|^2 \, d\xi \,,$$

we obtain, since $t-s < 0$:

$$I_2 \leq (1 + R^2)^{t-s} \|u_k - u_\ell\|_s^2 \leq 4(1 + R^2)^{t-s},$$

so that I_2 can be rendered arbitrarily small if R is chosen sufficiently large. For R fixed in this manner, I_1 can be rendered arbitrarily small if k, ℓ are sufficiently large, by virtue of (2.10.1).

PROPOSITION 2.11: *Let* s_1, s, s_2 *be such that* $s_1 < s < s_2$. *For any* $\varepsilon > 0$, *there exists a constant* C_ε *such that*

$$\|u\|_s^2 \leq \varepsilon \|u\|_{s_2}^2 + C_\varepsilon \|u\|_{s_1}^2 , \text{ for all } u \in H^{s_2} .$$

PROOF: We now show that

$$(1 + |\xi|^2)^s \leq \varepsilon(1 + |\xi|^2)^{s_2} + C_\varepsilon(1 + |\xi|^2)^{s_1} \quad \text{for all } \xi \in \mathbb{R}^n$$

with $C_\varepsilon = \varepsilon^{-(s-s_1)/(s_2-s)}$

i.e. by putting $\rho = 1 + |\xi|^2$, that

$$\rho^s \leq \varepsilon \rho^{s_2} + \varepsilon^{-(s-s_1)/(s_2-s)} \rho^{s_1}$$

i.e. that

$$1 \leq \varepsilon \rho^{s_2-s} + \varepsilon^{-(s-s_1)/(s_2-s)} \rho^{s_1-s} .$$

By putting $\lambda = \varepsilon^{1/s_2-s}$, this inequality can be written

$$(2.11.1.) \qquad 1 \leq (\lambda\rho)^{s_2-s} + (\lambda\rho)^{s_1-s}$$

If $\lambda\rho \geq 1$, (2.11.1) is satisfied since $(\lambda\rho)^{s_2 - s} \geq 1$

If $\lambda\rho \leq 1$, (2.11.1) is satisfied since $(\lambda\rho)^{s_1 - s} \geq 1$.

We now study distributions which behave locally like elements of H^s.

DEFINITION 2.12 : *Let* X *be an open subset of* $\mathbb{R}^n$ *and suppose* $s \in \mathbb{R}$. *We denote by* $H^s_{loc}(X)$ *the space of distributions* $u \in \mathcal{D}'(X)$ *such that* $\varphi u \in H^s$ *for all* $\varphi \in C^\infty_0(X)$. *We equip* $H^s_{loc}(X)$ *with the topology defined by the family of semi-norms* $u \longmapsto \|\varphi u\|_s$, φ *describing* $C^\infty_0(X)$.

If U is an open subset of X, the restriction operator is evidently continuous from $H^s_{loc}(X)$ into $H^s_{loc}(U)$. Furthermore, we have:

PROPOSITION 2.13 : $H^s_{loc}(X)$ *is a Fréchet space, and the injection of* $H^s_{loc}(X)$ *into* $\mathcal{D}'(X)$ *is continuous. When* $s > n/2 + k$, *we have* $H^s_{loc}(X) \subset C^k(X)$, *with continuous injection. We have* $\bigcap_{s \in \mathbb{R}} H^s_{loc}(X) = C^\infty(X)$. *A differential operator of degree* m *with* C^∞ *coefficients in* X *is continuous from* $H^s_{loc}(X)$ *into* $H^{s-m}_{loc}(X)$.

PROOF: Let K_j be an exhaustive sequence of compact subsets of X and suppose we have $\varphi_j \in C^\infty_0(X)$ equal to 1 in K_j. Then the sequence of semi-norms $u \longmapsto \|\varphi_j u\|_s$ suffices to define the topology of $H^s_{loc}(X)$: in fact, if $\varphi \in C_0^\infty(X)$, we have, for j sufficiently large,

$$\varphi = \varphi\varphi_j \quad \text{and} \quad \|\varphi u\|_s = \|\varphi\varphi_j u\|_s \leq C \|\varphi_j u\|_s$$

from Theorem 2.6, C only depending on s and φ. Thus $H^s_{loc}(X)$ is metrizable. Let (u_k) be a Cauchy sequence in $H^s_{loc}(X)$: for all $\varphi \in C_0^\infty(X)$, φu_k is a Cauchy sequence in H^s, thus φu_k possesses a limit u_φ in H^s, and a fortiori in $\mathcal{D}'$; it follows from this that u_k has a limit u in $\mathcal{D}'$, with $\varphi u = u_\varphi$ for all $\varphi \in C_0^\infty(X)$. Thus $u \in H^s_{loc}(X)$ and $u_k \to u$ in $H^s_{loc}(X)$. The other assertions of the proposition are immediate, by utilising in particular Theorem 2.5.

DEFINITION 2.14 : *We denote by* $H^s_{comp}(X)$ *the union of the spaces* H^s_K *when* K *describes the family of compact subsets of* X, *and we equip* $H^s_{comp}(X)$ *with the inductive limit topology of the topologies of the* H^s_K.

THEOREM 2.15 : $C_0^\infty(X)$ *is dense in* $H^s_{loc}(X)$ *and in* $H^s_{comp}(X)$. *The injections* $C^\infty(X) \longrightarrow H^s_{loc}(X)$ *and* $C_0^\infty(X) \longrightarrow H^s_{comp}(X)$ *are continuous and of dense image and allow us to identify the dual of* $H^s_{loc}(X)$ *with* $H^{-s}_{comp}(X)$ *and the dual of* $H^s_{comp}(X)$ *with* $H^{-s}_{loc}(X)$.

PROOF: The continuity of the injections follows immediately from the continuity of the injection $\mathcal{D} \longrightarrow H^s$.

Suppose we have functions φ_j as in the proof of Theorem 2.13. If $u \in H^s_{loc}(X)$, it is immediate that $\varphi_j u \to u$ in $H^s_{loc}(X)$; if ρ_k is a regularising sequence, we have seen (Theorem 2.7) that $\rho_k * (\varphi_j u) \to \varphi_j u$ in H^s when $k \to +\infty$; hence the density of $C^\infty_o(X)$ in $H^s_{loc}(X)$. Similarly, if $u \in H^s_{comp}(X)$, $\rho_k * u \to u$ in $H^s_{comp}(X)$ when $k \to +\infty$; hence the density of $C^\infty_o(X)$ in $H^s_{comp}(X)$.

We shall now show that each space $H^s_{loc}(X)$, $H^{-s}_{comp}(X)$ is identified with the dual of the other by means of the bilinear form on $H^s_{loc}(X) \times H^{-s}_{comp}(X)$ defined by $u, v \mapsto \langle u, v \rangle = \langle \chi u, v \rangle$, where $\chi \in C^\infty_o(X)$ is an arbitrary function having the value 1 in the neighbourhood of supp v, and where the last set of brackets are those of the pairing between $H^s(\mathbb{R}^n)$ and $H^{-s}(\mathbb{R}^n)$. First, it is immediately clear that $\langle u, v \rangle$ depends continuously on u for v fixed. This proves that $\langle u, v \rangle$ does not depend on the function χ chosen, and we thereby deduce that $\langle u, v \rangle$ depends continuously on v for u fixed.

Conversely, let ℓ be a continuous linear form on $H^s_{loc}(X)$, and let v be the distribution in $\mathcal{E}'(X)$ defined by $\langle \varphi, v \rangle = \ell(\varphi)$ for all $\varphi \in C^\infty(X)$; suppose we have $\chi \in C^\infty_o(X)$ equal to 1 in the neighbourhood of supp u. Then $\Phi \mapsto \langle \Phi, v \rangle = \langle \chi\Phi, v \rangle = \ell(\chi\Phi)$ depends continuously on $\Phi \in C^\infty_o(\mathbb{R}^n)$ for the topology induced by $H^s(\mathbb{R}^n)$. We therefore have $v \in H^{-s}(\mathbb{R}^n)$. In similar fashion, if ℓ is a continuous linear form on $H^s_{comp}(X)$, we consider the distribution $u \in \mathcal{D}'(X)$ defined by $< u, \psi > = \ell(\psi)$ for all $\psi \in C_o^\infty(X)$. If $\varphi \in C_o^\infty(X)$, then $\Phi \longmapsto < \varphi u, \Phi > = \ell(\varphi\Phi)$ depends continuously on $\Phi \in C_o^\infty(\mathbb{R}^n)$ for the topology induced by $H^{-s}(\mathbb{R}^n)$, thus $\varphi u \in H^s(\mathbb{R}^n)$ and finally $u \in H^s_{loc}(X)$.

REMARK 2.16 : The preceding identifications are algebraic. We can show that they are also topological when we equip each dual with the topology of uniform convergence on the bounded subdomains.

We can infer from the theory of pseudo-differential operators the invariance under diffeomorphism of the spaces $H^s_{loc}(X)$ (see Corollary 8.9 in Chapter IV; see also Exercise 7.2 for a direct proof):

THEOREM 2.17 : *Let* U, $\tilde{U}$ *be two open subsets of* $\mathbb{R}^n$ *and let* $\chi: U \to \tilde{U}$ *be a diffeomorphism. Then* χ_* *defines a topological vector isomorphism of* $H^s_{loc}(U)$ *onto* $H^s_{loc}(\tilde{U})$ *as well as of* $H^s_{comp}(U)$ *onto* $H^s_{comp}(\tilde{U})$.

This theorem allows us to generalise the definition of $H^s_{loc}(X)$ to the case where X is a manifold:

DEFINITION 2.18 : *Let X be a manifold. We denote by*

$H^s_{loc}(X)$ *the space of distributions* $u \in \mathcal{D}'(X)$ *such that* $\chi_* u \in H^s_{loc}(\tilde{U})$ *for any chart* $\chi : U \longrightarrow \tilde{U}$ *of* X. *We equip* $H^s_{loc}(X)$ *with the topology defined by the family of continuous semi-norms of* $\chi_* u$ *in* $H^s_{loc}(\tilde{U})$, χ *describing the family of charts of* X.

If U is an open subset of X, the operator of restriction to U is evidently continuous from $H^s_{loc}(X)$ into $H^s_{loc}(U)$.

THEOREM 2.19 : *Let* X *be a manifold.*

(i) (*Sheaf property of* $H^s_{loc}(X)$). *Let* $u \in \mathcal{D}'(X)$ *be such that, for all* $x \in X$, *there exists a neighbourhood* U *of* x *in* X *with* $u_{|U} \in H^s_{loc}(U)$. *Then* $u \in H^s_{loc}(X)$.

(ii) *Let* K *be a compact subset of* X. *Then the subspace* $H^s_K(X) = \{ u \in H^s_{loc}(X) \mid \text{supp } u \subset K \}$ *of* $H^s_{loc}(X)$ *is hilbertisable, and we denote by* $\|u\|_{s,K}$ *an admissible norm in* $H^s_K(X)$. *If the manifold* X *is compact, we put* $H^s_{loc}(X) = H^s(X)$.

(iii) *Proposition* 2.13 *and Theorem* 2.15 *remain valid. For the duality properties, we assume* X *to be endowed with a positive density.*

PROOF: For (i), consider an atlas $\chi_j : U_j \longrightarrow \tilde{U}_j$ of X such that $u \in H^s_{loc}(U_j)$ for any j belonging to the set of indices J. Let $\chi : U \longrightarrow \tilde{U}$ be an arbitrary chart of X; we show that $\chi_* u \in H^s_{loc}(U)$, i.e. that $\varphi \chi_* u \in H^s$ for

$\varphi \in C_0^\infty(U)$. The $\chi(U \cap U_j)$ constitute an open covering of $\tilde{U}$; thus there exists a finite subset F of J such that $\operatorname{supp} \varphi \subset \bigcup_{j \in F} \chi(U \cap U_j)$. Let there be functions $\varphi_j \in C_0^\infty(\chi(U \cap U_j))$ such that $\sum_{j \in F} \varphi_j = 1$ in the neighbourhood of supp φ. We have $\varphi \chi_* u = \sum_{j \in F} \varphi_j \varphi \chi_* u$. By the coordinate transformation $\chi_j \circ \chi^{-1}$, $\varphi_j \varphi \in C_0^\infty(\chi(U \cap U_j))$ is transported into a function $\psi_j \in C_0^\infty(\chi_j(U \cap U_j))$, and $\varphi_j \varphi \chi_* u$ is transported into $\psi_j \chi_{j*} u$ which belongs to H^s by hypothesis. Theorem 2.17 thus shows that $\varphi_j \varphi \chi_* u \in H^s$ for all $j \in F$, hence $\varphi \chi_* u \in H^s$. Furthermore, we see that, for φ fixed, there exists a constant C such that, for all $u \in H^s_{loc}(X)$, we have

$$\|\varphi \chi_* u\|_s \leq C \sum_{j \in F} \|\psi_j \chi_{j*} u\|_s .$$

Thus

(2.19.1.) *If* χ_j $(j \in J)$ *is an atlas of* X, *the semi-norms of* χ_j u *in* $H^s_{loc}(U_j)$ *are sufficient to define the topology of* $H^s_{loc}(X)$.

We now proceed to the proof of (ii) and (iii). We can clearly choose J to be countable in (2.19.1), which proves that $H^s_{loc}(X)$ is metrizable. Let (u_k) be a Cauchy sequence in $H^s_{loc}(X)$; for any chart $\chi : U \to \tilde{U}$ of X, $\chi_* u_k$ is a Cauchy sequence in $H^s_{loc}(\tilde{U})$, therefore $\chi_* u_k$ has a limit u_χ in $H^s_{loc}(\tilde{U})$; if χ' is another chart of X, we can see by proceeding to the limit that $u_{\chi'} = (\chi' \circ \chi^{-1})_* u_\chi$; thus there exists $u \in \mathcal{D}'(X)$ such that $u_\chi = \chi_* u$ for any chart χ of X. We have $u \in H^s_{loc}(X)$ and

$u_k \longrightarrow u$ in $H^s_{loc}(X)$. Therefore $H^s_{loc}(X)$ is a Fréchet space. Let K be a compact subset of X; $H^s_K(X)$ is closed in $H^s_{loc}(X)$, and therefore complete. Let there be a finite family of charts $\chi_j : U_j \longrightarrow \tilde{U}_j$ of X $(j \in F)$ such that $K \subset \bigcup_{j \in F} U_j$. Let there be functions $\varphi_j \in C_0^\infty(U_j)$ such that $\sum \varphi_j = 1$ in the neighbourhood of K. For $u \in H^s_K(X)$, we have $u = \sum_{j \in F} \varphi_j u$, and $u \to 0$ in $H^s_{loc}(X)$ if and only if $\chi_{j*}(\varphi_j u) \longrightarrow 0$ in H^s for every $j \in F$: this is clearly necessary, and it is sufficient from (2.19.1) by considering an atlas of X formed by the union of χ_j and an atlas of the complement of the neighbourhood of supp φ_j. Thus the topology in $H^s_K(X)$ can be defined by the norm associated with the inner product

$$(u, v)_{s,K} = \sum_{j \in F} (\chi_{j*}(\varphi_j u), \chi_{j*}(\varphi_j v))_s .$$

We leave to the reader the task of generalising the other properties stated in Proposition 2.13 and Theorem 2.15.

3. THE SPACES $H^s(\bar{X})$ AND $H^s_{loc}(\bar{X})$.

Let X be a regular open subset of $\mathbb{R}^n$ (see Definition 9.1, Chapter I) and let $s \in \mathbb{R}$. The restriction operator r: $H^s(\mathbb{R}^n) \longrightarrow \mathcal{D}'(X)$ defined by $ru = u_{|X}$, has for its kernel the closed subspace $H^s_{\mathbb{R}^n\backslash X}(\mathbb{R}^n)$ of $H^s(\mathbb{R}^n)$; we thus have an isomorphism of $H^s / H^s_{\mathbb{R}^n\backslash X}$ onto Im r. We equip $H^s / H^s_{\mathbb{R}^n\backslash X}$ with the Hilbert-quotient structure, which the preceding isomorphism allows us to transport into a Hilbert structure on Im r. This leads to:

DEFINITION 3.1 : *Let* X *be a regular open subset of* $\mathbb{R}^n$ *and let* $s \in \mathbb{R}$. *We denote by* $H^s(\bar{X})$ *the space of distributions* $u \in \mathcal{D}'(X)$ *admitting an extension* $\tilde{u}$ *in* $H^s(\mathbb{R}^n)$. *We equip* $H^s(\bar{X})$ *with the norm* $\|u\|_s = \text{Inf}\ \|\tilde{u}\|_s$, *where* $\tilde{u}$ *describes the family of extensions of* u *in* $H^s(\mathbb{R}^n)$.

The decomposition into an orthogonal direct sum

$$H^s(\mathbb{R}^n) = H^s_{\mathbb{R}^n\setminus X}(\mathbb{R}^n) \oplus (H^s_{\mathbb{R}^n\setminus X}(\mathbb{R}^n))^{\perp}$$

shows that, for all $u \in H^s(\bar{X})$, there exists a unique extension U of u such that $U \in (H^s_{\mathbb{R}^n\setminus X}(\mathbb{R}^n))^{\perp}$, and also that we have $\|u\|_s = \|U\|_s$. We have $H^0(\bar{X}) = L^2(X)$ and, more generally:

PROPOSITION 3.2 : *Suppose we have* $m \in \mathbb{N}$. *Then* $H^{-m}(\bar{X})$ *is the space of distributions* $u \in \mathcal{D}'(X)$ *of the form*

$$u = \sum_{|\alpha| \leq m} \partial^{\alpha} f_{\alpha}, \quad \text{with} \quad f_{\alpha} \in L^2(X). \tag{3.2.1.}$$

The norm in $H^{-m}(\bar{X})$ *is equivalent to the norm* $u \longmapsto |u|_{-m}$ *defined by*

$$|u|^2_{-m} = \operatorname{Inf} \sum_{|\alpha|\le m} \int_X |f_\alpha(x)|^2 \, dx$$

where the lower bound is taken for all the decompositions (3.2.1) *of* u.

PROOF: Suppose we have $u \in \mathcal{D}'(X)$ of the form (3.2.1). We put $\tilde{u} = \sum_{|\alpha|\le m} \partial^\alpha (f^0_\alpha)$ where $f^0_\alpha \in L^2(\mathbb{R}^n)$ is obtained by extending f_α by 0 outside of $\bar{X}$. The distribution $\tilde{u}$ is an extension of u and, from Corollary 2.9., we have $\tilde{u} \in H^{-m}(\mathbb{R}^n)$ with

$$|\tilde{u}|^2_{-m} \le \sum_{|\alpha|\le m} \|f^0_\alpha\|^2_0 = \sum_{|\alpha|\le m} \int_X |f_\alpha(x)|^2 \, dx \, .$$

Conversely, suppose $u \in H^{-m}(\bar{X})$; we consider $U \in H^{-m}(\mathbb{R}^n)$ such that $\|u\|_{-m} = \|U\|_{-m}$. From Corollary 2.9, there exists $F_\alpha \in L^2(\mathbb{R}^n)$ such that $|U|^2_{-m} = \sum_{|\alpha|\le m} \|F_\alpha\|^2_0$. If we put $f_\alpha = F_{\alpha_{|X}}$ we obtain (3.2.1) with

$$\sum_{|\alpha|\le m} \int_X |f_\alpha|^2 \, dx \le \sum_{|\alpha|\le m} \|F_\alpha\|^2_0 = |U|^2_{-m} \, .$$

We shall now see that $H^1(\bar{X})$ is indeed the space which appeared in the introduction. More generally:

PROPOSITION 3.3 : *We assume that* $X = \mathbb{R}^n_+$ *or that* ∂X *is compact. Then, for* $m \in \mathbb{N}$, $H^m(\bar{X})$ *is the space of distributions* $u \in \mathcal{D}'(X)$ *such that* $\partial^\alpha u \in L^2(X)$ *for* $|\alpha| \le m$.

The inner product in $H^m(\bar{X})$ *is equivalent to the inner product*

$$u, v \longmapsto \sum_{|\alpha| \leq m} \int_X \partial^\alpha u \, . \, \overline{\partial^\alpha v} \, dx \tag{3.3.1.}$$

PROOF: For the moment we shall denote by $\tilde{H}^m(\bar{X})$ the space of the $u \in \mathcal{D}'(X)$ such that $\partial^\alpha u \in L^2(X)$ for $|\alpha| \leq m$, equipped with the inner product (3.3.1). From Proposition 2.2, we have $H^m(\bar{X}) \subset \tilde{H}^m(\bar{X})$ with continuous injection. Conversely, we can show that $\tilde{H}^m(\bar{X}) \subset H^m(\bar{X})$ with continuous injection. This amounts to proving that there exists a continuous linear extension operator from $\tilde{H}^m(\bar{X})$ into $H^m(\mathbb{R}^n)$. By local charts and a partition of unity, it can easily be shown that it is sufficient to construct such an operator in the case $X = \mathbb{R}^n_+$. We consider the continuous linear extension operator $\rho : \mathcal{D}(\bar{\mathbb{R}}^n_+) \longrightarrow \mathcal{D}(\mathbb{R}^n)$ constructed in the proof of Proposition 9.2 Chapter I, or even simpler, the extension ρ corresponding to:

$$\lambda_k = 0 \text{ for } k \geq m+2 \, , \quad \sum_{k=1}^{m+1} \lambda_k \, 2^{k_j} = (-1)^j \text{ for } j = 0, \ldots, m$$

It is immediately evident that ρ is continuous with values in $H^m(\mathbb{R}^n)$ for the topology induced by $\tilde{H}^m(\bar{\mathbb{R}}^n_+)$ on $\mathcal{D}(\bar{\mathbb{R}}^n_+)$. In order to conclude the proof, it is thus sufficient to prove that $\mathcal{D}(\bar{\mathbb{R}}^n_+)$ is dense in $\tilde{H}^m(\bar{\mathbb{R}}^n_+)$, which we shall do by truncation and regularisation. Let there be functions χ_k and ρ_k as in the statement of Theorem 2.7, with in addition $\operatorname{supp} \rho \subset \{ x_n \leq 0 \}$. For $u \in \tilde{H}^m(\bar{\mathbb{R}}^n_+)$ Leibniz' formula shows that $\partial^\alpha(\chi_k u) = \chi_k \partial^\alpha u + v$, where v is a linear combination of terms of the form

$$k^{-|\alpha'|}(\partial^{\alpha'}\chi)(x/k)\,\partial^{\alpha''}u \qquad (\alpha' + \alpha'' = \alpha, \ \ |\alpha'| \geq 1) .$$

Thus $\partial^\alpha(\chi_k u) \longrightarrow \partial^\alpha u$ in $L^2(\mathbb{R}^n_+)$ for $|\alpha| \leq m$, i.e. $\chi_k u \to u$ in $\tilde{H}^m(\overline{\mathbb{R}}^n_+)$.

We now put $\varphi_k = (u^0 * \rho_k)_{|\mathbb{R}^n_+}$ where u^0 is the extension of u by 0 for $x_n < 0$. We have $\varphi_k \in C^\infty(\overline{\mathbb{R}}^n_+)$ and, for $|\alpha| \leq m$;

$$\partial^\alpha \varphi_k = (\partial^\alpha(u^0) * \rho_k)_{|\mathbb{R}^n_+} = ((\partial^\alpha u)^0 * \rho_k)_{|\mathbb{R}^n_+}$$

since, because $\partial^\alpha(u^0) - (\partial^\alpha u)^0$ is supported by $\{x_n = 0\}$, its convolution with ρ_k is supported by $\{x_n \leq 0\}$ in view of the choice of ρ. Since $(\partial^\alpha u)^0 * \rho_k \longrightarrow (\partial^\alpha u)^0$ in $L^2(\mathbb{R}^n)$, we can thereby deduce that $\partial^\alpha \varphi_k \longrightarrow \partial^\alpha u$ in $L^2(\mathbb{R}^n_+)$, and therefore that $\varphi_k \longrightarrow u$ in $\tilde{H}^m(\overline{\mathbb{R}}^n_+)$.

PROPOSITION 3.4: *Suppose* $s \in \mathbb{R}$. *Then the injection* $C_0^\infty(\overline{X}) \longrightarrow H^s(\overline{X})$ *is continuous and of dense image. Suppose we have* $s, t \in \mathbb{R}$ *with* $s > t$. *Then the injection* $H^s(\overline{X}) \longrightarrow H^t(\overline{X})$ *is compact when* $\overline{X}$ *is compact.*

PROOF: The first injection can be written $r \circ i \circ \rho$, where $\rho : C_0^\infty(\overline{X}) \longrightarrow C_0^\infty(\mathbb{R}^n)$ is a continuous linear extension operator, where $i : C_0^\infty(\mathbb{R}^n) \longrightarrow H^s(\mathbb{R}^n)$ is the injection, and where

$r : H^s(\mathbb{R}^n) \longrightarrow H^s(\overline{X})$ is the restriction. Its continuity thus follows from the continuity of i (see Proposition 2.3). The second injection can be written $r \circ j \circ \chi\rho_s$, where $\rho_s : H^s(\overline{X}) \longrightarrow H^s(\mathbb{R}^n)$ is a continuous linear extension operator, where $\chi \in C_0^\infty(\mathbb{R}^n)$ is a function equal to 1 in the neighbourhood of $\overline{X}$, and where $j : H^s_K \longrightarrow H^t$ is the injection ($K = \text{supp } \chi$). Its compactness thus follows from the compactness of j (see Theorem 2.10). Finally, suppose $u \in H^s(\overline{X})$ and suppose $\tilde{u} \in H^s(\mathbb{R}^n)$ is an extension of u ; we know (see Proposition 2.3) that there exists a sequence $\varphi_k \in C_0^\infty(\mathbb{R}^n)$ such that $\varphi_k \to \tilde{u}$ in $H^s(\mathbb{R}^n)$. We can therefore deduce that $\varphi_{k|X} \longrightarrow u$ in $H^s(\overline{X})$, and hence the density of $C_0^\infty(\overline{X})$ in $H^s(\overline{X})$.

PROPOSITION 3.5: *The dual of* $H^s(\overline{X})$ *(resp: of* $H^{-s}_{\overline{X}}(\mathbb{R}^n)$ *) is identified with the normed space* $H^{-s}_{\overline{X}}(\mathbb{R}^n)$ *(resp:* $H^s(\overline{X})$*) by the pairing*

$$< u, v > \; = \; < \tilde{u}, v > \quad \text{for } u \in H^s(\overline{X}) \;, \quad v \in H^{-s}_{\overline{X}}(\mathbb{R}^n) \;,$$

$\tilde{u}$ *being an arbitrary extension of* u *into* $H^s(\mathbb{R}^n)$.

PROOF: We begin by showing that $< \tilde{u}, v >$ does not depend on the extension $\tilde{u}$ chosen for u , i.e. that we have:

(3.5.1) LEMMA: *Let* $U \in H^s(\mathbb{R}^n)$, $v \in H^{-s}(\mathbb{R}^n)$ *be such that:* $U = 0$ *in* X, $v = 0$ *in* $\mathbb{R}^n \setminus \overline{X}$. *Then* $< U, v > \; = \; 0$.

In fact, by local charts and a partition of unity, we can

reduce to the case $X = \mathbb{R}^n_+$, U having compact support. Suppose we have a regularising sequence ρ_k (see Theorem 2.7) with $\operatorname{supp} \rho \subset \{x_n < 0\}$. We know that $\rho_k * U \longrightarrow U$ in H^s, so that $\langle U, v\rangle = \lim \langle \rho_k * U, v\rangle = 0$ since the distribution v is zero in the neighbourhood of the support of the test function $\rho_k * U$.

This lemma shows that $\langle u, v\rangle$ is well defined and that $|\langle u, v\rangle| \leq \|\tilde{u}\|_s \|v\|_{-s}$ for any extension $\tilde{u}$ of u in H^s, so that $|\langle u, v\rangle| \leq \|u\|_s \|v\|_{-s}$. Conversely, let ℓ be a continuous linear form on $H^s(\overline{X})$. Since the restriction operator $r : H^s(\mathbb{R}^n) \longrightarrow H^s(\overline{X})$ is continuous, $\ell \circ r$ is a continuous linear form on $H^s(\mathbb{R}^n)$; thus it is identified with a distribution $v \in H^{-s}(\mathbb{R}^n)$ satisfying $\langle U, v\rangle = \ell(U_{|X})$ for all $U \in H^s(\mathbb{R}^n)$, so that $v \in H^{-s}_{\overline{X}}(\mathbb{R}^n)$ and $\|v\|_{-s} = \|\ell \circ r\| \leq \|\ell\|$. Similarly let ℓ be a continuous linear form on $H^{-s}_{\overline{X}}(\mathbb{R}^n)$; from the Hahn-Banach theorem, it extends into a continuous linear form (with the same norm) on $H^{-s}(\mathbb{R}^n)$ which is identified with a distribution $U \in H^s(\mathbb{R}^n)$. If we put $u = U_{|X}$, we have $u \in H^s(\overline{X})$, $\langle u, v\rangle = \ell(v)$ for $v \in H^{-s}_{\overline{X}}(\mathbb{R}^n)$ and $\|u\|_s \leq \|U\|_s = \|\ell\|$.

DEFINITION 3.6: *Let $\overline{X}$ be a manifold with boundary imbedded in a manifold* M *(see Remark 9.6, Chapter* I*) and let* $s \in \mathbb{R}$. *We denote by* $H^s_{loc}(\overline{X})$ *the space of distributions* $u \in H^s_{loc}(X)$ *such that* $\varphi \chi_* u \in H^s(\overline{\mathbb{R}}^n_+)$ *for any chart* $\chi : U \longrightarrow \tilde{U}$ *of* $\overline{X}$ *at the* *boundary and for all* $\varphi \in C_o^\infty(\tilde{U})$. *We equip* $H^s_{loc}(\overline{X})$ *with the topology defined by the family of semi-norms* $u \longmapsto \|\varphi\, \chi_* u\|_s$ *and semi-norms which are continuous in* $H^s_{loc}(X)$.

As in the proof of Theorem 2.19, it can be shown that it is actually sufficient to consider an atlas of $\overline{X}$ in the above discussion. We can thereby deduce that $H^s_{loc}(\overline{X})$ is a Fréchet space, and that if K is a compact subset of $\overline{X}$, the subspace $H^s_K(\overline{X})$(formed by elements of $H^s_{loc}(\overline{X})$ which are zero in $X \setminus K$) of $H^s_{loc}(\overline{X})$ is hilbertisable; we denote by $\|u\|_{s,k}$ an admissible Hilbert norm in $H^s_K(\overline{X})$. It can immediately be shown that $H^s_{loc}(\overline{X}) \subset C^k(\overline{X})$, with continuous injection, when $s > n/2 + k$; $\bigcap_s H^s_{loc}(\overline{X}) = C^\infty(\overline{X})$; a differential operator of degree m with coefficients in $C^\infty(\overline{X})$ is continuous from $H^s_{loc}(\overline{X})$ into $H^{s-m}_{loc}(\overline{X})$. By a partition of unity, we can construct a continuous linear extension operator from $H^s_{loc}(\overline{X})$ into $H^s_{loc}(M)$, and we can thereby deduce the continuity of the injection $C^\infty(\overline{X}) \longrightarrow H^s_{loc}(\overline{X})$ together with the density of $C_o^\infty(\overline{X})$ in $H^s_{loc}(\overline{X})$. We denote by $H^s_{\overline{X},comp}(M)$ the union of the $H^s_K(M)$ when K describes the family of compact subsets of $\overline{X}$ and we equip $H^s_{\overline{X},comp}(M)$ with the inductive limit topology. When we assume that M is endowed with a positive density, arguments analogous to those of Theorem 2.15 and Proposition 3.5 show that the dual of $H^s_{loc}(\overline{X})$ (resp: of $H^{-s}_{\overline{X},comp}(M)$) is identified with $H^{-s}_{\overline{X},comp}(M)$ (resp: with $H^s_{loc}(\overline{X})$) by the pairing

$< u, v > = < \tilde{u}, v >$ of $u \in H^s_{loc}(\overline{X})$ and $v \in H^{-s}_{\overline{X},comp}(M)$, where $\tilde{u} \in H^s_{loc}(M)$ is an arbitrary extension of u and where $< \tilde{u}, v >$ is the pairing in $H^s_{loc}(M) \times H^{-s}_{comp}(M)$.

REMARK 3.7: In the case where X is a regular open subset of $\mathbb{R}^n$, we can show that $H^s_{loc}(\overline{X})$ is the space of the $u \in \mathcal{D}'(X)$ such that $\varphi u \in H^s(\overline{X})$ for all $\varphi \in C^\infty_0(X)$, and also that the semi-norms $u \longmapsto \|\varphi u\|_s$ define its topology.

REMARK 3.8: Let Y be a manifold (with or without boundary) and let $\mathcal{F}$ be a complex vector bundle of rank N above Y. We naturally define the space $H^s_{loc}(Y; \mathcal{F})$ in a manner analogous to Definitions 2.18 and 3.6: by using trivialisations of $\mathcal{F}$ above coordinate patches of Y, we arrive at $H^s(\mathbb{R}^n; \mathbb{C}^N) = (H^s(\mathbb{R}^n))^N$ or at $H^s(\overline{\mathbb{R}}^n_+; \mathbb{C}^N) = (H^s(\overline{\mathbb{R}}^n_+))^N$. We leave to the reader the task - of generalising to the case of spaces $H^s_{loc}(Y; \mathcal{F})$ the properties of the spaces $H^s_{loc}(Y)$ given earlier.

4. TRACE THEOREMS, SPACES $H^s_0(X)$.

We put $\mathbb{R}^n = \{ x = (x', x_n) \mid x' \in \mathbb{R}^{n-1}, x_n \in \mathbb{R} \}$. We shall see that we can define in a natural manner the trace on the hyperplane $\{ x_n = 0 \}$ of an element $u \in H^s(\mathbb{R}^n)$ under the condition $s > \frac{1}{2}$.

THEOREM 4.1: *Let* $s \in \mathbb{R}$ *be such that* $s > \frac{1}{2}$. *The trace operator* $\gamma : C_0^\infty(R^n) \longrightarrow C_0^\infty(R^{n-1})$ *defined by* $(\gamma u)(x') = u(x', 0)$ *extends in a unique manner into a continuous linear operator (again denoted by* γ*) from* $H^s(\mathbb{R}^n)$ *into* $H^{s-\frac{1}{2}}(\mathbb{R}^{n-1})$.

PROOF: For $u \in C_0^\infty(R^n)$, the Fourier inversion formula shows that $u(x', 0) = \int e^{ix'\xi'} \hat{u}(\xi', \xi_n)\, d\!\!\!/\xi'\, d\!\!\!/\xi_n$ so that

$$(4.1.1.) \qquad \widehat{\gamma u}(\xi') = \int \hat{u}(\xi', \xi_n)\, d\!\!\!/\xi_n .$$

We further observe that the change of variable $\xi_n = (1 + |\xi'|^2)^{\frac{1}{2}} \eta_n$ shows that, for $s > \frac{1}{2}$:

$$(4.1.2.) \qquad \int (1 + |\xi'|^2 + \xi_n^2)^{-s}\, d\!\!\!/\xi_n = C_s (1 + |\xi'|^2)^{-s+\frac{1}{2}}$$

where

$$C_s = \int_R (1 + \eta_n^2)^{-s}\, d\!\!\!/\eta_n .$$

From (4.1.1) and the Schwarz inequality, we have

$$\widehat{\gamma u}(\xi') = \int (1 + |\xi|^2)^{-s/2} (1 + |\xi|^2)^{s/2} \hat{u}(\xi)\, d\!\!\!/\xi_n ,$$

$$|\widehat{\gamma u}(\xi')|^2 \leq \int (1 + |\xi|^2)^{-s}\, d\!\!\!/\xi_n \int (1 + |\xi|^2)^s |\hat{u}(\xi)|^2\, d\!\!\!/\xi_n .$$

We therefore deduce, by virtue of (4.1.2), that

$$(1 + |\xi'|^2)^{s-\frac{1}{2}} |\widehat{\gamma u}(\xi')|^2 \leq C_s \int (1 + |\xi|^2)^s |\hat{u}(\xi)|^2\, d\!\!\!/\xi_n$$

By integrating both sides of this inequality with respect to ξ', we obtain

$$\|\gamma u\|^2_{s-\frac{1}{2}} \leq C_s \|u\|^2_s ,$$

and hence Theorem 4.1, since $C_0^\infty(\mathbb{R}^n)$ is dense in $H^s(\mathbb{R}^n)$.

REMARK 4.2: For a given in $\mathbb{R}$, we likewise define the trace operator γ^a on the hyperplane $\{x_n = a\}$; we have $\gamma^a = \gamma \circ \tau_h$, where $h = (0, \ldots, 0, a)$; thus Proposition 2.3 shows that the mapping $a \longmapsto \gamma^a u$ is, for $u \in H^s(\mathbb{R}^n)$ fixed ($s > \frac{1}{2}$), continuous from $\mathbb{R}$ into $H^{s-\frac{1}{2}}(\mathbb{R}^{n-1})$. The distribution associated with this function (see Section 1.5, Chapter I) coincides with u; in fact, the formula

$$< u, \varphi \otimes \psi > = \int < \gamma^{x_n} u, \varphi > \psi(x_n)\, dx_n$$

$$(\text{for } \varphi \in C_0^\infty(\mathbb{R}^{n-1}),\ \psi \in C_0^\infty(\mathbb{R}))$$

is obviously true for $u \in C_0^\infty(\mathbb{R}^n)$, and therefore also for all $u \in H^s$ since both sides of this equation depend continuously on u (the continuity of the right-hand side follows from the dominated convergence theorem by virtue of the upper bound

$$|< \gamma^{x_n} u, \varphi >| \leq \|\gamma^{x_n} u\|_{s-\frac{1}{2}} \|\varphi\|_{-s+\frac{1}{2}} \leq C_s \|u\|_s \|\varphi\|_{-s+\frac{1}{2}}) .$$

COROLLARY 4.2: *Let there by* $s \in \mathbb{R}$ *and* $j \in \mathbb{N}$ *such that* $s-j > \frac{1}{2}$. *Then the operator* $\gamma_j = \gamma \circ D^j_{x_n}$ *is continuous from* $H^s(\mathbb{R}^n)$ *into* $H^{s-j-\frac{1}{2}}(\mathbb{R}^{n-1})$. *If* $u \in H^s(\mathbb{R}^n)$, *then* u *is a function of class* C^k *with respect to* x_n *with values in* $H^{s-k-\frac{1}{2}}(\mathbb{R}^{n-1})$ *for* $0 \leq k \leq j$.

PROOF: The first part follows from the continuity of

$D_{x_n}^j : H^s(R^n) \longrightarrow H^{s-j}(R^n)$ and from Theorem 4.1.

We assume $j \geq 1$. The last part of Proposition 2.3 shows that, for $u \in H^s$, the function $x_n \longmapsto \gamma^{x_n} u$ is in $C^1(R, H^{s-1-\frac{1}{2}}(R^{n-1}))$ and also that its derivative is the function $x_n \longmapsto \gamma^{x_n} \partial_{x_n} u$; this implies the second part of the corollary.

COROLLARY 4.3: (Trace on a hypersurface). *Let* X *be a manifold,* Y *a hypersurface of* X, *and suppose* $s > \frac{1}{2}$. *Then the operator* $\gamma : C_0^\infty(X) \longrightarrow C_0^\infty(Y)$ *of restriction to* Y *extends uniquely into a continuous linear operator (again denoted by* γ*) from* $H^s_{loc}(X)$ *into* $H^{s-\frac{1}{2}}_{loc}(Y)$.

PROOF: For $x^0 \in Y$, let $\chi : U \longrightarrow \widetilde{U}$ be a diffeomorphism of a neighbourhood of x^0 in X onto an open subset of $\mathbb{R}^n$ such that $\chi(U \cap Y) = \{ x \in \widetilde{U} \mid x_n = 0 \}$; we put

$$\chi^0 = \chi_{|U \cap Y} \ , \ \widetilde{U}_0 = \chi(U \cap Y) \ .$$

For $\varphi_0 \in C_0^\infty(\widetilde{U}_0)$, we consider $\varphi \in C_0^\infty(\widetilde{U})$ such that $\varphi_{|x_n=0} = \varphi_0$. If $u \in C_0^\infty(X)$, we have $\varphi_0 \chi_*^0(\gamma u) = \gamma(\varphi \chi_* u)$. From Theorem 4.1, we have $\|\varphi_0 \chi_*^0(\gamma u)\|_{s-\frac{1}{2}} \leq C_s \|\varphi \chi_* u\|_s$; thus γ is continuous from $C_0^\infty(X)$ (equipped with the topology induced by $H^s_{loc}(X)$) into $H^{s-\frac{1}{2}}_{loc}(Y)$. Theorem 4.3 then follows from the density of $C_o^\infty(X)$ in $H_{loc}^s(X)$.

COROLLARY 4.4: *Suppose* $s > \frac{1}{2}$. *The trace operator* $\gamma : C_o^\infty(\overline{\mathbb{R}}_+^n) \longrightarrow C_o^\infty(\mathbb{R}^{n-1})$ *defined by* $(\gamma u)(x') = u(x', 0)$ *extends uniquely into a continuous linear operator* γ *from* $H^s(\overline{\mathbb{R}}_+^n)$ *into* $H^{s-\frac{1}{2}}(\mathbb{R}^{n-1})$.

PROOF: For $u \in H^s(\overline{\mathbb{R}}_+^n)$, we put $\gamma u = \gamma\tilde{u}$ where $\tilde{u}$ is an arbitrary extension of u into $H^s(\mathbb{R}^n)$. The fact that $\gamma\tilde{u}$ does not depend on the extension $\tilde{u}$ chosen follows for example from Remark 4.2. We have $\|\gamma u\|_{s-\frac{1}{2}} \leq C_s \|\tilde{u}\|_s$ for all $\tilde{u}$, thus $\|\gamma u\|_{s-\frac{1}{2}} \leq C_s \|u\|_s$. When $u \in C_o^\infty(\overline{\mathbb{R}}_+^n)$, we can choose $\tilde{u} \in C_o^\infty(\mathbb{R}^n)$; therefore $\gamma u = \tilde{u}|_{x_n=0} = u|_{x_n=0}$. We have thus proved the existence of an extension γ; its uniqueness follows from the density of $C_o^\infty(\overline{\mathbb{R}}_+^n)$ in $H^s(\overline{\mathbb{R}}_+^n)$.

From Corollary 4.4, we proceed exactly as in the proof of Corollary 4.3 in order to establish the following:

COROLLARY 4.5: (Trace at the boundary) *Let* $\overline{X}$ *be a manifold with boundary and suppose* $s > \frac{1}{2}$. *Then the operator* $\gamma : C_o^\infty(\overline{X}) \longrightarrow C_o^\infty(\partial X)$ *of restriction to* ∂X *extends uniquely into a continuous linear operator* γ *from* $H_{loc}^s(\overline{X})$ *into* $H_{loc}^{s-\frac{1}{2}}(\partial X)$.

We conclude this section by studying the closure of $C_o^\infty(X)$ in $H_{loc}^s(\overline{X})$. To do this, we shall need the following:

PROPOSITION 4.6: *Let* δ *be the Dirac measure at* 0 *in* $\mathbb{R}$, *and let* $j \in \mathbb{N}$ *and* $t \in \mathbb{R}$. *Then, for* $v \in \mathcal{D}'(\mathbb{R}^{n-1})$ *non-zero,* $v \otimes D^j \delta$ *is in* $H^t(\mathbb{R}^n)$ *if and only if* $t+j < -\frac{1}{2}$ *and if* $v \in H^{t+j+\frac{1}{2}}(\mathbb{R}^{n-1})$. *We then have*

$$\text{(4.6.1)} \qquad \|v \otimes D^j \delta\|_t = C_{t,j} \|v\|_{t+j+\frac{1}{2}},$$

$$\text{with } C_{t,j}^2 = \int_{\mathbb{R}} (1+\eta_n^2)^t \, \eta_n^{2j} \, d\!\!\!/\eta_n .$$

PROOF: We put $u = v \otimes D^j \delta$; we have $\hat{u}(\xi) = \hat{v}(\xi') \, \xi_n^j$. By using, as in (4.1.2), the change of variable $\xi_n = (1 + |\xi'|^2)^{\frac{1}{2}} \eta_n$, we thus obtain:

$$\|u\|_t^2 = \int (1 + \eta_n^2)^t \, \eta_n^{2j} \, d\!\!\!/\eta_n \int (1 + |\xi'|^2)^{t+j+\frac{1}{2}} |\hat{v}(\xi')|^2 \, d\!\!\!/\xi'$$

hence the conclusion.

REMARK 4.7: Proposition 4.6 shows in particular that, for $t < -\frac{1}{2}$ the operator $v \longmapsto v \otimes \delta$ is continuous from $H^{t+\frac{1}{2}}(\mathbb{R}^{n-1})$ into $H^t(\mathbb{R}^n)$. By transposition we obtain, by putting $-t = s$, a continuous operator A from $H^s(\mathbb{R}^n)$ into $H^{s-\frac{1}{2}}(\mathbb{R}^{n-1})$ and it is immediately obvious that $A\varphi = \gamma\varphi$ for $\varphi \in C_o^\infty(\mathbb{R}^n)$. We thus obtain a fresh proof of the Trace Theorem 4.1.

PROPOSITION 4.8: *We put* $X = \mathbb{R}_+^n$. *For* $t \in \mathbb{R}$, $H^t_{\partial X,\text{comp}}(\mathbb{R}^n)$ *(resp:* $H^t_{\partial X}(\mathbb{R}^n)$*) is the space of distributions* u *of the form* $u = \sum_{0 \leq j < -t-\frac{1}{2}} v_j \otimes D^j \delta$, with $v_j \in H^{t+j+\frac{1}{2}}_{comp}(\mathbb{R}^{n-1})$ (resp : $v_j \in H^{t+j+\frac{1}{2}}(\mathbb{R}^{n-1})$)

(for $t \geq -\frac{1}{2}$, we have $H^t_{\partial X}(\mathbb{R}^n) = \{ 0 \}$).

PROOF: If u is of the above form, then $u \in H^t_{\partial X, comp}(\mathbb{R}^n)$ (resp: $u \in H^t_{\partial X}(\mathbb{R}^n)$) from Proposition 4.6. Conversely, suppose $u \in H^t_{\partial X, comp}(\mathbb{R}^n)$. Since $u \in \mathcal{E}'(\mathbb{R}^n)$ is supported by the hyperplane $\{ x_n = 0 \}$, we know (SCHWARTZ [1]) that u (which we assume non-zero) can be written uniquely in the form

$$u = \sum_{j=0}^{N} v_j \otimes D^j \delta \quad , \text{ with } v_j \in \mathcal{E}'(\mathbb{R}^{n-1}) \text{ and } v_N \neq 0.$$

By again using the change of variable $\xi_n = (1 + |\xi'|^2)^{\frac{1}{2}} \eta_n$, we obtain, since $\hat{u}(\xi) = \sum_{j=0}^{N} \hat{v}_j(\xi') \xi_n^j$:

$$\|u\|_t^2 = \iint (1 + \eta_n^2)^t \left|\sum_{j=0}^{N} (1 + |\xi'|^2)^{\frac{t}{2}+\frac{j}{2}+\frac{1}{4}} \hat{v}_j(\xi') \eta_n^j\right|^2 d\xi' \, d\eta_n < +\infty \; .$$

Thus the integral with respect to η_n is convergent for almost all ξ' ; by choosing such a ξ' with $\hat{v}_N(\xi') \neq 0$, we therefore deduce that $\int (1 + \eta_n^2)^t \eta_n^{2N} d\eta_n < +\infty$, so that $N < -t - \frac{1}{2}$.

In the same manner, the integral with respect to ξ' is convergent for almost all η_n; by giving (N + 1) such values to η_n, we see by solving a Cramer system that the N + 1 functions $(1 + |\xi'|^2)^{\frac{t}{2}+\frac{j}{2}+\frac{1}{4}} \hat{v}_j$ are in $L^2(\mathbb{R}^{n-1})$, i.e. $v_j \in H^{t+j+\frac{1}{2}}$ for $0 \leq j \leq N$. Finally suppose $u \in H^t_{\partial X}(\mathbb{R}^n)$. Since $u \in \mathcal{D}'(\mathbb{R}^n)$ is supported by the hyperplane $\{ x_n = 0 \}$, we know (SCHWARTZ [1] that u can be written uniquely in the form

$u = \sum_{j\geq 0} v_j \otimes D^j \delta$, where the sum is locally finite. By truncation (see Theorem 2.7), u is the limit of a sequence $u^{(k)} \in H^t_{\partial X, comp}(\mathbb{R}^n)$ and we have seen that

$u^{(k)} = \sum_{j\geq 0} v_j^{(k)} \otimes D^j \delta$ with $v_j^{(k)} = 0$ if $j \geq -t-\frac{1}{2}$.

Suppose we have $\psi_j \in C_o^\infty(\mathbb{R}^n)$ such that $< D^j \delta, \psi > = 1$, $< D^\ell \delta, \psi > = 0$ for $\ell \neq j$. For $\varphi \in C_o^\infty(\mathbb{R}^{n-1})$, we have

$$< v_j, \varphi > = < u, \varphi \otimes \psi_j >$$

$$= \lim_{k\to+\infty} < u^{(k)}, \varphi \otimes \psi_j > = \lim_{k\to+\infty} < v_j^{(k)}, \varphi >,$$

thus $v_j = 0$ if $j \geq -t-\frac{1}{2}$. We then proceed exactly as in the case $u \in H^t_{\partial X, comp}(\mathbb{R}^n)$ to prove that $v_j \in H^{t+j+\frac{1}{2}}(\mathbb{R}^{n-1})$.

THEOREM 4.9: *Let X be a regular open subset of* $\mathbb{R}^n$. *Then the closure* $H^s_{o,loc}(X)$ *(resp:* $H^s_o(X)$*) of* $C^\infty_o(X)$ *in* $H^s_{loc}(\overline{X})$ *(resp:* $H^s(\overline{X})$*) is the space of the* $u \in H^s_{loc}(\overline{X})$ *(resp:* $H^s(\overline{X})$*) such that* $\gamma\, \partial^\alpha u = 0$ *for all* α *satisfying* $|\alpha| < s-\frac{1}{2}$ *(*γ *denoting the trace operator on* ∂X*). In particular,* $C^\infty_o(X)$ *is dense in* $H^s_{loc}(\overline{X})$ *(resp:* $H^s(\overline{X})$*) if and only if* $s \leq \frac{1}{2}$.

PROOF: We can consider the space $H^s_{o,loc}(X)$ as being the bipolar space of $C^\infty_o(X)$ in $H^s_{loc}(X)$. Now, according to the end of Section 3, the space orthogonal to $C^\infty_o(X)$ is identified with the space of the $v \in H^{-s}_{\overline{X},comp}(\mathbb{R}^n)$ such that $< v, \varphi > = 0$ for all $\varphi \in C^\infty_o(X)$, i.e. with the space $H^{-s}_{\partial X, comp}(\mathbb{R}^n)$. Thus $u \in H^s_{loc}(\overline{X})$ is in $H^s_o(X)$ if and only if

(4.9.1.) $\quad < u, v > = 0 \quad$ for all $\quad v \in H^{-s}_{\partial X, comp}(\mathbb{R}^n)$.

By partition of unity, we see that, in (4.9.1), we can force v to have support in an open coordinate patch of $\bar{X}$ at the boundary, and this allows us to reduce to the case $X = \mathbb{R}^n_+$.

Proposition 4.8 then shows that (4.9.1) is equivalent to

$$< u, \sum_{0 \le j < s-\frac{1}{2}} v_j \otimes D^j \delta > = 0 \text{ for all } v_j \in H^{-s+j+\frac{1}{2}}_{comp}(\mathbb{R}^{n-1}) \ ,$$

i.e. to $\sum_{0 \le j < s-\frac{1}{2}} < \gamma_j u, (-1)^j v_j > = 0$ for all $v_j \in H^{-s+j+\frac{1}{2}}_{comp}(\mathbb{R}^{n-1})$, i.e. finally to $\gamma_j u = 0$ for $0 < j < s - \frac{1}{2}$. Theorem 4.9 is thus proved in the case of $H^s_{loc}(\bar{X})$. In the case of $H^s(\bar{X})$ we proceed in the same manner.

Let us consider more particularly the space $H^s_o(X)$ (i.e. the closure of $C^\infty_o(X)$ in $H^s(\bar{X})$) and let us equip it with the topology induced by $H^s(\bar{X})$. Since the injection of $C^\infty_o(X)$ into $H^s_o(X)$ is continuous and of dense image, the dual of $H^s_o(X)$ is naturally identified with a subspace of $\mathcal{D}'(X)$ and we shall limit ourselves here to studying the very simple case where $s \in \mathbb{N}$ (see LIONS-MAGENES [1] for the case $s \in \mathbb{R}$).

THEOREM 4.10 : *For* $m \in \mathbb{N}$, *the dual of* $H^m_o(X)$ *is identified topologically with* $H^{-m}(\bar{X})$.

PROOF: Let ℓ be a continuous linear form on $H^m_o(X)$; we identify this with the distribution $u \in \mathcal{D}'(X)$ defined by

$< u, \varphi > = \ell(\varphi)$ for all $\varphi \in C_0^\infty(X)$. From the Hahn-Banach Theorem and Proposition 3.5, there exists $U \in H_{\bar{X}}^{-m}(\mathbb{R}^n)$ such that $< u, \varphi > = < U, \varphi >$ for all $\varphi \in C_0^\infty(X)$. We thus have $u = U_{|X} \in H^{-m}(\bar{X})$. Conversely, suppose $u \in H^{-m}(\bar{X})$. From Proposition 3.2., u is of the form $u = \sum_{|\alpha| \leq m} \partial^\alpha f_\alpha$, with $f_\alpha \in L^2(X)$. We put $U = \sum_{|\alpha| \leq m} \partial^\alpha(f_\alpha^0)$, where $f_\alpha^o \in L^2(\mathbb{R}^n)$ is the extension of f_α by 0 outside of $\bar{X}$. For $\varphi \in C_0^\infty(X)$, we have $< u, \varphi > = < U, \varphi >$, which depends continuously on φ for the topology induced by $H^m(\bar{X})$, in view of Proposition 3.5 since $U \in H_{\bar{X}}^{-m}(\mathbb{R}^n)$. This proves that the dual of $H_o^m(X)$ is algebraically identified with $H^{-m}(\bar{X})$; the topological identification is obtained in similar manner.

REMARK 4.11 : Suppose $s \in \mathbb{R}$ is such that $s > \frac{1}{2}$. Propositions 4.6 and 4.8 show that the linear mapping

$$\prod_{0 \leq j < s-\frac{1}{2}} H^{-s+j+\frac{1}{2}}(\mathbb{R}^{n-1}) \longrightarrow H^{-s}(\mathbb{R}^n)$$

$$(v_j) \longmapsto \sum_{0 \leq j < s-\frac{1}{2}} v_j \otimes D^j \delta$$

is continuous, injective and of closed image.

Its transpose, which is actually the restriction mapping

$$H^s(\mathbb{R}^n) \longrightarrow \prod_{0 \leq j < s-\frac{1}{2}} H^{s-j-\frac{1}{2}}(\mathbb{R}^{n-1})$$

$$u \longmapsto (\gamma_j u)$$

is thus surjective.

By partition of unity and local charts at the boundary, we can show that this surjectivity result remains valid when $\mathbb{R}^n$ is replaced by $\bar{X}$, $\mathbb{R}^{n-1}$ by ∂X, and $\gamma_j u$ by $\gamma\left(\frac{1}{i}\frac{\partial}{\partial\nu}\right)^j u$, where X is a regular open subset of $\mathbb{R}^n$ with compact boundary ∂X, and where ν is a field of vectors on $\mathbb{R}^n$ transversal to ∂X.

REMARK 4.12 : Let $\mathcal{H}$ be a complex Hilbert space. We know that the properties of Fourier transformation generalise without difficulty to the case of functions or distributions with values in $\mathcal{H}$; for example, we have Parseval's formula

$$\int (u(x), v(x))_{\mathcal{H}}\, dx = \int (\hat{u}(\xi), \hat{v}(\xi))_{\mathcal{H}}\, đ\xi$$

for $u, v \in \mathcal{S}(\mathbb{R}^n; \mathcal{H})$. This allows us to generalise word for word all the preceding results and proofs to the case of Sobolev spaces of distributions with values in $\mathcal{H}$.

5. APPLICATION TO THE DIRICHLET PROBLEM

We return to the example in the introduction, in a rather more general form. Let X be a regular open subset of $\mathbb{R}^n$ with compact boundary ∂X and let P be the elliptic differential operator $P = -\Delta + V(x)$, where $V \in C^\infty(\bar{X})$ is such that there exist constants c and C satisfying $0 < c \leq V(x) \leq C$ for all $x \in \bar{X}$.

THEOREM 5.1 : *Suppose we have* $f \in H^{-1}(\bar{X})$ *and* $g \in H^{\frac{1}{2}}(\partial X)$. *Then there exists a unique* $u \in H^1(\bar{X})$ *such that*

$$\left\{\begin{array}{ll} Pu = f & \text{(in the sense of distributions in X)} \quad (5.1.1.) \\ u_{|\partial X} = g & \quad (5.1.2.) \end{array}\right.$$

PROOF: Equation (5.1.1.) is equivalent to

$< Pu, \varphi > = < f, \varphi >$ for all $\varphi \in C_0^\infty(X)$ i.e.

$$((u, \bar{\varphi})) = < f, \varphi > \quad \text{for all} \quad \varphi \in C_0^\infty(X) , \tag{5.1.3.}$$

where we have put

$$((u, v)) = \sum_{j=1}^{n} \int_X \partial_j u \cdot \overline{\partial_j v} \, dx + \int_X V u\bar{v} \, dx$$

for $u, v \in H^1(\bar{X})$. From the assumptions made concerning V(x), ((u, v)) is an inner product equivalent to the inner product of $H^1(\bar{X})$.

We first prove Theorem 5.1 in the particular case f = 0. Since $C_0^\infty(X)$ is dense in $H_0^1(X)$, (5.1.3) is then equivalent to

$$((u, v)) = 0 \quad \text{for all} \quad v \in H_0^1(X) .$$

We put $\sum = \{ u \in H^1(\bar{X}) \mid \gamma u = g \}$. Theorems 4.5, 4.9 and Remark 4.11 show that $\sum$ is a closed affine variety in $H^1(\bar{X})$ parallel to $H_0^1(X)$, and therefore that

$u \in H^1(\bar{X})$ satisfies (5.1.1), (5.1.2) if and only if u is the orthogonal projection (relative to the inner product $((u, v))$) of 0 onto Σ, i.e. if and only if u is the unique element of Σ which minimises the energy $E(u) = ((u, u))$.

It remains for us to prove Theorem 5.1 in the case $g = 0$. From Theorem 4.9, conditions (5.1.1), (5.1.2) are then equivalent to

$$((u, \bar{\varphi})) = \langle f, \varphi \rangle \quad \text{for all } \varphi \in C_0^\infty(X) \quad \text{and} \quad u \in H_0^1(X) .$$

Since $f \in H^{-1}(\bar{X})$, Theorem 4.10 shows that f is identified with a continuous linear form on $H_0^1(X)$. This means, from Riesz's theorem, that there exists a unique $u \in H_0^1(X)$ such that $((u, \bar{\varphi})) = \langle f, \varphi \rangle$ for all $\varphi \in C_0^\infty(X)$; this concludes the proof of Theorem 5.1.

The method used in the preceding proof, called the variational method, allows considerable generalisations (see LIONS [1]). For example, suppose X is a regular bounded open subset of $\mathbb{R}^n$ and $P(x, D_x)$ is a differential operator of degree $2m$ with coefficients in $C^\infty(\bar{X})$, which we can always assume to be of the form $Pu = \sum_{|\alpha|, |\beta| \leq m} D^\beta(a_{\alpha\beta} D^\alpha u)$ with $a_{\alpha\beta} \in C^\infty(\bar{X})$.

We assume that P is strongly elliptic in $\bar{X}$, i.e. that there exists $c > 0$ such that

$$\operatorname{Re} P_{2m}(x, \xi) \geq c\,|\xi|^{2m} \quad \text{for} \quad x \in \bar{X}, \quad \xi \in \mathbb{R}^n .$$

Let us consider the sesquilinear form with respect to $u, v \in C_0^\infty(X)$:

$$u, v \longmapsto \langle Pu, \bar{v} \rangle = \sum_{|\alpha| \le m, |\beta| \le m} \langle a_{\alpha\beta} D^\alpha u, \overline{D^\beta v} \rangle .$$

This extends into a continuous sesquilinear form

$u, v \longmapsto a(u, v)$ on $H^m(\bar{X}) \times H^m(\bar{X})$.

Gårding's inequality (see Theorem 8.6, Chapter IV) implies that there exists λ_0 such that

$$\text{(5.1.4.)} \quad \mathrm{Re}\; a(u, u) \ge \frac{c}{2} \|u\|_m^2 - \lambda_0 \|u\|_0^2 \quad \text{for } u \in H_0^m(X).$$

We can then state the following:

THEOREM 5.2 : *Suppose we have* $f \in H^{-m}(\bar{X})$ *and* $g_j \in H^{m-j-\frac{1}{2}}(\partial X)$ *for* $j = 0, \ldots, m-1$. *Then for* $\lambda \in \mathbb{C}$ *given such that* $\mathrm{Re}\,\lambda \ge \lambda_0$, *there exists a unique* $u \in H^m(\bar{X})$ *which is a solution of the Dirichlet problem:*

$$\begin{cases} (P + \lambda)u = f & \text{(in the sense of distributions in } X \text{)} \\ \gamma_j u = g_j & (j = 0, \ldots, m-1) \end{cases} ,$$

where we have put $\gamma_j u = \gamma\left(\frac{1}{i} \frac{\partial u}{\partial Y}\right)^j$, γ *being a field of vectors on* $\mathbb{R}^n$ *transversal to* ∂X.

The uniqueness of u clearly follows from (5.1.4). We do not give in detail the proof of the existence of u (see Lax-Milgram theorem, Exercise 7.4), as we shall be establishing some more general results in Chapter VI.

6. SOBOLEV SPACES AND REGULARISATION

Occasionally we need to prove a regularity result for solutions of an equation Pu = f from an a priori inequality of the type

$$\|v\|_{s_1} \leq C_K(\|Pv\|_{s_2} + \|v\|_{s_3})$$

for all $v \in \mathcal{D}_K$, with appropriate indices s_j. Since we cannot apply this inequality directly to a distribution u , we are led to regularise u and to estimate Sobolev norms of $\rho_\epsilon * u$ and of the commutator $[P , \rho_\epsilon *]u$.

For technical reasons, which we shall discuss in Chapter VII, section 6, we begin by introducing a family of equivalent norms on H^s which depend on a parameter $0 < \delta \leq 1$.

We follow closely the methods of HÖRMANDER [1], [2].

DEFINITION 6.1 : *If we have* $0 < \delta \leq 1$, *we put for* $u \in H^s(\mathbb{R}^n)$:

$$\|u\|^2_{s,\delta} = \int (1 + |\xi|^2)^{s+1}(1 + |\delta\xi|^2)^{-1} |\hat{u}(\xi)|^2 d\xi \quad . \tag{6.1.1}$$

It is clear that for δ fixed, this is a norm which is equivalent to $\|u\|^2_s$ and furthermore, we have:

$$\|u\|_{s,\delta} \xrightarrow[\delta \searrow 0]{} \|u\|_{s+1} \qquad \text{for} \quad u \in H^{s+1} \quad .$$

The usefulness of these norms is to show that if a distribution $u \in H^s$ satisfies $\|u\|_{s,\delta} \le C$ with C independent of δ, then this distribution is in H^{s+1} and $\|u\|_{s+1} \le C$.

We give below another expression for this norm

PROPOSITION 6.2 : *If we have* $s \in \mathbb{R}$, *there exist* C_1 *and* $C_2 > 0$ *such that*

$$(6.2.1)\quad C_2 \|u\|^2_{s,\delta} \le \inf_{\substack{u = u'+u'' \\ u',u'' \in H^{-\infty}}} \left(\frac{1}{\delta^2} \|u'\|^2_s + \|u''\|^2_{s+1} \right) \le C_1 \|u\|^2_{s,\delta}$$

for $0 < \delta \le 1$ *and* $u \in H^s$.

PROOF of the 2nd inequality. If we have $u \in H^s$, we utilise the decomposition $u = u' + u''$ where the Fourier transform of u' is zero for $|\xi| < 1/\delta$ and that of u" is zero for $|\xi| \ge 1/\delta$. Then, we can show that

$$\frac{1}{\delta^2} \|u''\|^2_s + \|u''\|^2_{s+1} \le 2\|u\|^2_{s,\delta}$$

by observing that for $|\delta\xi| \ge 1$:

$$\frac{1}{\delta^2}(1 + |\xi|^2)^s \le 2\,(1 + |\xi|^2)^{s+1}\,(1 + |\delta\xi|^2)^{-1}$$

and for $|\delta\xi| < 1$:

$$(1 + |\xi|^2)^{s+1} \le 2\,(1 + |\xi|^2)^{s+1}\,(1 + |\delta\xi|^2)^{-1} .$$

In order to prove the first inequality in (6.2.1), it is sufficient to note that we always have

$$\|u\|^2_{s,\delta} \le \frac{1}{\delta^2} \|u\|^2_s \quad \text{and} \quad \|u\|^2_{s,\delta} \le \|u\|^2_{s+1} .$$

This equivalent expression for the norm $\| \ \|_{s,\delta}$ will allow us to prove the following:

PROPOSITION 6.3:

a) *Let* $P(x,D)$ *be a differential operator of degree* m *with* C^∞ *coefficients which are constant outside a compact domain. Then for all* $s \in \mathbb{R}$ *there exists* C *such that*

$$\|Pu\|_{s,\delta} \le C \|u\|_{s+m,\delta} \text{ for } 0 < \delta \le 1 \text{ and } u \in H^{s+m}. \tag{6.3.1}$$

b) *Let* χ *be a diffeomorphism between two open subsets* Ω_1 *and* Ω_2 *of* $\mathbb{R}^n$*, and let* K *be a compact subset of* Ω_2*; then for all* $s \in \mathbb{R}$*, there exists* C *such that*

$$\|\chi^* u\|_{s,\delta} \le C \|u\|_{s,\delta} \text{ for } 0 < \delta \le 1 \text{ and } u \in H^s(\Omega_2) \cap \mathcal{E}'_K. \tag{6.3.2}$$

PROOF OF a). If $u = u' + u''$ is a decomposition of $u \in H^{s+m}$, from (6.2.1) we have

$$\|Pu\|^2_{s,\delta} \le \frac{1}{C_2} \left(\frac{\|Pu'\|^2_s}{\delta^2} + \|Pu''\|^2_{s+1} \right) .$$

By utilising the continuity of P in Sobolev spaces, we thereby deduce that there exists a constant C which depends only on s and P such that

$$\|Pu\|^2_{s,\delta} \le C \left(\frac{\|u'\|^2_{s+m}}{\delta^2} + \|u''\|^2_{s+m+1} \right)$$

which implies (6.3.1).

The proof of b) is identical and is based on the invariance of the norms $\| \ \|_s$ under diffeomorphism.

We now give an expression for $\|u\|_{s,\delta}$ using regularisations of u.

If we have $\rho \in C_0^\infty(\mathbb{R}^n)$, we assume that for an integer $h \geq 0$, we have

$$\hat{\rho}(\xi) = O(|\xi|^h) \quad , \quad \xi \to 0 \quad . \tag{6.3.3}$$

We assume also that:

$$\hat{\rho}(t\xi) = 0 \text{ for all } t \in \mathbb{R} \text{ implies } \xi = 0 \text{ if } \xi \in \mathbb{R}^n. \tag{6.3.4}$$

We can easily construct such functions, by putting $\rho = \Delta^h\theta$ where $\theta \in C_0^\infty$ and $\hat{\theta}(0) \neq 0$.

THEOREM 6.4 : *Suppose we have* $s \in \mathbb{R}$ *and a function* ρ *with* $h > s+1$. *Then, there exist* C_2 *and* $C_1 > 0$ *such that*

$$C_1\|u\|_{s,\delta}^2 \leq \|u\|_s^2 + \int_0^1 \|u * \rho_\epsilon\|_0^2 \, \epsilon^{-2(s+1)}\left(1 + \frac{\delta^2}{\epsilon^2}\right)^{-1} \frac{d\epsilon}{\epsilon} \leq C_2\|u\|_{s,\delta}^2 \tag{6.4.1}$$

for $u \in H^s$ *and* $0 < \delta \leq 1$. *We have put* $\rho_\epsilon(x) = \epsilon^{-n}\rho(x/\epsilon)$.

(The second inequality in (6.4.1) *does not make use of assumption* (6.3.4)*).*

PROOF: We have

$$\int_0^1 \|u * \rho_\epsilon\|_0^2 \ \epsilon^{-2(s+1)}(1 + \delta^2/\epsilon^2)^{-1} \ \frac{d\epsilon}{\epsilon} = \int F(\xi,\delta)\,|\hat{u}(\xi)|^2 d\xi$$

with

$$F(\xi,\delta) = \int_0^1 |\hat{\rho}(\epsilon\xi)|^2 \ \epsilon^{-2s-2}\,(1 + \delta^2/\epsilon^2)^{-1} \ \frac{d\epsilon}{\epsilon} . \tag{6.4.2}$$

Since $h > s+1$, F extends into a continuous function on $\mathbb{R}^n \times [0,1]$, thus, it is sufficient to show that there exist constants c_2 and $c_1 > 0$ such that

$$c_1 |\xi|^{2s+2}(1 + |\delta\xi|^2)^{-1} \le F(\xi,\delta) \le c_2 |\xi|^{2s+2}(1 + |\delta\xi|^2)^{-1} \tag{6.4.3}$$

for $|\xi| \ge 1$ and $\delta \in [0,1]$.

If we put $\eta = \dfrac{\xi}{|\xi|}$ and $t = \epsilon|\xi|$ in the integral (6.4.2), we obtain

$$F(\xi,\delta) = |\xi|^{2s+2} \int_0^{|\xi|} |\hat{\rho}(t\eta)|^2 \ t^{-2s-2}\left(1 + \frac{|\delta\xi|^2}{t^2}\right)^{-1} \ \frac{dt}{t} .$$

We therefore deduce $F(\xi,\delta) \le$

$$|\xi|^{2s+2}(1 + |\delta\xi|^2)^{-1}\left[\int_0^1 |\hat{\rho}(t\eta)|^2 \ t^{-2s-2}\,\frac{dt}{t} + \int_1^\infty |\hat{\rho}(t\eta)|^2 t^{-2s-2}\,\frac{dt}{t}\right]$$

and since $\hat{\rho} \in \mathcal{S}$, we thus obtain the second inequality in (6.4.3).

In order to prove the first inequality in (6.4.3), we observe that

$$F(\xi,\delta) \ge \tfrac{1}{4} \ |\xi|^{2s+2} \ (1 + |\hat{\rho}\xi|^2)^{-1} \int_{\frac{1}{2}}^1 \ |\hat{\rho}(t\eta)|^2 \ t^{-2s-2} \ \frac{dt}{t} , \ |\xi| \ge 1$$

and we take

$$c_1 = \inf_{|\eta|=1} \ \tfrac{1}{4} \int_{\frac{1}{2}}^1 \ |\hat{\rho}(t\eta)|^2 \ t^{-2s-2} \ \frac{dt}{t} .$$

We have $c_1 > 0$, since the integral depends continuously on η and $\hat{\rho}$ satisfies the condition (6.3.4).

The following theorem will be proved in Chapter IV in the more general case where P is a pseudo-differential operator.

THEOREM 6.5 : *Let* $P(x,D)$ *be a differential operator of degree* m *with* C^∞ *coefficients which are constant outside a compact domain. If we have* $s \in \mathbb{R}$, *then there exists* C *such that*

$$(6.5.1.)\quad \int_0^1 \|[P\,,\,\rho_\varepsilon *]u\|_0^2 \ \varepsilon^{-2s-2}(1+\delta^2/\varepsilon^2)^{-1}\,\frac{d\varepsilon}{\varepsilon} \le C\|u\|^2_{s+m-1,\,\delta}$$

for $u \in H^{s+m-1}$ *and* $0 < \delta \le 1$.

REMARK 6.6 : The upper bound (6.5.1) is still valid if the operator P depends continuously on a parameter varying within a compact space.

7. ADDITIONAL NOTES AND EXERCISES.

7.1 POINCARE'S INEQUALITY.

a) Suppose $\varphi \in C_0^\infty(R^n)$ is such that

$$\operatorname{supp}\varphi \subset \{\, x \mid a \le x_1 \le a+L \,\}\,.$$

Show that $$\int_{R^n} |\varphi(x)|^2\,dx \le L^2 \int_{R^n} \Big|\frac{\partial\varphi}{\partial x_1}(x)\Big|^2\,dx$$

[express φ as the integral of $\frac{\partial\varphi}{\partial x_1}$ and majorise the integral using the Cauchy-Schwarz inequality].

b) Let m be an integer ≥ 1. Deduce from the preceding question that there exists a function $]0, +\infty[\ni L \longrightarrow C(L) \in [0, +\infty[$ tending to 0 with L and such that

$$\|\varphi\|_{m-1} \leq C(L) \left(\sum_{|\alpha|=m} \|D^\alpha \varphi\|_0 \right) \text{ for all } \varphi \in C_0^\infty(\{|x| < L\}).$$

c) Let X be a regular open subset of $\mathbb{R}^n$ such that $X \subset \{|x| < L\}$. Prove that $\|u\|_{m-1} \leq C(L) \|u\|_m$ for all $u \in H_0^m(X)$ and show that $\sum_{|\alpha|=m} \|D^\alpha u\|_0$ is a norm equivalent to $\|u\|_m$ on $H_0^m(X)$.

d) Thus deduce that in Theorem 5.1 we can replace the assumption $0 < c \leq V(x) \leq C$ by $0 \leq V(x) \leq C$.

7.2 INVARIANCE OF H^s UNDER DIFFEOMORPHISM

a) With $s \in]0, 1[$ and $v \in C_0^\infty(\mathbb{R}^n)$, use Parseval's equality to prove the equality

$$\iint_{\mathbb{R}^n \times \mathbb{R}^n} \frac{|v(x) - v(y)|^2}{|x - y|^{n+s}} \, dx \, dy = \int F_s(\xi) \, |\hat{v}(\xi)|^2 \, d\!\!\!/\xi$$

where we have put $F_s(\xi) = \int_{\mathbb{R}^n} |e^{ix\xi} - 1|^2 \, |x|^{-n-2s} \, dx$.

b) Show that F_s is a homogeneous function of degree $2s$ with respect to ξ. Deduce from this that the function $v \longrightarrow [v]_s$ defined by

$$[v]_s^2 = \int |v(x)|^2 \, dx + \iint |v(x) - v(y)|^2 \, |x - y|^{-n-2s} \, dx \, dy$$

is equivalent to the norm $\|v\|_s$ on $H^s(\mathbb{R}^n)$.

c) Use the expression $[\, v \,]_s$ to prove Theorem 2.17 when $0 \leq s < 1$. Deduce from this the case $s \geq 0$, then finally the case $s < 0$ by making use of the duality of Sobolev spaces.

7.3 Show that the following inclusions are strict:

$$\mathcal{S} \subset H^{+\infty} \subset H^{-\infty} \subset \mathcal{S}'$$

[use for example the distributions $(1 + x^2)^{-1}$ and δ].

7.4 LAX-MILGRAM THEOREM

Let V be a Hilbert space and let a be a continuous sesquilinear form. We assume a to be coercive; this means that there exists $\alpha > 0$ such that $|a(u, u)| \geq \alpha \|u\|_V^2$ for all $u \in V$.

a) Prove that for any continuous antilinear form T on V, there exists a unique $u \in V$ such that

(7.4.1.) $$a(u, v) = T(v) \qquad \text{for all} \qquad v \in V.$$

[Let $A \in \mathcal{L}(V, V)$ be the operator such that $a(u, v) = (Au, v)_V$ for all u,v. Show that A is injective and that its image is both dense and closed. If f is the element of V such that $T(v) = (f, v)_V$, then the solution of (7.4.1) is given by $u = A^{-1}. f$].

b) Apply this result with $V = H_0^m(X)$ to prove Theorem 4.2, after first reducing to the case where the g_j are zero.

c) If we further assume that a is Hermitian, then prove that u is a solution of (7.4.1) if and only if u minimises the functional

$$v \in V \longrightarrow a(v, v) - 2 \operatorname{Re} T(v) \in \mathbb{R} \quad .$$

7.5 HÖRMANDER'S INEQUALITIES

Let $P = p(x, D)$ be a differential operator of degree m with coefficients which are C^∞ in a neighbourhood Ω of 0 in $\mathbb{R}^n$.

a) We put $P_1 = \dfrac{\partial p}{\partial \xi_1}(x, D)$.

By using formula (1.1.3) of Chapter I, show that

$$P_1 \varphi = i\, P(x_1 \varphi) - i\, x_1 P\varphi \qquad \text{for} \qquad \varphi \in C_0^\infty(\Omega) \quad .$$

b) Show that

$$|(x_1 P\varphi, P_1 \varphi)| \leq \ell\, \|P\varphi\|_0 \, \|P_1 \varphi\|_0 \leq \frac{\ell}{2}\left(\|P\varphi\|_0^2 + \|P_1 \varphi\|_0^2\right),$$

where $\ell = \sup\limits_{x \in \operatorname{supp} \varphi} |x_1|$.

c) By utilising Poincaré's inequality (Exercise 7.1), prove that there exists a function $L \longmapsto C(L)$ tending to 0 when $L \to 0$, such that

$$\|x_1 \varphi\|_{m-1} \leq C(L)\, \|\varphi\|_{m-1} \qquad \text{for} \qquad \varphi \in C_0^\infty(\{|x| < L\}),$$

$$|(x_1 \varphi, Q\varphi)| \leq C(L)\, \|\varphi\|_{m-1}^2 \qquad \text{for} \qquad \varphi \in C_0^\infty(\{|x| < L\}),$$

where $Q = [P^*, P_1]$ (note that Q is of degree $2m - 2$).

d) Deduce from this that there exist $L_0 > 0$, $C \geq 0$ and two functions: $L \longmapsto C(L)$, $L \longmapsto \varepsilon(L)$ from $]0, L_0]$ into $[0, +\infty[$, tending to 0 when $L \to 0$, such that

$$\text{(7.5.1.)} \quad \sum_{j=1}^{n} \left\| \frac{\partial p}{\partial \xi_j}(x, D)\, \varphi \right\|_0^2 \leq C(L)\left(\|P\varphi\|_0^2 + \|P^* \varphi\|_0^2\right) + \varepsilon(L)\, \|\varphi\|_{m-1}^2$$

for $\varphi \in C_0^\infty(\{|x| < L\})$, $L \leq L_0$.

e) In the case where P has constant coefficients and where Ω is a bounded open subset of $\mathbb{R}^n$, prove that $\|P\varphi\|_0 = \|P^* \varphi\|_0$, and that there exists $C \geq 0$ such that

$$\sum_{j=1}^{n} \| \frac{\partial p}{\partial \xi_j} (D) \varphi\|_0^2 \leq C \|P^* \varphi\|_0^2 \quad \text{for all} \quad \varphi \in C_0^\infty(\Omega) .$$

Thus deduce that if P is non-zero, there exists C such that

$$\|\varphi\|_0^2 \leq C \|P^* \varphi\|_0^2 \quad \text{for all} \quad \varphi \in C_0^\infty(\Omega) .$$

7.6 LOCAL SOLVABILITY FOR DIFFERENTIAL OPERATORS OF PRINCIPAL TYPE WITH REAL PRINCIPAL SYMBOL

Suppose Ω is an open subset of $\mathbb{R}^n$; let $P = P(x, D)$ be a differential operator of degree m with C^∞ coefficients in Ω. We suppose that P is of principal type, i.e. that:

$$d_\xi p_m(x, \xi) \neq 0 \quad \text{for} \quad x \in \Omega , \quad \xi \in \mathbb{R}^n \setminus 0 ,$$

where P_m is the principal symbol (homogeneous of degree m) of P.

With $a \in \Omega$, we denote by $B(a, L)$ the open ball with centre a and radius L.

a) Prove that there exists $c > 0$ such that

$$\sum_{j=1}^{n} | \frac{\partial p_m}{\partial \xi_j} (a, \xi) |^2 \geq c |\xi|^{2(m-1)} \quad \text{for} \quad \xi \in \mathbb{R}^n .$$

b) Suppose we have $L_0 > 0$ such that $B(a, L_0) \subset \Omega$.

Deduce from the previous question and from section 7.1 that

there exists $C \geq 0$ such that

$$\|\varphi\|^2_{m-1} \leq C \sum_{j=1}^{m} \left\| \frac{\partial p_m}{\partial \xi_j} (a, D) \varphi \right\|^2_0 \quad \text{for} \quad \varphi \in C^\infty_0(B(a, L)).$$

Deduce from this, again by using section 7.1, that there exist $C_1, C \geq 0$ such that:

$$(7.6.1.) \quad \left.\begin{aligned} &\|\varphi\|^2_{m-1} \leq C \sum_{j=1}^{n} \left\| \frac{\partial p}{\partial \xi_j} (a, D) \varphi \right\|^2_0 \\ &\text{and} \\ &\|\varphi\|^2_{m-1} \leq C \sum_{j=1}^{n} \left\| \frac{\partial p}{\partial \xi_j} (x, D) \varphi \right\|^2_0 \end{aligned}\right\} \text{for } \varphi \in C^\infty_0(B(a, L_0)) \text{ with } L_0 \text{ sufficiently small}.$$

We shall actually give later (see section 11.3 of Chapter IV) a simpler proof of (7.6.1).

c) Prove, with the aid of (7.5.1) and (7.6.1) that there exists a function $L \to M(L)$ from $]0, L_0]$ into $[o, +\infty[$, tending to 0 when $L \to 0$, such that:

$$(7.6.2.) \quad \|\varphi\|^2_{m-1} \leq M(L) \left(\|P\varphi\|^2_0 + \|P^* \varphi\|^2_0 \right) \text{ for } \varphi \in C^\infty_0(B(a, L)).$$

d) Henceforth we make the supplementary assumption that the coefficients of the polynomial p_m are real-valued.

Prove that there exists $M_0 \geq 0$ such that

$$(7.6.3.) \quad \|P\varphi\|^2_0 \leq 2 \|P^* \varphi\|^2_0 + M_0 \|\varphi\|^2_{m-1}$$

for $\varphi \in C^\infty_0(B(a, L)$.

e) Deduce from (7.6.2) and (7.6.3) that there exist $L > 0$ and $M \geq 0$ such that

(7.6.4.) $\|\varphi\|^2_{m-1} \leq M \|P^* \varphi\|^2_0$ for $\varphi \in C^\infty_0(B(a, L))$.

f) With the number L chosen in this manner, suppose $f \in H^{-(m-1)}(R^n)$. We put $E = \{ P^* \varphi \mid \varphi \in C^\infty_0(B(a, L))\}$.

Deduce from (7.6.4) that

$$\ell \;:\; E \longrightarrow C$$
$$\psi = P^* \varphi \longrightarrow (f, \varphi)$$

is a well-defined antilinear form on E, and is continuous when we equip E with the topology induced by L^2 (B(a,L)). Infer from this that there exists $u \in L^2(B(a, L))$ such that $Pu = f$ in B(a, L).

CHAPTER 3

SYMBOLS, OSCILLATORY INTEGRALS AND STATIONARY-PHASE THEOREMS

1. INTRODUCTION.

Let P(x, D) be a differential operator of degree m with C^∞ coefficients in an open subset X of $\mathbb{R}^n$. By Fourier transformation, we can express it in the form:

$$(1.1.)\quad (Pu)(x) = \int_{\mathbb{R}^n} e^{ix.\xi}\, P(x, \xi)\, \hat{u}(\xi)\, đ\xi, \qquad u \in C_0^\infty(X)$$

If we assume that P - P(D) is elliptic with constant coefficients, we have seen (see Chapter I, Proposition 6.7) that we can define a parametrix F of P in the form $F = (\mathcal{F}^{-1})\left(\frac{\theta(\xi)}{P(\xi)}\right)$.

Letting Q be the operator of convolution with F, we likewise say that this is a parametrix of P because the operators P o Q - I and Q o P - I are regularising. The operator Q is expressed by

$$(Qu)(x) = \int e^{ix\xi}\, \frac{\theta(\xi)}{P(\xi)}\, \hat{u}(\xi)\, đ\xi \quad , \qquad u \in C_0^\infty$$

and we have proved that the function $q = \frac{\theta}{P}$ behaves for $|\xi| \to +\infty$ like a homogeneous function of degree -m; more precisely, for any multi-index α, we have

$$|\partial^\alpha q(\xi)| = O((1+|\xi|)^{-m-|\alpha|}) .$$

When P(x, D) is elliptic with variable coefficients, we are thus led to try to construct a parametrix Q of P in the form

$$(1.2.)\qquad (Qu)(x) = \int e^{ix\cdot\xi}\, q(x,\xi)\, \hat{u}(\xi)\, đ\xi \quad , \qquad u \in C_0^\infty(X)$$

where $q(x, \xi)$ behaves for $|\xi| \to +\infty$ like the function $q(\xi)$. For example, as a first approximation, we may use the function

$$\frac{\theta(\xi)}{P_m(x,\xi)} .$$

We are thus led to define and study a class of such functions q, termed symbols; the operators associated with the formula (1.2) will be called pseudo-differential operators.

2. SYMBOLS

Let X be an open subset of $\mathbb{R}^n$ $(n \geq 0)$ and let N be an integer ≥ 1. We define the space $S^m(X \times \mathbb{R}^N)$ of symbols of degree $m \in \mathbb{R}$.

DEFINITION 2.1 : *We denote by* $S^m(X \times \mathbb{R}^N)$ *the set of complex-valued functions* $a \in C^\infty(X \times \mathbb{R}^N)$ *such that, for any compact subset* K *of* X *and any multi-indices* α, β *there exists* C *such that*

$$(2.1.1.)\qquad |\partial_x^\beta \partial_\theta^\alpha a(x,\theta)| \leq C(1+|\theta|)^{m-|\alpha|} \quad \text{for } x \in K,\ \theta \in \mathbb{R}^N .$$

We say that θ *is the frequency variable.*

EXAMPLE 2.2 : A polynomial $p(x, \xi) = \sum_{|\alpha| \leq m} a_\alpha(x)\xi^\alpha$ with coefficients $a_\alpha \in C^\infty(X)$ is in $S^m(X \times \mathbb{R}^n)$.

For $m \in \mathbb{R}$, the function $(x, \xi) \longrightarrow (1 + |\xi|^2)^{m/2}$ is in $S^m(X \times \mathbb{R}^n)$.

In order to generalise Definition 2.1 to certain open subsets of $X \times \mathbb{R}^N$, we next introduce a number of concepts. The multiplicative group $\mathbb{R}^+$ acts on $X \times \mathbb{R}^N$ by: $t \in \mathbb{R}^+$, $(x, \theta) \longmapsto (x, t\theta)$. A subset of $X \times \mathbb{R}^N$ is said to be conic if it is invariant under this action.

We define the conic support of a function $a \in C^\infty(X \times \mathbb{R}^N)$ to be the smallest closed conic subset outside of which a is zero. If $\Gamma \subset X \times \mathbb{R}^N$, we put $\Gamma \setminus 0 = \{ (x, \theta) \in \Gamma \mid \theta \neq 0 \}$, and $\Gamma^c = \{ (x, t\theta) | (x, \theta) \in \Gamma, t \geq 1 \}$. We also put $\mathbb{R}^N_0 = \mathbb{R}^N \setminus 0$.

DEFINITION 2.3 : *Let* Γ *be an open conic subset of* $X \times \mathbb{R}^N$. *We denote by* $S^m(\Gamma)$ *the set of complex-valued functions* $a \in C^\infty(\Gamma)$ *such that, for any compact subset* K *of* Γ *and for any multi-indices* α, β, *there exists* C *such that:*

(2.3.1.) $$|\partial_x^\beta \partial_\theta^\alpha a(x, \theta)| \leq C(1 + |\theta|)^{m-|\alpha|} \text{ for } (x, \theta) \subset K^c$$

We put

$$S^{+\infty}(\Gamma) = \bigcup_m S^m(\Gamma) \ , \quad S^{-\infty}(\Gamma) = \bigcap_m S^m(\Gamma) \ .$$

$S^m(\Gamma)$ is obviously a vector sub-space of $C^\infty(\Gamma)$; a function $a \in C^\infty(\Gamma)$ which is *homogeneous* of degree m with respect to θ, (i.e. satisfying $a(x,t\theta) = t^m\, a(x,\theta)$ for $t > 0$, $(x,\theta) \in \Gamma$) is in $S^m(\Gamma)$. A function $a \in C^\infty(\Gamma)$ which is zero for $|\theta|$ sufficiently large is in $S^{-\infty}(\Gamma)$. We can immediately show the following:

PROPOSITION 2.4 : *We have* $S^{-\infty}(\Gamma) \subset S^m(\Gamma) \subset S^{m'}(\Gamma) \subset S^{+\infty}(\Gamma)$ *for* $m \le m'$.

Suppose $a \in S^m(\Gamma)$, $b \in S^{m'}(\Gamma)$; *then* $ab \in S^{m+m'}(\Gamma)$, $\partial_x^\beta \partial_\theta^\alpha a \in S^{m-|\alpha|}(\Gamma)$.

Suppose $a \in C^\infty(\Gamma)$; *then* $a \in S^m(\Gamma)$ *if and only if* $a \in S^m(\Gamma \setminus 0)$.

In particular, $S^{+\infty}(\Gamma)$ is a commutative algebra, and $S^{-\infty}(\Gamma)$ is an ideal of this; we write $a \sim b$ when $a,b \in S^{+\infty}(\Gamma)$, with $a - b \in S^{-\infty}(\Gamma)$. Being a symbol is a local property in the sense of open conic sets:

PROPOSITION 2.5 : *Let* Γ *be an open conic subset of* $X \times \mathbb{R}^N$. *Suppose* $a \in C^\infty(\Gamma)$. *If, for any point* $(x_0, \theta_0) \in \Gamma \setminus 0$, *there exists a conic neighbourhood* Γ_0 *of* (x_0,θ_0) *in* $\Gamma \setminus 0$ *such that* $a_{|\Gamma_0} \in S^m(\Gamma_0)$, *then* $a \in S^m(\Gamma)$.

PROOF: According to the final part of Proposition 2.4, we can assume $\Gamma = \Gamma \setminus 0$. Let K be a compact subset of Γ; we put $\epsilon = \inf_{(x,\theta) \in K} |\theta|$; we have $\epsilon > 0$. By hypothesis, for any $(x_o, \theta_o) \in \Gamma$, there exists a conic neighbourhood V_o of (x_o, θ_o) in Γ such that the upper bounds (2.3.1) are satisfied for $(x, \theta) \in V$, $|\theta| \geq \varepsilon$. By covering K by a finite number of such neighbourhoods V_o, we see that we obtain the upper bounds (2.3.1) for $(x, \theta) \in K^c$.

We shall now see that symbols of degree m are preserved under certain changes of variable. More precisely:

PROPOSITION 2.6 : *Let* Γ_j *be an open conic subset of* $R^{n_j} \times R_o^{N_j}$, $(j = 1, 2)$.

Let $f : \Gamma_1 \longrightarrow \Gamma_2$ *be* C^∞ *and homogeneous of degree 1 (i.e. commuting with the action of* $\mathbb{R}^+$*).*

$$(x, \theta) \longmapsto (y, w)$$

If $b \in S^m(\Gamma_2)$, *then* $a = b \circ f \in S^m(\Gamma_1)$.

PROOF: Let K_1 be a compact subset of Γ_1; we have $f(K_1^c) = K_2^c$, with $K_2 = f(K_1)$ a compact subset of Γ_2. We put $f(x,\theta) = (g(x,\theta), h(x,\theta))$; g is homogeneous of degree 0 with respect to θ, and h is homogeneous of degree 1 with respect to θ; since $g(K_1)$ is compact in $\mathbb{R}_o^{N_2}$, there exist c, c' > 0 such that $c|\theta| \leq |h(x, \theta)| \leq c'|\theta|$ for $(x, \theta) \in K_1^c$. We thus obtain $|a(x, \theta)| \leq C(1 + |h(x, \theta)|)^m \leq C'(1 + |\theta|)^m$ for $(x, \theta) \in K_1^c$. We now establish the upper bounds (2.3.1) for $(x, \theta) \in K_1^c$ by recurrence on $|\alpha| + |\beta|$. We assume them to be satisfied (for any b and any m) when $|\alpha| + |\beta| \leq p-1$. Now, we have

$$\begin{cases} \dfrac{\partial a}{\partial x_j} = \displaystyle\sum_{k=1}^{n} \left(\frac{\partial b}{\partial y_k} \circ f\right) \frac{\partial g_k}{\partial x_j} + \sum_{k=1}^{N} \left(\frac{\partial b}{\partial \theta_k} \circ f\right) \frac{\partial h_k}{\partial x_j} \\ \dfrac{\partial a}{\partial \theta_j} = \displaystyle\sum_{k=1}^{n} \left(\frac{\partial b}{\partial y_k} \circ f\right) \frac{\partial g_k}{\partial \theta_j} + \sum_{k=1}^{N} \left(\frac{\partial b}{\partial \theta_k} \circ f\right) \frac{\partial h_k}{\partial \theta_j} \end{cases}$$

with $\dfrac{\partial g_k}{\partial x_j} \in S^0(\Gamma_1)$, $\dfrac{\partial h_k}{\partial x_j} \in S^1(\Gamma_1)$, $\dfrac{\partial g_k}{\partial \theta_j} \in S^{-1}(\Gamma_1)$.

Since $\dfrac{\partial b}{\partial y_k} \circ f$, $\dfrac{\partial b}{\partial \theta_k} \circ f$, satisfy upper bounds of the type (2.3.1) for $|\alpha| + |\beta| \leq p-1$ (for degrees respectively m, m-1), Leibniz's formula proves that the same is also true of $\dfrac{\partial a}{\partial x_j}$, $\dfrac{\partial a}{\partial \theta_j}$ (for degrees respectively m, m-1). We thus obtain for a the upper bounds (2.3.1) when $|\alpha| + |\beta| \leq p$.

The following is a useful special case of this proposition. Suppose Γ_1 is an open conic subset of $X \times \mathbb{R}^N_0$ and suppose f is the mapping corresponding to passage into polar coordinates in $\mathbb{R}^N_0$: $X \times \mathbb{R}^N_0 \ni (x, \theta) \overset{f}{\longmapsto} (x, \frac{\theta}{|\theta|}, |\theta|) \in X \times S_{N-1} \times \mathbb{R}^+$. Then in this case f is a diffeomorphism, homogeneous of degree 1, of Γ_1 onto the open conic subset $\Gamma_2 = f(\Gamma_1) \subset X \times S_{N-1} \times \mathbb{R}^+$, and Proposition 2.6 shows that :

(2.6.1.) $a \in S^m(\Gamma_1)$ if and only if $a \circ f^{-1} \in S^m(\Gamma_2)$.

The two preceding propositions allow us to generalise the

definition of $S^m(\Gamma)$.

DEFINITION 2.7 : *We define a conic manifold to be a manifold* M *equipped with a* C^∞ *action of the multiplicative group* $\mathbb{R}^+$ *such that:*

(2.7.1.) *For any* $\lambda \in M$, *there exists a diffeomorphism* χ, *homogeneous of degree* 1, *of an open conic neighbourhood* U *of* λ *in* M *onto an open conic subset* Γ *of* $\mathbb{R}^n \times \mathbb{R}^N_o$.

DEFINITION 2.8 : *Suppose we have a conic manifold* M *and* $m \in \mathbb{R}$. $S^m(M)$ *denotes the space of functions* $a \in C^\infty(M)$ *such that, for any* $\lambda \in M$, *there exists a diffeomorphism* χ *satisfying the conditions of* (2.7.1) with $a \circ \chi^{-1} \in S^m(\Gamma)$.

EXAMPLE 2.9 : If E is a vector bundle on a C^∞ manifold, we denote by $E \setminus 0$, the bundle E deprived of its 0-section. For $t \in \mathbb{R}^+$, we take as action multiplication by t in each fibre. Then $E \setminus 0$, and any open conic subset of $E \setminus 0$, is a conic manifold. If M is a conic manifold, any C^∞ function, homogeneous of degree m in M, is in $S^m(M)$.

PROPOSITION 2.10 :

i) *Let* M_1, M_2 *be two conic manifolds and let* $f : M_1 \longrightarrow M_2$ *be a* C^∞ *mapping of degree* 1. *If* $b \in S^m(M_2)$, *then* $a = b \circ f \in S^m(M_1)$.

ii) *Let* M *be a conic manifold, and let* $(U_j)_{j \in J}$ *be a covering of* M *by open conic sets. Suppose we have* $m \in \mathbb{R}$ *and* m' *with* $-\infty \leq m' < m$. *Let there be* $a_j \in S^m(U_j)$ *such that, in* $U_j \cap U_k$, *we have*

$a_j \equiv a_k \pmod{S^{m'}(U_j \cap U_k)}$ *for all* $j, k \in J$. *Then there exists* $a \in S^m(M)$ *unique modulo* $S^{m'}(M)$ *such that*

$a_{|U_j} \equiv a_j \pmod{S^{m'}(U_j)}$ *for all* $j \in J$.

PROOF: Part (i) is an immediate consequence of Proposition 2.6. In order to establish (ii), we first observe that the quotient space $M' = M / \mathbb{R}^+$ is a manifold and that the projection $p: M \to M'$ is C^∞; we put $U'_j = p(U_j)$; since U_j is conic, we have $U_j = p^{-1}(U'_j)$; since the U'_j form an open covering of M', we know that we can therefore find a locally finite sub-covering, and hence by inverse image under p a sub-covering of the covering of M by the U_j which is locally finite in the conic sense (i.e. such that for any compact K of M, there is only a finite number of indices corresponding to a non-empty intersection with $\mathbb{R}^+ K$). It will therefore be sufficient to prove (ii) when the covering is of this form. By a partition of unity, we know that there exist $\alpha'_j \in C^\infty(M')$ such that $\operatorname{Supp} \alpha'_j \subset U'_j$, $\sum_j \alpha'_j = 1$ in M'; by putting $\alpha_j = \alpha'_j \circ p$, we obtain functions $\alpha_j \in C^\infty(M)$, homogeneous of degree 0, with $\operatorname{Supp} \alpha_j \subset U_j$, $\sum_j \alpha_j = 1$ in M.

We put $a = \sum_j \alpha_j a_j$ (where $\alpha_j a_j$ denotes the extension by 0 outside U_j of $\alpha_{j|U_j} \cdot a_j$) . We have $a \in S^m(M)$, and, for $k \in J$ we have: $a_{|U_k} = \sum_j (\alpha_j a_j)_{|U_k} \equiv \sum_j \alpha_{j|U_k} a_k$ (mod $S^{m'}(U_k)$) and therefore $a_{|U_k} \equiv a_k$ (mod $S^{m'}(U_k)$); hence the existence of a. The uniqueness of a modulo $S^{m'}(M)$ is obvious.

We now consider a conic manifold N, and a conic sub-manifold M of N (i.e. a sub-manifold M of the manifold N which is stable under the action of $\mathbb{R}^+$ in M). If $a \in S^m(N)$, part (i) of the preceding proposition shows that $a_{|M} \in S^m(M)$. Conversely, we have:

PROPOSITION 2.11 : *Let* M *be a locally closed conic sub-manifold of a conic manifold* N; *for* $a \in S^m(M)$, *there exists* $\tilde{a} \in S^m(N)$ *extending* a.

PROOF: We first show that the proposition is true when we restrict ourselves to a sufficiently small conic neighbourhood (in N) of any point of M. By using the diffeomorphism χ of (2.7.1), then by taking polar coordinates in $\mathbb{R}^N_o$ (see (2.6.1)), we can reduce to $N = Y \times \mathbb{R}^+$, where Y is a manifold. The projection on Y of the sub-manifold M of N is a locally closed sub-manifold X of Y and we have $M = X \times \mathbb{R}^+$ since M is conic.

By taking local coordinates in Y we reduce to the case where $N = (U \times V) \times \mathbb{R}^+$, $M = (U \times \{0\}) \times \mathbb{R}^+$ where U, V are open neighbourhoods of 0 in $\mathbb{R}^p$, $\mathbb{R}^q$. If $a = a(y ; t) \in S^m(M)$, we define $\tilde{a} = \tilde{a}(y, z ; t) \in S^m(N)$ by $\tilde{a}(y, z ; t) = a(y, t)$.

In order to prove that the proposition is true globally, we consider a covering of M by open conic subsets W_j of N which are locally finite (in the conic sense) and, as in the proof of Proposition 2.10, we construct functions $\alpha_j, \beta_j \in C^\infty(N)$, homogeneous of degree 0, with $\operatorname{supp} \alpha_j \subset \operatorname{supp} \beta_j \subset W_j$, $\beta_j = 1$ in $\operatorname{supp} \alpha_j$ and $\sum \alpha_j = 1$ in the neighbourhood of M. We have $a = \sum a_j$, with $a_j = \alpha_j . a$; now, we have seen that there exist $\tilde{a}_j \in S^m(W_j)$ such that $\tilde{a}_{j|W_j \cap M} = a_{j|W_j \cap M}$. If we put $\tilde{a} = \sum \beta_j \tilde{a}_j$, then $\tilde{a} \in S^m(N)$ and $\tilde{a}_{|M} = \sum (\beta_j . \tilde{a}_j)_{|M} = \sum a_j = a$.

In applications, we encounter symbols which possess an expansion into homogeneous components in the following sense:

DEFINITION 2.12 : *Let* M *be a conic manifold. We say that* $a \in S^m(M)$ *is a classical symbol of degree* m *if there exists for any* $j \in \mathbb{N}$ *a function* $a_j \in C^\infty(M)$, *homogeneous of degree* m-j, *such that* $\left(a - \sum_{j=0}^{k-1} a_j\right) \in S^{m-k}(M)$ *for any integer* $k \geq 1$. *We denote by* $S^m_{cl}(M)$ *the space of classical symbols of degree* m.

In the case of the space $X \times \mathbb{R}^N$, we denote by $S^m_{cl}(X \times \mathbb{R}^N)$ the space of the $a \in C^\infty(X \times \mathbb{R}^N)$ such that

$$a_{|X \times \mathbb{R}^N_o} \in S^m_{cl}(X \times \mathbb{R}^N_o) .$$

It is clear that the functions a_j are uniquely determined for $a \in S^m_{cl}(M)$; the function a_o is called the principal part, homogeneous of degree m, of a.

EXAMPLE 2.13 : We can show that the symbols considered in Example 2.2 are classical, and that they admit respectively as principal part of degree m:

$$\sum_{|\alpha| = m} a_\alpha(x)\xi^\alpha \quad , \quad |\xi|^m \quad .$$

3. ELLIPTIC SYMBOLS

DEFINITION 3.1 : *Suppose* M *is a conic manifold and* $m \in \mathbb{R}$. *We say that a symbol* $a \in S^m(M)$ *is elliptic of degree* m *if there exists* $b \in S^{-m}(M)$ *with* $a.b \equiv 1$ (modulo $S^{-1}(M)$) .
Note that this definition only depends on the class of a in $S^m(M)/S^{m-1}(M)$.

From Proposition 2.10, ellipticity is a local property in the sense of conic neighbourhoods; in order to study this, we can therefore reduce to the case where M is an open conic subset of $X \times \mathbb{R}^N_o$.

PROPOSITION 3.2: *Let* Γ *be an open conic subset of* $X \times \mathbb{R}^N_o$; *a symbol* $a \in S^m(\Gamma)$ *is elliptic of degree* m, *if and only if, for any compact subset* K *of* Γ, *there exists* $c > 0$ *such that*

$$(3.2.1.) \qquad |a(x, \theta)| \geq c\,|\theta|^m$$

$$\text{for } (x, \theta) \in K^c \text{ and } |\theta| \geq \frac{1}{c} .$$

PROOF: We assume that (3.2.1) is satisfied; we let $K_1 \subset K_2 \subset \ldots \subset K_j \subset \ldots$ be an exhaustive sequence of compact subsets of Γ and we let $c_1 \geq c_2 \geq \ldots \geq c_j \geq \ldots$ be a sequence of associated constants, in accordance with (3.2.1). We suppose that $\varepsilon \in \mathbb{R}^+$ and that F is a subset of Γ, and we put

$$F^\varepsilon = F \cap \{ (x, \theta) \mid |\theta| \geq \frac{1}{\varepsilon} \} \quad ; \text{ then the set}$$

$$A = \left(\bigcup_{j\geq 1} (K_j^c)^{1/2\, c_j} \right) \text{is included in the interior } B \text{ of the set}$$

$$\left(\bigcup_{j\geq 1} (K_{j+1}^c)^{c_j} \right) \text{since } K_j \subset\subset K_{j+1} .$$

It can easily be shown that A is closed, and we take a function $\rho \in C^\infty(\Gamma)$ equal to 1 on A and with support in B. Since a does not vanish on B, we define $b = \frac{\rho}{a} \in C^\infty(\Gamma)$. We put $r = ab - 1$; to prove that $r \in S^{-\infty}(\Gamma)$, we observe that any point $(x, \theta) \in \Gamma$ admits a conic neighbourhood Γ_o such that for $\varepsilon > 0$, sufficiently small, we have $\Gamma_o^\varepsilon \subset A$, and since r is zero in the neighbourhood of A it suffices to apply Proposition 2.5. Likewise, in order to prove that $b \in S^{-m}(\Gamma)$ it is sufficient to majorize the derivatives of b on A. On A we have

$\partial^{\alpha}_{x,\theta} b = \partial^{\alpha}_{x,\theta}(\frac{1}{a})$ and we have already observed in Chapter I (6.8.2) that this is a finite linear combination of terms of the form

$$\frac{\partial^{\beta_1} a \dots \partial^{\beta_j} a}{a^{j+1}} \quad \text{with} \quad |\beta_1| + \dots + |\beta_j| = |\alpha| \quad ;$$

we therefore immediately deduce the desired upper bound of $\partial^{\alpha}_{x,\theta} b$ in any domain $(K^c)^{\varepsilon}$, with K a compact subset of Γ_o.

Conversely, if we take a compact subset K of Γ and put $r = ab - 1$, since $r \in S^{-1}(\Gamma)$, there exists $\varepsilon > 0$ such that

$|r(x, \theta)| \leq \frac{1}{2}$ for $(x, \theta) \in (K^c)^{\varepsilon}$ and

$|b(x, \theta)| \leq \frac{1}{\varepsilon}(1 + |\theta|)^{-m}$ for $(x,\theta) \in K^c$. Consequently we have $|a.b| \geq \frac{1}{2}$ in $(K^c)^{\varepsilon}$ and finally $\frac{1}{\varepsilon}(1 + |\theta|)^{-m} \; |a(x,\theta)| \geq \frac{1}{2}$ in $(K^c)^{\varepsilon}$ which proves that a satisfies (3.2.1).

DEFINITION 3.3: *Suppose* M *is a conic manifold, and* $m \in \mathbb{R}$. *We say that a symbol* $a \in S^{+\infty}(M)$ *is elliptic of degree* m *at a point of* M *if it is an elliptic symbol of degree* m *in an open conic neighbourhood of this point. We denote by* Ell(a) *the set of points of* M *at which* a *is elliptic; this is an open conic set whose complement is a closed conic set called the characteristic set of* a *and denoted by* char(a).

In the case of a symbol $a \in S^m(X \times \mathbb{R}^N)$, we define the ellipticity, Ell(a) and char(a) by considering $a_{|X \times \mathbb{R}^N_o}$.

EXAMPLE 3.4: If a is a classical symbol with principal part a_o, homogeneous of degree m, we have $\text{char}(a) = a_o^{-1}(0)$.

4. ASYMPTOTIC EXPANSIONS OF SYMBOLS

We can meaningfully consider a formal series of symbols (a_j) with degrees m_j tending to $-\infty$, when M is a conic manifold. We state:

DEFINITION 4.1: *Let* $(a_j)_{j \in J}$ *be a countable family of symbols* $a_j \in S^{m_j}(M)$ *; we assume that for any real* r *, there is only a finite number of indices* j *such that* $m_j \geq r$. *We say that a function* $a \in C^\infty(M)$ *satisfies* $a \sim \sum_j a_j$ *if there exists a sequence of real numbers* (μ_k) *tending to* $-\infty$ *such that for all* k:

$$\Big(a - \sum_{m_j \geq -k} a_j\Big) \in S^{\mu_k}(M) .$$

For example, with the notation of Definition 2.12, we have

$$a \sim \sum_{j \geq 0} a_j .$$

THEOREM 4.2: *Let* $(a_j)_{j \in J}$ *be a family of symbols as in*

Definition 4.1. *Then there exists* $a \in S^m(M)$ *unique modulo* $S^{-\infty}(M)$, *such that* $a \sim \sum a_j$, *with* $m = \max_j m_j$.

PROOF: The uniqueness of a modulo $S^{-\infty}$ is immediate. To prove the existence of a , we will not restrict the generality, bearing in mind Proposition 2.10 (ii), by assuming that $M = \Gamma$ is an open conic subset of $X \times \mathbb{R}^N$. We begin with the particular case where $J = \mathbb{N}$ and where the sequence (m_j) is strictly decreasing (and tends to $-\infty$). Suppose we have $\chi \in C^\infty(\mathbb{R}^N)$ such that $\chi(\theta) = 0$ if $|\theta| \leq 1/2$, and $\chi(\theta) = 1$ if $|\theta| \geq 1$. Let (K_j) be an exhaustive sequence of compact subsets of Γ. We can find an increasing sequence (t_j) such that;

$$(4.2.1.) \qquad |\partial_x^\beta \partial_\theta^\alpha(\chi(\tfrac{\theta}{t_j})\, a_j(x, \theta))| \leq 2^{-j}(1 + |\theta|)^{m_{j-1}-|\alpha|}$$

for $(x, \theta) \in K_i^c$, $|\alpha| + |\beta| + i \leq j$

(we can easily construct the t_j in stepwise fashion by observing that for j fixed, there is only a finite number of upper bounds (4.2.1) to be realised, and that $\dfrac{(1+|\theta|)^{m_{j-1}}}{(1+|\theta|)^{m_j}} \longrightarrow 0$

if $|\theta| \to \infty$) . We then put

$a(x, \theta) = \sum_{j=0}^{+\infty} \chi(\frac{\theta}{t_j})\, a_j(x, \theta)$ which is clearly in $C^\infty(\Gamma)$ since the sum is locally finite; the convergence of the series $\sum 2^{-j}$ allows us to deduce from (4.2.1) that

$a - \sum_{j=0}^{k-1} a_j \in S^{m_k}(\Gamma)$ for any integer $k \geq 1$.

We now pass on to the existence of a in the general case; we denote by $(n_k)_{k \in N}$ the strictly decreasing sequence of the possible values of m_j when j ranges over J. We put

$b_k = \sum_{m_j = n_k} a_j \in S^{n_k}(\Gamma)$, and we consider $a \in C^\infty(\Gamma)$ such that $a \sim \sum_{k \in \mathbb{N}} b_k$. Then for all $\ell \in \mathbb{N}$ we have:

$a - \sum_{m_j \geq -\ell} a_j = a - \sum_{k=0}^{r} b_k \in S^{n_{r+1}}(\Gamma)$, where r is determined by $n_r \geq -\ell > n_{r+1}$. Finally, the symbol a meets the requirements since $n_{r+1} \xrightarrow[\ell \to +\infty]{} -\infty$.

REMARK 4.3: We can confirm directly from the definitions that we have the following associativity property. Let $(J_h)_{h \in H}$ be an arbitrary partition of the set of indices J; for $h \in H$, suppose $a_h \sim \sum_{j \in J_h} a_j$. Then we have $a \sim \sum_{h \in H} a_h$

(note that $a_h \in S^{m_h}(M)$, where $m_h = \max_{j \in H} m_j$).

REMARK 4.4: In the case where $J = \mathbb{N}$ and where the sequence (m_j) is decreasing, we have $a \sim \sum_{j \geq 0} a_j$ if and only if for any integer k, we have $(a - \sum_{j=0}^{k} a_j) \in S^{m_{k+1}}(M)$. In the case where $M = \Gamma$, this condition is equivalent to the following: for any integer k, and for any compact subset K of Γ, there exists C such that $|\partial_\theta^\alpha \partial_x^\beta (a - (\sum_{j=0}^{k} a_j)(x, \theta)| \leq C(1 + |\theta|)^{m_{k+1} - |\alpha|}$

for $(x, \theta) \in K^c$ and $|\alpha| + |\beta| \leq k$.

In fact, in order to majorise the derivatives of order $k' > k$, it suffices to write

$$a - \sum_{j=0}^{k} a_j = (a - \sum_{j=0}^{k'} a_j) + (\sum_{j=k+1}^{k'} a_j)$$

and to note that

$$\sum_{j=k+1}^{k'} a_j \in S^{m_{k+1}}(\Gamma) .$$

5. <u>TOPOLOGY ON THE SYMBOL SPACES</u>

Let Γ be an open conic subset of $X \times \mathbb{R}^N$. With any compact subset $K \subset \Gamma$ and any integer j we associate a semi-norm on $S^m(\Gamma)$ by putting

$$C_{K,j,m}(a) = \sup_{\substack{|\alpha|+|\beta| \leq j \\ (x,\theta) \in K^c}} |\partial_\theta^\alpha \partial_x^\beta a(x, \theta)/(1 + |\theta|)^{m-|\alpha|}| ,$$

these semi-norms define on $S^m(\Gamma)$ a metrizable topological vector space structure, because it suffices to consider an exhaustive sequence of compact subsets (K_h).

We have the approximation result:

PROPOSITION 5.1: *If* $a \in S^m(\Gamma)$, *we take* $\rho \in C_o^\infty(\mathbb{R}^N)$ *identical to* 1 *in the neighbourhood of* 0 *and we put* $\rho_k(\theta) = \rho(\frac{\theta}{k})$.
Then $a_k = \rho_k \cdot a \in S^{-\infty}(\Gamma)$ *and for all* $m' > m$, *the sequence* a_k

converges to a *for the topology of* $S^{m'}(\Gamma)$.

PROOF: If $C_{K,j,m'}$ is a semi-norm on $S^{m'}(\Gamma)$, since $a \in S^m(\Gamma)$ there exists C such that

$$\frac{|\partial_\theta^{\alpha'}(1 - \rho_k(\theta)).\partial_\theta^{\alpha''}\partial_x^\beta\, a(x, \theta)|}{(1 + |\theta|)^{m'-|\alpha|}} \leq C\,\frac{(1 + |\theta|)^{m-|\alpha''|}}{(1 + |\theta|)^{m'-|\alpha|}} \,.\, k^{-|\alpha'|}$$

for $(x, \theta) \in K^c$, $\alpha = \alpha' + \alpha''$ and $|\alpha| + |\beta| \leq j$.

For $\alpha' = 0$, the value of the left-hand side is $O(k^{m-m'})$ because $|\theta| \geq k$ for $\theta \in \text{supp.}(1 - \rho_k)$; if $\alpha' \neq 0$ the left-hand side is again $O(k^{m-m'})$ because we have $k \leq |\theta| \leq 2k$ for $\theta \in \text{supp}\, \partial_\theta^{\alpha'}(1 - \rho_k)$. Consequently, by utilising Leibniz's formula, we deduce that

$$C_{K,j,m'}(a - a_k) = O(k^{m-m'})$$

which proves the proposition.

We can thus immediately deduce the following result concerning extension by continuity:

PROPOSITION 5.2: *Let* ℓ *be a linear mapping defined on the subspace of* $S^{-\infty}(\Gamma)$, *formed by the functions which vanish for* $|\theta|$ *large, and with values in a Frechét space* F. *We assume that for all* $m \in \mathbb{R}$, ℓ *is continuous for the topology induced by* $S^m(\Gamma)$. *Then there exists a unique extension of* $\ell : S^{+\infty}(\Gamma) \to F$ *which is continuous on each* $S^m(\Gamma)$.

6. VARIOUS GENERALISATIONS

(6.1.) Up to now, we have considered symbols with values in $\mathbb{C}$. We are now going to define the symbol-sections of a complex vector bundle above a conic manifold, under suitable assumptions.

Let M be a conic manifold and let F be a vector bundle on M; we specify a C^∞ action of $\mathbb{R}^+$ on F which commutes, via the projection of F onto M, with that of $\mathbb{R}^+$ in M

$$\begin{array}{ccc} F & \xrightarrow{(t)} & F \\ \downarrow & & \downarrow \\ M & \xrightarrow{(t)} & M \end{array}$$

We can show that there exists, above a conic neighbourhood of any point of M, a trivialisation of F commuting with the actions of $\mathbb{R}^+$ allowing a reduction to $M = \Gamma$ (an open conic subset of $\mathbb{R}^n \times \mathbb{R}^N_o$), i.e. $F = \Gamma \times \mathbb{C}^d$; we put in this case $S^m(M, F) = (S^m(\Gamma))^d$, hence the general definition of $S^m(M, F)$ by restriction and transport.

If $F = \mathcal{L}(F_1, F_2)$, where F_1, F_2 are two bundles on M of the preceding type, we define $S^m(M; \mathcal{L}(F_1, F_2))$ accordingly.

We say that $a \in S^m(M; \mathcal{L}(F_1, F_2))$ is right elliptic of degree m (resp: left elliptic) if there exists $b \in S^{-m}(M; \mathcal{L}(F_2, F_1))$ such that $ab - 1_{F_2} \in S^{-1}(M; \mathcal{L}(F_2, F_2))$ (resp : $ba - 1_{F_1} \in S^{-1}(M; \mathcal{L}(F_1, F_1))$. We can show that, in the case of a classical symbol a with homogeneous principal part a_o of degree m, the right ellipticity (resp: left) of a is equivalent to the right invertibility (resp: left) of a_o.

(6.2.) In fact, there is not just one single type of symbols but rather we can say that there are almost as many of them as there are types of problems, and we need to be able to adapt one to the other. For example, if P(D) is a hypoelliptic operator then in order to include functions of the type $\xi \to \frac{\theta(\xi)}{P(\xi)}$ the upper bounds (6.9.2) in Chapter I lead to the following generalisation of the spaces $S^m(\Gamma)$. With $0 < \rho \leq 1$, we define the space $S^m_\rho(\Gamma)$ of symbols of degree m and of type ρ by replacing the upper bounds (2.3.1) by

$$|\partial_x^\beta \, \partial_\theta^\alpha \, a(x, \theta)| \quad \leq \quad C \, (1 + |\theta|)^{m-\rho|\alpha|+\delta|\beta|}$$

where $\delta = 1 - \rho$. We can prove that all the preceding results in this chapter generalise, with only minor modifications, to symbols of type ρ (see HÖRMANDER [3], [5] for symbols which are still more general).

7. OSCILLATORY INTEGRALS

7.1. INTRODUCTION: When $\hat{u}(\xi)$ is defined explicitly, expression (1.1) is written formally as

$$(7.1.1.)\quad (Pu)(x) = \iint e^{i(x-y)\cdot\xi}\, p(x, \xi)\, u(y)\, dy\, đ\xi \; ;$$

however this integral is not absolutely convergent with respect to ξ.

The objective of this section is to prove that we can in fact work with "integrals" of the type (7.1.1) and with integrals which are even more general, as if they were absolutely convergent. We shall explain the idea underlying the method in the particular case where $P(x, D)$ is the identity operator in one dimension. In order to give a meaning to the integral

$$u(x) = \iint e^{i(x-y)\xi}\, u(y)\, dy\, đ\xi \quad ,$$

we can clearly write it as an iterated integral

$$u(x) = \int_R e^{ix\xi} \left(\int_R e^{-iy\xi}\, u(y)\, dy \right) đ\xi \quad ;$$

however this method is not very easy to handle. Since the difficulty exists for $|\xi|$ large, it is sufficient, after truncation with respect to ξ, to only consider the case where we integrate for $|\xi| \geq 1$. We note that:

$$e^{i(x-y)\xi} = -\frac{1}{\xi^2}\,\partial_y^2\left(e^{i(x-y)\xi}\right) ,$$

which leads us, after formal integration by parts with respect to y, to put

$$(7.1.2.)\quad \iint_{\substack{|\xi|\geq 1\\ y\in\mathbb{R}}} e^{i(x-y)\xi}\,u(y)\,dy\,đ\xi = $$

$$= \iint_{\substack{|\xi|\geq 1\\ y\in\mathbb{R}}} e^{i(x-y)\xi}\left(-\frac{1}{\xi^2}.\right)\partial_y^2\,u(y)\,dy\,đ\xi$$

and to define the left-hand side by the absolutely convergent integral appearing in the right-hand side, this being legitimate because it can easily be shown that the right-hand side of (7.1.2.) coincides with the iterated integral

$$\int_{|\xi|\geq 1} e^{ix\xi}\left(\int_{\mathbb{R}} e^{-iy\xi}\,u(y)\,dy\right) đ\xi \quad .$$

This type of generalised integral has been studied systematically by HÖRMANDER [5] under the name of oscillatory integral.

7.2. DEFINITION OF OSCILLATORY INTEGRALS: Let X be an open subset of $\mathbb{R}^n$, $(n \geq 0)$, let $\varphi \in C^\infty(X \times \mathbb{R}^N \setminus 0)$ be a real-valued function which is homogeneous of degree one with respect to $\theta \in \mathbb{R}^N \setminus 0$ and let $a \in S^m(X \times \mathbb{R}^N)$ be a symbol; we consider the integral

$$(7.2.1.) \quad I_\varphi(au) = \iint e^{i\varphi(x,\theta)}\, a(x,\theta)\, u(x)\, dx\, d\theta, \quad u \in C_0^\infty(X).$$

It is clear that this integral is absolutely convergent when $m < -N$, and in this case it depends continuously on $a \in S^m$; we shall prove that we can, using integrations by parts, make it meaningful for arbitrary m, if φ has no critical points.

We observe that the mapping ℓ :

$$(7.2.2.) \qquad a \in S^{-\infty}(X \times \mathbb{R}^N) \longrightarrow \ell(a) = I_\varphi(au) \in \mathbb{C}$$

is a linear form; in order to demonstrate the continuity of ℓ in the sense of Proposition 5.2, we establish the following:

PROPOSITION 7.3 : *If we assume that* $d_{x,\theta}\,\varphi(x,\theta) \neq 0$ *for* $(x,\theta) \in X \times \mathbb{R}^N \setminus 0$, *then there exists a first-order differential operator*

$$(7.3.1.) \quad L = \sum_1^N a_j(x,\theta)\, D_{\theta_j} + \sum_1^n b_j(x,\theta)\, D_{x_j} + c(x,\theta)$$

such that $L(e^{i\varphi}) = e^{i\varphi}$ *and with* $a_j \in S^0$, $b_j, c \in S^{-1}$.

PROOF: For an operator of the form (7.3.1), we have

$$e^{-i\varphi} L(e^{i\varphi}) = \sum_j a_j\, \partial_{\theta_j}\varphi + \sum_j b_j\, \partial_{x_j}\varphi + c\ .$$

We make the right-hand side identical to 1 by putting

$$a_j = (1-\rho)\, D\, |\theta|^2\, \partial_{\theta_j}\varphi\ , \quad b_j = (1-\rho)\, D\, \partial_{x_j}\varphi\ , \quad c = \rho$$

where $\rho \in C_0^\infty(R^N)$ is identical to 1 for $|\theta| \leq 1$ and where

$$D = (|\theta|^2 \cdot \sum_j |\partial_{\theta_j} \varphi|^2 + \sum_j |\partial_{x_j} \varphi|^2)^{-1} .$$

Since φ has no critical points, it is clear that $(1 - \rho) D \in S^{-2}(X \times R^N)$ and consequently the coefficients a_j, b_j, c are in the spaces indicated.

It is useful for later work to note that the transpose ${}^tL$ of L has coefficients a_j', b_j', c' in the same symbol spaces.

This lemma leads us to state:

DEFINITION 7.4 : *If Γ is an open conic subset of $X \times (R^N \setminus 0)$, we say that a function $\varphi \in C^\infty(\Gamma)$ is a phase function if it is homogeneous of degree 1 with respect to θ, with real values and such that $d_{x,\theta}\, \varphi(x, \theta) \neq 0$ on Γ.*

PROPOSITION 7.5 : *If φ is a phase function on $X \times (R^N \setminus 0)$, then:*

i) *the linear form ℓ in (7.2.2) extends uniquely into a linear form ℓ on $S^{+\infty}(X \times R^N)$, which is continuous on any S^m. This extension is termed an oscillatory integral and is denoted by*

$$(7.5.1.) \qquad \ell(a) = \tilde{\iint} e^{i\varphi(x, \theta)}\, a(x, \theta)\, u(x)\, dx\, d\theta .$$

For $a \in S^m$*, we have the expression*

$$\ell(a) = \iint e^{i\varphi(x,\theta)}({}^tL)^k(a(x,\theta)\,u(x))\,dx\,d\theta \tag{7.5.2.}$$

when $k > m + N$*, where* L *is taken as in* 7.3.

ii) *If* ρ_k *is defined as in Proposition* 5.1*, then*

$$\ell(a) = \lim_{k\to\infty} \iint e^{i\varphi}\,\rho_k\,au\,dx\,d\theta \tag{7.5.3.}$$

iii) *For* a *fixed in* S^m*, the mapping*

$$u \in C_0^\infty(X) \longrightarrow I_\varphi(au) \in \mathbb{C}$$

is a distribution of order k *(for* $k > m+N$*) which we denote by*

$$\tilde{\int} e^{i\varphi(x,\theta)}a(x,\theta)\,d\theta \tag{7.5.4.}$$

and which we call the integral distribution associated with the phase φ *and the amplitude* a

PROOF OF i) : If $a \in S^{-\infty}$, we can, by virtue of Lemma 7.3, replace $e^{i\varphi}$ by $L^k(e^{i\varphi})$ in (7.2.1), and by integration by parts we obtain

$$\ell(a) = I_\varphi(au) = \iint e^{i\varphi(x,\theta)}({}^tL)^k(a(x,\theta)a(x))dx\,d\theta\,. \tag{7.5.5.}$$

Furthermore, the properties of the coefficients of ${}^tL$ imply that ${}^tL$ maps S^p continuously into S^{p-1} for any real value p. Consequently, if $a \in S^m$ and if we choose k such that $k > m+N$, the integral which occurs on the right-hand side of (7.5.5) depends continuously on $a \in S^m$; thus, it suffices to apply

Proposition 5.2.

In order to prove ii), we utilise the continuity of ℓ on any subspace $S^{m'}$ and we conclude by using Proposition 5.1.

To prove iii), we note that the representation (7.5.2) allows us to majorise $I_\varphi(au)$ by

$$|I_\varphi(au)| \leq C_K \sup_{|\alpha|\leq k} |\partial_x^\alpha u(x)| \quad , \quad u \in \mathcal{D}_K(X)$$

when $k > m+N$.

Oscillatory integrals which depend on one parameter behave like ordinary integrals, in the sense that we can, under certain assumptions, differentiate or integrate under the integral sign $\tilde{\iint}$.

PROPOSITION 7.6 : *Let* Y *be an open subset of* $\mathbb{R}^p$ *and let* φ *be a phase function on* $X \times Y \times (\mathbb{R}^N \setminus 0)$ *such that* $d_{x,\theta}\,\varphi(x, y, \theta) \neq 0$. *If* $a \in S^m(X \times Y \times \mathbb{R}^N)$ *and* $u \in C_0^\infty(X \times Y)$, *we put for* $y \in Y$

$$F(y) = \tilde{\iint} e^{i\varphi(x, y, \theta)}\, a(x, y, \theta)\, u(x, y)\, dx\, d\theta \ .$$

Then

i) *We have* $F \in C_0^\infty(Y)$ *and we can differentiate under the integral sign* $\widetilde{\iint}$.

ii) *We have* $\displaystyle\int F(y)\,dy = \widetilde{\iiint} e^{i\varphi}\, au\; dx\, dy\, d\theta$.

PROOF OF ii): As in Lemma 7.3 we can again undertake with the parameter y the construction of an operator $L(x, y, \theta, \partial_\theta, \partial_x)$ with coefficients $a_j \in S^0(X \times Y \times \mathbb{R}^N)$, $b_j, c \in S^{-1}(X \times Y \times \mathbb{R}^N)$. If $k > m+N$, then (7.5.2) shows that we have

$F(y) = \displaystyle\iint e^{i\varphi}({}^tL^k)(au)\; dx\, d\theta$ and since this integral is absolutely convergent, the usual Fubini theorem allows us to draw the required conclusion.

PROOF OF i): By truncation, we can reduce to the case where the symbol a is zero for $|\theta| \le 1$. If we have $\rho_k(\theta)$ as in Proposition 5.1, we denote by $F_k(y)$ the function obtained when a is replaced by $a_k = a.\rho_k$.

The operator L above allows us to adapt the proof of Proposition 7.5 i) and ii) to the case where we have a parameter, in order to prove that $F(y) = \lim_{k\to\infty} F_k(y)$ uniformly with respect to y. Furthermore, it is clear that $F_k \in C_0^\infty(Y)$; if α is a multi-index, we immediately obtain the expression

$$\partial_y^\alpha F_k(y) = \iint e^{i\varphi(x,y,\theta)}\, b_\alpha(x, y, \theta)\, \rho_k(\theta)\; dx\, d\theta$$

with

$$b_\alpha = e^{-i\varphi} \partial_y^\alpha (e^{i\varphi} au) \in S^{m+|\alpha|}(X \times Y \times \mathbb{R}^N) \quad .$$

Consequently, the argument used for F_k can still be used to prove that

$$\partial_y^\alpha F(y) = \lim_{k\, \infty} \partial_y^\alpha F_k(y) = \widetilde{\iint} \partial_y^\alpha (e^{i\varphi} au)\, dx\, d\theta$$

uniformly with respect to y.

Henceforth, we shall omit the sign $\sim$ on an oscillatory integral since Proposition 7.6 allows us to manipulate it like an ordinary integral.

We can give an inclusion for the singular support of a distribution of the type (7.5.4). We put $\mathbb{R}_0^N = \mathbb{R}^N \setminus 0$.

PROPOSITION 7.7: *If* φ *is a phase function on* $X \times \mathbb{R}_0^N$ *and if we have* $a \in S^m(X \times \mathbb{R}^N)$ *then the distribution* $A = \int e^{i\varphi(x,\,\theta)} a(x,\,\theta) d\theta$ *satisfies*

(7.7.1.) $$\text{sing supp } A \subset \{x \in X \mid \text{there exists } \theta \neq 0,\ \varphi'_\theta(x,\,\theta) = 0\}$$

PROOF: Suppose we have $x_0 \in X$ such that $\varphi'_\theta(x_0,\,\theta) \neq 0$ for all $\theta \in \mathbb{R}_0^N$; the homogeneity and the continuity of φ show that there exists an open neighbourhood V of x_0 such that

$\varphi'_\theta(x, \theta) \neq 0$ for $(x, \theta) \in V \times R^N_0$. Furthermore, Proposition 7.6 shows that the mapping

$$V \ni x \longrightarrow A(x) = \int e^{i\varphi(x, \theta)} a(x, \theta) d\theta \in C$$

is C^∞ and since we have

$$< A, u > = \int A(x)\, u(x)\, dx \qquad \text{for} \qquad u \in C_0^\infty(V) \quad ,$$

we therefore deduce that $A_{|V} \in C^\infty(V)$, which demonstrates the inclusion (7.7.1).

8. INTEGRAL OPERATORS ASSOCIATED WITH A PHASE AND AN AMPLITUDE

CONSTRUCTION 8.1: Let X, Y be open subsets of $\mathbb{R}^n$, $\mathbb{R}^p$; let φ be a phase function in $X \times Y \times (\mathbb{R}^N \setminus 0)$ and let $a \in S^m(X \times Y \times \mathbb{R}^N)$. We consider the integral distribution $K \in \mathcal{D}'(X \times Y)$ associated with the phase φ and the amplitude a, i.e.

$$(8.1.1.) \quad < K, w > = \int e^{i\varphi(x,y,\theta)} a(x, y, \theta)\, w(x, y)\, dx\, dy\, d\theta,$$

$$w \in C_0^\infty(X \times Y) \quad .$$

Let $A : C_0^\infty(Y) \longrightarrow \mathcal{D}'(X)$ be the operator with kernel K. For $v \in C_0^\infty(Y)$, Av is thus the distribution in X defined by

$$(8.1.2.)\quad \langle Av, u\rangle = \int e^{i\varphi(x,y,\theta)}\, a(x, y, \theta)\, u(x)\, v(y)\, dx\, dy\, d\theta, \qquad u \in C_0^\infty(X)$$

and A is termed the integral operator associated with the phase φ and the amplitude a.

PROPOSITION 8.2: i) *If* K *is the kernel of* A, *we have*

$$(8.2.1.)\quad \text{sing supp } K \subset \{(x, y) \in X \times Y \mid \text{ there exists } \theta \in R_0^N \text{ with } \varphi'_\theta(x, y, \theta) = 0\}.$$

ii) *If* $d_{y,\theta}\, \varphi(x, y, \theta) \neq 0$ *for* $(x, y, \theta) \in X \times Y \times R_0^N$, A *is continuous from* $C_0^\infty(Y)$ *into* $C^\infty(X)$.

iii) *If* $d_{x,\theta}\, \varphi(x, y, \theta) \neq 0$ *for* $(x, y, \theta) \in X \times Y \times R_0^N$, *then* A *extends into an operator which is continuous from* $\mathcal{E}'(Y)$ *into* $\mathcal{D}'(X)$.

PROOF: i) follows by application of (7.7.1) to K.

For ii), Proposition 7.6 and (8.1.2) show that if $v \in C_0^\infty(Y)$, then Av is the C^∞ function in X defined by

$$(8.2.2.)\quad (Av)(x) = \int e^{i\varphi(x,y,\theta)}\, a(x, y, \theta)\, v(y)\, dy\, d\theta .$$

The continuity of A is obtained by majorising a semi-norm $p_{K,j}(A)$ after having replaced (8.2.2) by an integral of the type (7.5.2) with $k > m+N+j$.

For iii), we note that the operator ${}^tA$ is obtained by permuting the roles of x,y in (8.1.2), but i) shows that ${}^tA$ is continuous from $C_0^\infty(X)$ into $C^\infty(Y)$ and consequently Proposition 4.5 of Chapter I shows that A extends into an operator which is continuous from $\mathcal{E}'(Y)$ into $\mathcal{D}'(X)$.

EXAMPLE 8.3: The identity operator for $C_0^\infty(X)$; for $u \in C_0^\infty(X)$, we have

$$(8.3.1.)\quad u(x) = \iint e^{i(x-y)\xi}\, u(y)\, dy\, đ\xi \qquad x \in X .$$

This formula is deduced from the Fourier inversion formula by utilising the passage to the limit (7.5.3)

EXAMPLE 8.4: Let P(x, D) be a differential operator with C^∞ coefficients on X. By differentiating in the oscillating integral (8.3.1) we obtain

$$(8.4.1.)\quad (Pu)(x) = \iint e^{i(x-y)\xi}\, P(x, \xi)\, u(y)\, dy\, đ\xi , \qquad u \in C_0^\infty(X) , \quad x \in X .$$

EXAMPLE 8.5: Composition with a C^∞ mapping f from X into Y. We define the operator $f^* : C^\infty(Y) \longrightarrow C^\infty(X)$ by $f^*v = v \circ f$. For $x \in X$, (8.3.1) shows that

$$(8.5.1.)\quad (f^* v)(x) = \iint e^{i(f(x)-y).\xi}\, v(y)\, dy\, đ\xi , \qquad v \in C_0^\infty(Y) .$$

EXAMPLE 8.6: Trace Operator. Suppose we have $\mathbb{R}^p = \mathbb{R}^n \times \mathbb{R}^{p-n}$ and suppose Y is an open subset of $\mathbb{R}^p$. Let X be the open subset of $\mathbb{R}^n$ defined by

$$Y \cap (\mathbb{R}^n \times \{0\}) = X \times \{0\} \quad (\text{ we assume } X \neq \emptyset).$$

Let γ be the trace operator on X:

$$C_0^\infty(Y) \ni v \longrightarrow \gamma v \in C_0^\infty(X), \quad (\gamma v)(x) = v(x, 0) \quad ;$$

then we have $\gamma = f^*$ with $f(x) = (x, 0)$. Therefore Example 8.5 shows that

$$(\gamma v)(x) = \iint e^{i(x-y)\xi} v(y)\, dy\, đ\xi \qquad v \in C_0^\infty(Y) \quad , \quad x \in X$$

EXAMPLE 8.7: Pseudo-differential operators. These are the operators P defined by (8.4.1) when we replace the polynomial with respect to ξ, $P(x, \xi)$, by an arbitrary amplitude $p(x, y, \xi) \in S^m(X \times X \times \mathbb{R}^n)$. These will be studied in detail in Chapter IV.

8.8 GENERALISATION TO THE CASE WHERE THE PHASE FUNCTION φ IS DEFINED IN AN OPEN CONIC SET. Let Γ be an open conic subset of $X \times (\mathbb{R}^N \setminus 0)$ and let φ be a phase function in Γ. The methods of Section 7 allow us, with minor modifications, to define and study the oscillating integral

$$\iint e^{i\varphi(x, \theta)} a(x, \theta)\, u(x)\, dx\, d\theta \quad , \qquad u \in C_0^\infty(X)$$

when $a \in S^m(X \times \mathbb{R}^N)$ and when a satisfies conic supp $a \subset \Gamma \cup (X \times \{0\})$.

If φ is a phase function in an open conic set $\Gamma \subset X \times Y \times (\mathbb{R}^N \setminus 0)$ and $a \in S^m(X \times Y \times \mathbb{R}^N)$ is such that conic supp $a \subset \Gamma \cup (X \times Y \times \{0\})$, then, we define the integral operator $A : C_0^\infty(Y) \longrightarrow \mathcal{D}'(X)$ associated with the phase φ and the amplitude a as in Section 8.1, and we obtain analogous properties.

8.9 GENERALISATION TO THE CASE OF SYMBOLS OF TYPE $\rho > 0$ (see Section 6). All the preceding constructions and properties remain valid, with minor modifications, when the amplitudes a are symbols from S^m_ρ.

9. STATIONARY PHASE THEOREM

9.1 CASE WHERE THE PHASE IS NEVER STATIONARY. Let X, Y be open subsets of $\mathbb{R}^p$, $\mathbb{R}^q$; we often have occasion to study the behaviour for $\tau \to +\infty$ of integrals of the type

$$(9.1.1.) \qquad I(x, \tau) = \int_Y e^{i\tau\varphi(x, y)}\, a(x, y, \tau)\, dy$$

where $\varphi \in C^\infty(X \times Y)$ is real-valued and where $a \in S^m(X \times Y \times \mathbb{R}^+)$. In order to avoid any convergence problems, we assume that a is zero for y outside a compact set $K \subset Y$.

If the function φ has no critical points with respect to the integration variable y, then I is of rapid decrease with respect to τ; more precisely, we have:

PROPOSITION 9.2: *We assume that* $d_y \varphi(x, y) \neq 0$ *on* $X \times Y$, *and in this case* $I(x, \tau) \in S^{-\infty}(X \times R^+)$, *where* τ *is the frequency variable.*

PROOF: The first-order operator

$$L(x, y, D_y) = \frac{1}{\tau} |d_y \varphi(x, y)|^{-2} \left(\sum_1^q \partial_{y_j} \varphi(x, y) D_{y_j}\right)$$

satisfies for any integer $k \geq 1$

$$L^k(e^{i\tau\varphi}) = e^{i\tau\varphi} .$$

Substituting back into (9.1.1) we obtain, after integration by parts,

$$I(x, \tau) = \int e^{i\tau\varphi} ({}^tL)^k a \, dy .$$

Since $({}^tL)^k a \in S^{m-k}(X \times Y \times R^+)$, we then deduce that for any k and for any compact set $H \subset X \times R^+$ we have the upper bound

$$|I(x, \tau)| \leq C \tau^{m-k} \quad \text{for} \quad (x, \tau) \in H^c .$$

The derivatives of I are defined by integrals of the same type as I; thus we majorise them in the same manner, and this proves the proposition.

9.3 CASE WHERE φ HAS A QUADRATIC CRITICAL POINT. The above proposition leads us to study the behaviour of I when φ possesses a critical point. We begin with the simplest case, i.e. that in which φ is a quadratic form which is non-degenerate with respect to y. More precisely, we assume that

$$\varphi(x, y) = 1/2 < Q(x)y, y >$$

where Q(x) is an invertible symmetric matrix which has C^∞ dependence on $x \in X$, and we take $Y = \mathbb{R}^q$. In this case, we have:

THEOREM 9.3: *The integral*

$$\text{(9.3.1.)} \qquad I(x, \tau) = \int \exp(i \tfrac{\tau}{2} < Q(x)y, y >) \; a(x, y, \tau) \, dy$$

satisfies

$$\text{(9.3.2.)} \qquad I(x, \tau) \in S^{m-q/2} (X \times \mathbb{R}^+)$$

and this symbol admits an asymptotic expansion

$$\text{(9.3.3.)} \quad I(x, \tau) \sim$$

$$\left(\frac{2\pi}{\tau}\right)^{q/2} |\det Q(x)|^{-1/2} e^{i \pi/4 \operatorname{sign} Q(x)} \Big(\sum_{k \geq 0} \frac{\tau^{-k}}{k!} R^k a(x, y, \tau)_{|y=0} \Big)$$

$$\text{(9.3.4.) where } R(x, \partial_y) = \frac{i}{2} < Q^{-1}(x) \partial_y, \partial_y >$$

and sign Q(x) = (*number of positive eigenvalues of* Q(x))-

(*number of negative eigenvalues of* Q(x)).

We begin with a Fourier transform calculation.

LEMMA 9.4: *Let Q be an invertible symmetric matrix in* $\mathbb{R}^q$. *Then the Fourier transform of the function* $\exp(\frac{i}{2} < Qy, y >)$ *is given by the function*

$$(9.4.1.) \quad \eta \in \mathbb{R}^q \longrightarrow (2\pi)^{q/2} \, |\det Q|^{-1/2} \, e^{i\,\pi/4 \text{ sign } Q} \, . \exp(-\tfrac{i}{2} < Q^{-1}\eta, \eta >).$$

PROOF OF THE LEMMA: Let M be an orthogonal matrix such that ${}^tM\,Q\,M$ is diagonal; the change of variable $y = Mx$ in the calculation of the Fourier transform takes us to the case where Q is diagonal. If we thus put $< Qy, y > = \sum_1^q \lambda_j y_j^2$, we have

$$\exp(\tfrac{i}{2} < Qy, y >) = \prod_{j=1}^{q} \exp(\tfrac{i}{2} \lambda_j y_j^2) .$$

We are thus led to calculate $\mathcal{F}(e^{i/2\,\lambda\, y^2})$ and we have seen in Exercise 10.3 of Chapter I that we obtain the function

$$\eta \in \mathbb{R} \longrightarrow (2\pi)^{1/2} \, |\lambda|^{-1/2} \, e^{i\,\pi/4 \text{ sign } \lambda} \exp(-\tfrac{i}{2} \tfrac{\eta^2}{\lambda}) .$$

We hence obtain the lemma, by observing that

$$\prod_j |\lambda_j|^{-1/2} = |\det Q|^{-1/2} , \quad \sum_j \text{sign } \lambda_j = \text{sign } Q$$

$$\sum_j \frac{1}{\lambda_j} \eta_j^2 = < Q^{-1}\eta, \eta > .$$

We now return to the proof of the theorem. We apply Parseval's formula in the integral (9.3.1), and we obtain

$$(9.4.1.) \; I(x,\tau) = (2\pi\,\tau)^{-q/2} \, |\det Q(x)|^{-1/2} \, e^{i\,\pi/4 \text{ sign } Q(x)} \, J(x, \tau)$$

with

$$(9.4.2.) \qquad J(x, \tau) = \int_{R^q} e^{-i/2\tau < Q^{-1}(x)\eta,\ \eta >} \hat{a}(x, \eta, \tau)\, d\eta$$

(where ^ indicates that we have taken the Fourier transform with respect to the variable y). The advantage of expression (9.4.2) lies in the fact that τ appears in the exponential via the function $\frac{1}{\tau}$; thus, we can obtain the asymptotic expansion of J by utilising the Taylor expansion of the exponential function.

LEMMA 9.5: *For any integer* $N \geq 1$, *we have*

$$(9.5.1.) \qquad e^{is} = \sum_{0}^{N-1} \frac{(is)^k}{k!} + r_N(s) \qquad s \in R$$

where the remainder r_N *satisfies the upper bounds*

$$(9.5.2.) \qquad \left|\left(\frac{d}{ds}\right)^j r_N(s)\right| \leq \frac{s^{N-j}}{(N-j)!} \quad \text{for} \quad 0 \leq j \leq N.$$

PROOF OF THE LEMMA: For $j = 0$, the upper bound (9.5.2) is in fact just the usual upper bound for the remainder in the Taylor formula. We obtain the upper bounds (9.5.2) for $N \geq j > 0$ from the case $j = 0$ by differentiating the equality (9.5.1) j times.

9.6 CONCLUSION OF THE PROOF OF THE THEOREM: We apply this lemma to the exponential in the integral (9.4.2), replacing s by $-\frac{1}{2\tau} < Q^{-1}(x)\eta,\ \eta >$. We obtain,

$$(9.6.1.) \qquad J(x, \tau) = \sum_{0}^{N-1} J_k(x, \tau) + R_N(x, \tau)$$

where

$$J_k(x, \tau) = \frac{1}{k!} \int (-\frac{i}{2\tau} < Q^{-1}(x)\eta, \eta >)^k \, a(x, \hat{\eta}, \tau) \, d\eta$$

$$(9.6.2.) \qquad = (2\pi)^q \frac{\tau^{-k}}{k!} (R(x, \partial_y))^k \, a(x, y, \tau)_{|y=0}$$

and

$$R_N(x, \tau) = \int r_N(-\frac{1}{2\tau} < Q^{-1}(x)\eta, \eta >) \, a(x, \hat{\eta}, \tau) \, d\eta \quad .$$

We note that $J_k(x, \tau) \in S^{m-k}(X \times \mathbb{R}^+)$; thus, in view of Remark 4.4, it suffices to show that the remainder R_N satisfies the upper bounds

$$(9.6.3.) \qquad \partial_\tau^\alpha \, \partial_x^\beta \, R_N(x, \tau) = O(\tau^{m-N-|\alpha|})$$

uniformly for x within a compact subset $K \subset X$ and for $|\alpha| + |\beta| \leq N$. We begin by proving this with $\alpha = \beta = 0$. Utilising (9.5.2), we obtain:

$$(9.6.4.) \qquad |R_N(x, \tau)| \leq C \, \tau^{-N} \int |\eta|^{2N} \, |a(x, \hat{\eta}, \tau)| \, d\eta \; .$$

Furthermore, since a is a symbol of degree m with compact support with respect to y , we have for any integer $\mu \geq 0$

$$(9.6.5.) \qquad \sup_\eta \, (1 + |\eta|^2)^\mu \, |a(x, \hat{\eta}, \tau)| = O(\tau^m)$$

uniformly for $x \in K$. By taking μ sufficiently large, we deduce from (9.6.4), and (9.6.5) the inequality

$$\sup_{x \in K} |R_N(x, \tau)| = O(\tau^{m-N}) \; .$$

In the general case, Leibniz's formula shows us that $\partial_\tau^\alpha \partial_x^\beta R_N(x, \tau)$ is a finite sum of terms of the form

$$\int \partial_\tau^{\alpha'} \partial_x^{\beta'} (r_N(-\frac{1}{2\tau} < Q^{-1}(x)\eta, \eta >)) . \partial_\tau^{\alpha''} \partial_x^{\beta''} a(x, \hat{\eta}, \tau)\, d\eta$$

with

$$\alpha = \alpha' + \alpha'' \quad , \quad \beta = \beta' + \beta'' \quad .$$

We can show by recurrence on $|\alpha'| + |\beta'|$ that $\partial_\tau^{\alpha'} \partial_x^{\beta'} r_N(\ldots)$ is a finite sum of terms of the form

$$r_N^{(j)} (-\frac{1}{2\tau} < Q^{-1}(x)\eta, \eta >) \prod_{\ell=1}^{j} \partial_\tau^{\gamma_\ell} \partial_x^{\delta_\ell} (-\frac{1}{2\tau} < Q^{-1}(x)\eta, \eta >)$$

where

$$j \leq \alpha' + |\beta'| \, , \quad \sum_\ell \gamma_\ell = \alpha' \quad , \quad \sum_\ell \delta_\ell = |\beta'| \quad .$$

By combining Lemma 9.5 and the immediate upper bound

$$\left| \prod_{\ell=1}^{j} \partial_\tau^{\gamma_\ell} \partial_x^{\delta_\ell} (\ldots) \right| = O(\tau^{-\alpha'} (\frac{|\eta|^2}{\tau})^j) ,$$

we easily obtain the upper bound (9.6.3).

EXAMPLE 9.7: Suppose we have $X \subset R^n$, $Y = R^n \times R^n$; a point of Y is denoted by $y = (z, \zeta)$ and we define $Q(x) = Q$ by $< Qy, y > = < z, \zeta >$. We have det Q = 1, sign Q = 0 and $R(x, \partial_y) = \frac{i}{2} \sum_1^n \frac{\partial^2}{\partial z_j \partial \zeta_j}$. In this case Theorem 9.3 gives

$$\int e^{i\tau z.\zeta} a(x, z, \zeta, \tau) dz\, d\zeta \sim (\frac{2\pi}{\tau})^n \sum_{k \geq 0} \frac{\tau^{-k}}{k!} R^k a|_{z=\zeta=0}$$

(9.7.1.)

$$\sim (\frac{2\pi}{\tau})^n \sum_{\alpha \in \mathbb{N}^n} \frac{\tau^{-|\alpha|}}{\alpha!} \partial_z^\alpha D_\zeta^\alpha a(x, z, \zeta, \tau)_{|z=\zeta=0} .$$

9.8 CASE WHERE φ HAS A NON-DEGENERATE CRITICAL POINT WITH RESPECT TO y: If we have $\varphi \in C^\infty(X \times Y)$, we say that (x°, y°) is a non-degenerate critical point of φ with respect to y if $\partial_y \varphi(x^\circ, y^\circ) = 0$ and if the Hessian matrix $(\partial^2_{y_j y_k} \varphi(x^\circ, y^\circ))_{jk} = Q$ is invertible.

In this case, the implicit function theorem shows that there exist neighbourhoods U, V of x°, y° such that for any $x \in U$, there exists a unique $y(x) \in V$ satisfying $\partial_y \varphi(x, y(x)) = 0$ and $y(x^\circ) = y^\circ$.

We put $Q(x) = \partial^2_{yy} \varphi(x, y)_{|y=y(x)}$, this being an invertible matrix when U is small. By a change of variables with respect to y, we can reduce φ to the form (9.3.1), this being expressed in the following theorem:

THEOREM 9.9: (or Morse's lemma with parameter). *We assume that* (x°, y°) *is a non-degenerate critical point of* φ *with respect to* y . *Then, after appropriately shrinking the open sets* U, V, *there exists a* C^∞ *function* $h : U \times V \longrightarrow \mathbb{R}^q$ *such that if we put* $z = h(x, y)$, *we have*

$$(9.9.1.) \qquad \varphi(x, y) - \varphi(x, y(x)) = \frac{1}{2}\, {}^t z . Q(x) . z$$

$$\text{for} \quad (x, y) \in U \times V .$$

Moreover h *is a diffeomorphism with respect to* y *for* x *fixed, and satisfies*

(9.9.2.) $h(x, y(x)) = 0 \ , \quad \partial_y h(x, y)_{|y=y(x)} = I \ , \qquad x \in U .$

(h *is termed the "Morse diffeomorphism"*).

PROOF: Taylor's formula with integral remainder allows us to write

$$\varphi(x, y) - \varphi(x, y(x)) = \frac{1}{2}\,{}^t(y - y(x)).Q(x, y).(y - y(x)) \ , \qquad (x, y) \in U \times V$$

with $Q(x,y(x)) = Q(x)$.

Let us for the moment assume the existence of a C^∞ function

$U \times V \ni (x, y) \xrightarrow{R} R(x, y) \in \mathcal{L}(R^q, R^q)$ such that

(9.9.3.) $${}^tR(x, y).Q(x).R(x, y) = Q(x, y) \ ,$$
$$R(x, y(x)) = I \quad \text{for} \quad (x, y) \in U \times V \ .$$

Then, the function $h(x,y) = R(x,y).(y - y(x))$ satisfies the conditions of the theorem, after appropriately shrinking U and V so as to have a diffeomorphism with respect to y for x fixed.

In order to prove the existence of the function R, we consider the mapping Φ, dependent on the parameter $x \in U$, defined by

$$Q^{-1}.\mathcal{S} \ni R \xrightarrow{\Phi} \Phi(R) = {}^tR\, Q(x)\, R$$

where $\mathcal{S}$ denotes the space of symmetric $q \times q$ matrices. For $x = x^0$, the derivative of Φ at the point $R = Id$ is given by the linear mapping

$$Q^{-1}\mathcal{S} \ni T \xmapsto{d\Phi(Id)} {}^tT\,Q + Q\,T \in \mathcal{S} \quad ;$$

this linear mapping is surjective, and since every $A \in \mathcal{S}$ is the image of $\frac{1}{2} Q^{-1} A \in Q^{-1}\mathcal{S}$, it is therefore bijective.

Consequently, Φ is a local diffeomorphism in the neighbourhood of the identity operator, and the function $R(x, y) = \Phi^{-1}(Q(x, y))$ is defined for $(x, y) \in U \times V$ and satisfies (9.9.3) when U and V are small enough.

We are now in a position to state the theorem of stationary phase with parameter.

THEOREM 9.10: *We adopt the assumptions of Theorem* 9.9, *and we assume that the symbol* $a \in S^m(X \times Y \times \mathbb{R}^+)$ *is zero for* (x, y) *outside a compact subset of* U x V. *We put*

$Q(x) = \partial^2_{yy}\, \varphi(x, y)_{|y=y(x)}$ *and we denote by*

$k : U \times W \ni (x, z) \longmapsto y = k(x, z) \in V$ *the inverse of the Morse diffeomorphism for* x *fixed in* U. *Then the integral* (9.1.1) *satisfies*

(9.10.1.) i) $e^{-i\tau\varphi(x, y(x))}\, I(x, \tau) \in S^{m-\frac{1}{2}q}(X \times \mathbb{R}^+)$

ii) *This symbol admits an asymptotic expansion in* $U \times \mathbb{R}^+$ *of the form*

$$(9.10.2.)\quad \left(\frac{2\pi}{\tau}\right)^{q/2} |\det Q(x)|^{-1/2} e^{i\,\pi/4\,\mathrm{sign}\,Q(x)} \cdot \left[\sum_{k \geq 0} \frac{\tau^{-k}}{k!} R^k(x, z, \partial_z)\, \tilde{a}(x, z, \tau)_{|z=0} \right]$$

where

$$\tilde{a}(x, z, \tau) = a(x, k(x, z), \tau) \cdot \left| \frac{Dy}{Dz}(x, z) \right|$$

and

$$R(x, z, \partial_z) = \frac{i}{2} < Q^{-1}(x)\, \partial_z, \partial_z > \quad .$$

iii) *In particular, the leading term in this expansion is given by*

$$(9.10.3.)\quad \left(\frac{2\pi}{\tau}\right)^{q/2} |\det Q(x)|^{-1/2} e^{i\,\pi/4\,\mathrm{sign}\,Q(x)} \cdot \left[a(x, y(x), \tau) \right] \mathrm{mod}\ S^{m-\frac{1}{2}q-1} \ .$$

PROOF: The assumption concerning the support of a allows us to write the phase function φ in the integral (9.1.1) in the form (9.9.1), and we have $I(x, \tau) =$

$$e^{i\tau\,\varphi(x, y(x))} \int e^{i\,\tau/2 < Q(x).h(x, y), h(x, y) >} a(x, y, \tau) dy \ .$$

Then, we use the Morse diffeomorphism in order to effect the change of variables $y \to z = h(x, y)$, and we find

$$e^{-i\tau\,\varphi(x, y(x))}\, I(x, \tau) = \int e^{i\,\tau/2\,< Q(x)z, z >} \tilde{a}(x, z, \tau) dz$$

and consequently Theorem 9.10 follows from Theorem 9.3.

9.11 VARIOUS GENERALISATIONS: All the results generalize word for word when a is a symbol with values in a finite-dimensional vector space F: $a \in S^m(X \times Y \times R^+ ; F)$. All the results likewise generalise, with minor modifications, to the case where a is a symbol of type ρ with $\rho > 1/2$.

10. ADDITIONAL NOTES AND EXERCISES

(*) 10.1 SYMBOLS WITH UNIFORM UPPER BOUNDS W.R.T. x IN $\mathbb{R}^n$:

We say that $a \in BS^m(\mathbb{R}^n \times \mathbb{R}^N)$ if $a \in S^m(\mathbb{R}^n \times \mathbb{R}^N)$ and if a satisfies (2.1.1) with C independent of the compact set K.

We put $BS^{+\infty} = \bigcup_m BS^m$, $BS^{-\infty} = \bigcap_m BS^m$.

a) Show that $BS^{+\infty}$ is a graded sub-algebra of $S^{+\infty}$.

b) Supposing $a \in BS^m$, prove that there exists $b \in BS^{-m}$ such that $a.b-1 \in BS^{-1}$ if and only if there exists $c > 0$ such that

$$|a(x,\theta)| \geq c(1+|\theta|)^m \qquad \text{for} \qquad |\theta| \geq \frac{1}{c} .$$

We say that a is elliptic of degree m .

c) Suppose we have a sequence of symbols $(a_j)_{j \leq 0}$ with $a_j \in BS^{m_j}$ and $m_j \searrow -\infty$. Prove that there exists $a \in BS^{m_o}$, unique modulo $BS^{-\infty}$, such that for any k we have

$$(a - \sum_{j=0}^{k} a_j) \in BS^{m_{k+1}} .$$

Under these conditions we write

$$a \underset{B}{\sim} \sum_{j\geq 0} a_j \qquad .$$

10.2. OSCILLATORY INTEGRALS AND DECOMPOSITION INTO PLANE WAVES :

a) Suppose we have a symbol $b \in S^m(\mathbb{R} \times [0, +\infty[\,)$ and consider the distribution $v \in \mathcal{D}'(\mathbb{R})$ defined by the oscillating integral

$$v(y) = \int_0^{+\infty} e^{iy\xi} b(y,\xi)d\xi \quad .$$

Let there be $f : \mathbb{R}^n \longrightarrow \mathbb{R}$ such that $f'(x) \neq 0$ when $f(x) = 0$. Consider the distribution $u = f^* v \in \mathcal{D}'(\mathbb{R}^n)$ defined in Exercise 10.9 of Chapter I. Prove that

$$u(x) = \int_0^{+\infty} e^{if(x)\xi} \, b(f(x) \, , \, \xi)d\xi \quad .$$

[Approximate b by symbols with compact support in ξ]

b) Let $a = a(x, \theta)$ be a symbol in $\mathbb{R}^n \times \mathbb{R}^N$ and let $\varphi = \varphi(x \, , \, \theta)$ be a phase function in $\mathbb{R}^n \times (\mathbb{R}^N \setminus 0)$.

Consider the distribution $u \in \mathcal{D}'(\mathbb{R}^n)$ defined by

$$u(x) = \int e^{i\varphi(x,\theta)} a(x,\theta) \, d\theta$$

and assume that $\varphi'_x(x \, , \, \theta) \neq 0$ when $\varphi(x \, , \, \theta) = 0$.

Prove (by changing into spherical coordinates) that

$$u = \int_{S_{N-1}} u_w \, dw \quad , \text{ where } u_w \in \mathcal{D}'(\mathbb{R}^n) \text{ is defined by}$$

$$u_w(x) = \int_0^{+\infty} e^{ir\varphi(x,w)} \, a(x \, , \, rw) \, r^{n-1} dr \quad ,$$

where S_{N-1} is the unit sphere of $\mathbb{R}^N$ and where dw is the Riemannian density on S_{N-1} [Equip $\mathbb{R}^N$ with its canonical Euclidian structure].

c) By applying the above in the case

$$u(x) = \delta(x) = \int_{R^n} e^{ix.\theta}\, d\!\!/\theta \ , \quad \text{show that}$$

$$\delta(x) = (2\pi)^{-n} \int_{S_{n-1}} v(w \,.\, x)\, dw \ , \qquad \text{where}$$

$$v(y) = \int_0^{+\infty} e^{iry}\, r^{n-1}\, dr = D^{n-1}\, 2\pi(\mathcal{F}^{-1}Y)$$

$$= i\, D^{n-1}\left(\operatorname{pv} \frac{1}{y} - i\pi\, \delta(y)\right) \quad (\text{cf. } 10.2.1, I)$$

Deduce from this the following decomposition into plane waves of the Dirac distribution in $\mathbb{R}^n$ $(n \geq 2)$:

$$\delta(x) = i(2\pi)^{-n} \int_{S_{n-1}} D^{n-1}\left(\operatorname{pv} \frac{1}{y}\right)\Big|_{y=x.w} dw \qquad \text{if } n \text{ is even}$$

$$\delta(x) = \pi(2\pi)^{-n} \int_{S_{n-1}} (D^{n-1}\delta)(x.w)\, dw \qquad \text{if } n \text{ is odd} \quad .$$

10.3 LACUNAS OF THE FUNDAMENTAL SOLUTION OF THE WAVE OPERATOR:

We again adopt the notation of Exercise 10.8 in Chapter I, and we suppose that $n \geq 2$. We recall that the distribution $E_1 \in \mathcal{D}'(R^n)$ is defined by

$$E_1(x) = \mathcal{F}^{-1}\frac{\sin|\xi|}{|\xi|} = \int \left(e^{i(x.\xi+|\xi|)} - e^{i(x.\xi-|\xi|)}\right)\frac{1}{2i|\xi|}\, d\!\!\!/\xi \, .$$

a) Reasoning as in the above exercise, establish the decomposition into plane waves:

$$E_1(x) = \tfrac{1}{2}(2\pi)^{-n}\int_{S_{n-1}} [v(x.\omega+1) - v(x.\omega-1)]\, d\omega$$

where

$$v(y) = D^{n-2}\left(\mathrm{pv}\,\frac{1}{y} - i\pi\,\delta(y)\right) \in \mathcal{D}'(\mathbb{R}) \, .$$

Deduce from this that

(10.3.1) $$E_1(x) = (2\pi)^{-n}\int_{S_{n-1}} w(x.\,\omega+1)\, d\omega$$

where $w \in \mathcal{D}'(\mathbb{R})$ is defined by

(10.3.2) $$w(y) = D^{n-2}\,\mathrm{pv}\,\frac{1}{y} \quad \text{if } n \text{ is even},$$

(10.3.3) $$w(y) = -i\pi\, D^{n-2}\,\delta(y) \quad \text{if } n \text{ is odd}.$$

b) From (10.3.1), (10.3.3) and from (10.9.1) of Chapter I, deduce that when n is odd, we have $E_1 = 0$ in $\{x \in \mathbb{R}^n \mid |x| < 1\}$, and therefore (see section 10.8 of Chapter I) that $\mathrm{supp}\, E_1 = \{x \in \mathbb{R}^n \mid |x| = 1\}$. Infer from this (see (10.8.1), Chapter I) that the support of the fundamental solution E of the wave operator is equal, when n is odd and ≥ 3, to the cone

$$\{(t,x) \in \mathbb{R}^{n+1} \mid t = |x|\} \, .$$

c) Suppose $f : \mathbb{R} \to \mathbb{C}$ is continuous. Prove that we have, for $x \in \mathbb{R}^n$:

(10.3.4) $$\int_{\omega\in S_{n-1}} f(x.\omega)\,d\omega = \sigma_{n-2}\int_{-1}^{1} f(s|x|)(1-s^2)^{\frac{n-3}{2}}\, ds \, ,$$

where σ_{n-2} is the area of the unit sphere S_{n-2} of $\mathbb{R}^{n-1}$, with $\sigma_0 = 2$ [Cut S_{n-1} up into zones $x.\omega$ = constant, and put $s = \frac{x}{|x|}.\omega$).

d) We shall henceforth assume that n is even, and we propose to study the restriction of E_1 to $\{|x| < 1\}$.

Deduce from (10.3.1), (10.3.2) and (10.3.4) that this restriction is the C^∞ function

$$x \longmapsto F_n(|x|^2)$$

where, for $0 \leq r < 1$, we have put

$$F_n(r) = C_n \int_{-1}^{1} (1 + s/r)^{1-n}(1-s^2)^{\frac{n-3}{2}} \, ds \quad ,$$

and $C_n = -(2i\pi)^{-n}(n-2)!\,\sigma_{n-2}$.

Prove, by integration by parts, that

$$F'_n = \frac{n}{2}\,\frac{C_n}{C_{n+2}}\,F_{n+2} \quad .$$

Deduce from this that

$$(10.3.5) \qquad F_{2k} = \frac{(-1)^{k-1}}{\pi^{k-1}}\,F_2^{(k-1)} \quad .$$

[Recall that $\dfrac{\sigma_{2k-2}}{\sigma_{2k-4}} = \dfrac{\pi}{k-3/2}$).

Calculate directly the integral $\displaystyle\int_{\omega\in S_1} \frac{1}{x.\omega+1}\,d\omega$ when $x \in \mathbb{R}^2$, $|x| < 1$.

Deduce from this that

$$F_2(r) = \frac{1}{2\pi}(1-r)^{-\frac{1}{2}}$$

and thus, by utilising (10.3.5), that the restriction of E_1 to $\{x \in R^n \,|\, |x| < 1\}$ is equal to

$$K_n(1 - |x|^2)^{-\frac{n-1}{2}} \quad ; \quad \text{with}$$

$$K_n = \frac{(-1)^{k-1}}{(2\pi)^k} 1.3,\ldots(2k - 3) \quad \text{for} \quad n = 2k \quad .$$

Infer from this that, when n is even, the restriction of the fundamental solution E of the wave operator (see (10.8.1) Chapter I) to the cone $\{(t,x) \in R \times R^n \mid |x| < t\}$ is given by

$$K_n(t^2 - |x|^2)^{-\frac{n-1}{2}} \quad .$$

10.4 FUNDAMENTAL SOLUTION OF THE WAVE OPERATOR FOR SPACE DIMENSION EQUAL TO 1, 2 OR 3:

We once more adopt the notation of Exercise 10.8. of Chapter I.

a) In the case $n = 1$, prove that E_1 is the characteristic function of the interval $[-1, 1]$, and thus that E is the characteristic function of $\{(t,x) \in R^2 \mid |x| \leq t\}$.

b) In the case $n = 3$, consider (cf. section 7.2, Chapter I) the distribution $\delta(1 - |x|^2)$. Calculate its Fourier transform by utilising section 7.2 of Chapter I and (for example) the formula (10.3.4). Thus deduce that

$$\langle E , \varphi \rangle = \frac{1}{4\pi} \int_{R^3} \frac{\varphi(|x|, x)}{|x|} dx \quad \text{for} \quad \varphi \in C_0^\infty(R^4) \quad .$$

c) In the case n = 2, we express E on the basis of the case n = 3 by using the "method of descent". Let $\tilde{E} = \tilde{E}(t, x_1, x_2, x_3)$ denote the fundamental solution corresponding to the case n = 3. Prove that the mapping

$$\mathcal{D}(\mathbb{R}^3) \ni \varphi \longmapsto \langle \tilde{E}(t, x_1, x_2, x_3), \varphi(t, x_1, x_2) \otimes 1_{x_3} \rangle \in \mathbb{C}$$

defines a distribution on $\mathbb{R}^3$ with support in $\{t \geq 0\}$, which is a fundamental solution of $\dfrac{\partial^2}{\partial t^2} - \sum_{j=1}^{2} \dfrac{\partial^2}{\partial x_j^2}$. Prove that this distribution coincides with the fundamental solution E. Thus deduce, by utilising the last part of b), that E is the function $\frac{1}{2\pi} Y(t)\, Y(t - |x|)\, (t^2 - |x|^2)^{-1/2}$.

10.5 STATIONARY PHASE THEOREM FOR A SURFACE INTEGRAL.

Let S be a hypersurface of $\mathbb{R}^n$ and suppose $a \in C_0^\infty(S)$. We equip S with the Riemannian structure induced by $\mathbb{R}^n$ and we denote the associated density by $d\sigma$. We assume that the total curvature K(y) of S does not vanish at any point y of the support of a.

Supposing $x \in \mathbb{R}^n \setminus 0$, let $y_j(x)$ denote the points $y \in S \cap \operatorname{supp} a$ where the normal to S is parallel to x ; in this case the integral

$$I(\tau,x) = \int_S e^{i\tau\, x.y}\, a(y)\, d\sigma(y)$$

exhibits the following behaviour as $\tau \to +\infty$:

(10.5.1) $I(\tau,x) =$

$$\left(\frac{2\pi}{\tau\,|x|}\right)^{\frac{n-1}{2}} \sum_j a(y_j(x))\ |K(y_j(x))|^{-\frac{1}{2}}\, e^{i(\pi/4)\sigma_j} . e^{i\tau x.y_j(x)} + O(\tau^{-\frac{n+1}{2}})$$

where σ_j is equal to the number of positive principal curvatures minus the number of negative principal curvatures when the normal to S at $y_j(x)$ is orientated positively in the direction of x.

a) Prove that we can reduce to the case where a has its support in the domain of a local chart containing at most one point $y_j(x)$ and of the form $y' \in R^{n-1} \to (y', f(y')) \in S$. Show that in such a chart we have $I(\tau,x) =$

$$\int_{R^{n-1}} e^{i\tau(x'.y' + x_n f(y'))}\, a(y', f(y'))(1 + |f'(y')|^2)^{\frac{1}{2}}\, dy' .$$

b) Deduce (10.5.1) from the stationary phase theorem by observing that the modulus of the total curvature is expressed in this chart by

$$|K(y', f(y'))| = (1 + |f'(y')|^2)^{\frac{n+1}{2}}\ |\det(f''(y'))| .$$

(see Dieudonné [1], Chapter XX, Problem 14.1).

CHAPTER 4

PSEUDO DIFFERENTIAL OPERATORS

1. DEFINITION

1.1 INTRODUCTION : In order to invert elliptic operators, we saw in (1.2) of Chapter III that we were led to generalise the differential operators by operators of the form

$$(1.1.1) \qquad Pu(x) = \int e^{ix\xi}\, p(x,\xi)\,\hat{u}(\xi)\, đ\xi \quad , \quad u \in C_0^\infty(X)$$

where p is in a space $S^m(X \times \mathbb{R}^n)$. Writing out the Fourier transform, we obtain an expression using an oscillatory integral depending on the parameter $x \in X$:

$$(1.1.2) \qquad (Pu)(x) = \iint e^{i(x-y).\xi}\, p(x,\xi)\, u(y)\,dy\, đ\xi \quad .$$

Moreover, we also want the operators of the form (1.1.2) to possess the stability properties of the differential operators for the operations of addition, of composition, of transposition,....

For example, if we compose P with the operator of multiplication by $g \in C^{\infty}(X)$, we obtain for the composition the expression

$$(P \cdot gu)(x) = \iint e^{i(x-y)\xi} \, p(x,\xi) \, g(y) \, u(y) \, dy \, đ\xi$$

where the amplitude $p(x,\xi)g(y)$ is an element of $S^m(X \times X \times \mathbb{R}^n)$. This leads us to state the following :

DEFINITION 1.2 : *Let* X *be an open subset of* $\mathbb{R}^n$, *and let* $m \in \mathbb{R} \cup \{-\infty\}$. *We define a pseudo-differential operator (p.d.o.) of degree* m *in* X *to be an integral operator* P *associated with the phase function*

$$\varphi(x,y,\xi) = (x-y).\xi \qquad (x,y,\xi) \in X \times X \times \mathbb{R}^n$$

and with an amplitude

$$p(x,y,\xi) \in S^m(X \times X \times \mathbb{R}^n) \quad , \quad (\xi \text{ is the frequency variable}).$$

We denote by $L^m(X)$ *the space of p.d.o.'s of degree* m *in* X.

By definition, a p.d.o. P is written:

$$(1.2.1) \qquad (Pu)(x) = \iint e^{i(x-y)\xi} \, p(x,y,\xi) \, u(y) \, dy \, đ\xi \; , \quad u \in C_0^{\infty}(X).$$

If we carry out the integration with respect to ξ in (1.1.2) we formally obtain.

$$(1.2.2) \qquad (Pu)(x) = \int K(x,x-y) \, u(y) \, dy$$

where

$$K(x,z) = \int e^{iz.\xi} \, p(x,\xi) \, đ\xi \quad .$$

In the particular case where p is homogeneous with respect to ξ, the kernel K is a distribution which is homogeneous with respect to z and which has C^{∞} dependence on x.

It was in the form (1.2.2), termed a singular integral, that CALDERON-ZYGMUND [1] studied these operators and developed the first symbolic calculus. Subsequently, numerous other authors expanded and generalised the theory, replacing (1.2.2) by the more tractable expression (1.2.1); we cite the names of KUMANO-GO, KOHN-NIRENBERG, UNTERBERGER-BOKOBZA, HORMANDER. We shall primarily follow the approach of HORMANDER [5], but putting the emphasis more on the role of the stationary phase theorem.

1.3 IMMEDIATE PROPERTIES. From Proposition 8.2 of Chapter III, we know that any pseudo-differential operator P is continuous from $C_0^{\infty}(X)$ into $C^{\infty}(X)$ and extends into a continuous operator from $\mathcal{E}'(X)$ into $\mathcal{D}'(X)$. Suppose K_p is the kernel distribution of P; Proposition 7.7 of Chapter III, shows that

$$\text{(1.3.1)} \qquad \text{sing supp } K_p \subset \text{Diag}(X \times X) ,$$

consequently Proposition 5.4 of Chapter I shows that we have

$$\text{(1.3.2)} \qquad \text{sing supp } Pu \subset \text{sing supp } u \quad , \quad u \in \mathcal{E}'(X) .$$

In the particular case where p does not depend on y we have the representation (1.1.1) of Pu for $u \in C_0^{\infty}(X)$, and we therefore deduce by continuity that

$$(1.3.3) \qquad \langle Pu , v \rangle = \int \hat{u}(\xi) \left(\int e^{ix\xi} p(x,\xi) v(x) \, dx \right) đ\xi$$

for $u \in \mathcal{E}'(X)$, $v \in C_0^\infty(X)$, since the function

$\xi \to \int e^{ix\xi} p(x,\xi) v(x) dx$ is of rapid decrease.

The regularising operators are pseudo-differential operators:

PROPOSITION 1.4 : *The set of regularising operators is equal to* $L^{-\infty}(X)$ *which is itself equal to* $\bigcap_m L^m(X)$.

PROOF: Supposing $P \in \bigcap L^m(X)$, then for all $m \in \mathbb{R}$ there exists $a \in S^m$ such that P can be written in the form (1.2.1).

Its kernel distribution

$$(1.4.1) \qquad K_p(x,y) = \int e^{i(x-y)\xi} p(x,y,\xi) \, đ\xi$$

is in $C^k(X \times X)$ when $m < -n-k$. Consequently $K_p \in C^\infty(X \times X)$.

We now assume that K_p is C^∞; we can write P in the form (1.2.1) by taking

$$p(x,y,\xi) = e^{-i(x-y)\xi} \rho(\xi) K_p(x,y) \quad \text{with} \quad \rho \in C_0^\infty(\mathbb{R}^n)$$

and $\int \rho(\xi) \, đ\xi = 1$. We thus have $p \in S^{-\infty}$ and $P \in L^{-\infty}$.

Finally, it is clear that $L^{-\infty}(X) \subset \bigcap_m L^m(X)$.

We can always reduce to the case of a proper p.d.o. if we argue modulo a regularising operator; this gives us

PROPOSITION 1.5 : *If we have* $P \in L^m(X)$, *then we can write*

(1.5.1) $$P = P' + R$$

where R *is a regularising operator and* $P' \in L^m(X)$ *is a proper operator.*

PROOF: We prove that there exists a function $\rho \in C^\infty(X \times X)$ such that the projections

$$p_1|_{\text{supp}\,\rho} \ni (x,y) \mapsto x \in X \;, \quad p_2|_{\text{supp}\,\rho} \ni (x,y) \overset{p_2}{\mapsto} y \in X$$

are proper. Let $(\mathcal{V}_j)_{j \in J}$ be a locally finite covering of X by relatively compact open subsets; then $U = \bigcup_j \bar{\mathcal{V}}_j \times \bar{\mathcal{V}}_j$ is a neighbourhood of $\Delta = \text{Diag}\,(X \times X)$ such that $p_1|_U$ and $p_2|_U$ are proper. If V is a neighbourhood of Δ such that $\bar{V} \subset U$, then it suffices to take for ρ a $C^\infty(X \times X)$ function identical to 1 on V and with support in U.

We define the operators P' and R by their kernels

$$K_{P'} = \rho K_P \;, \quad K_R = (1-\rho)K_P \qquad .$$

By construction the operator P' is proper and the inclusion (1.3.1) shows that R is regularising. Finally, $P' \in L^m(X)$, since if p is an amplitude for P, then ρp is an amplitude for P'.

REMARK 1.6 : Proposition 5.3 of Chapter I shows that if P is a proper p.d.o. then it is a continuous operator on each of the following spaces:

$$C_0^\infty(X) \quad , \quad C^\infty(X) \quad , \quad \mathcal{E}'(X) \quad , \quad \mathcal{D}'(X) \quad .$$

2. A CHARACTERISATION OF P.D.O.'s.

We are now going to generalise to p.d.o.'s the characterisation (see Proposition 7.3 of Chapter I) of differential operators. The result in question is somewhat technical, but it contains practically all the analysis necessary to the theory of p.d.o.'s; in particular, the symbolic calculus and invariance under diffeomorphism are easily deduced from it.

THEOREM 2.1 : *Suppose* $P \in L^m(X)$; *we take a symbol* $a \in S^q(X \times \mathbb{R}^N)$ *and a real-valued function* $\psi \in C^\infty(X \times \mathbb{R}^N_0)$ *such that*

$$(2.1.1) \quad \begin{cases} d_x \psi(x,\theta) \neq 0 \text{ for } (x,\theta) \in \text{conic supp of } a \text{ and } \theta \neq 0, \\ \psi(x,\theta) \text{ is homogeneous of degree 1 with respect to } \theta \end{cases}$$

We assume that P *is proper or that* a *has compact support with respect to* y, *and we put*

(2.1.2) $b(x,\theta) = e^{-i\psi(x,\theta)}\, P(a\, e^{i\psi})(x,\theta)$. *Then:*

(2.1.3) i) $b \in S^{m+q}(X \times R^N_o)$

ii) *we have an expression for the asymptotic expansion of* b :

(2.1.4) $b(x,\theta) \sim$

$$\sum_{\alpha,\beta} \frac{1}{\alpha!\ \beta!} (\partial_\xi^{\alpha+\beta} D_y^\beta p)(x,x,d_x\psi(x,\theta)) D_y^\alpha (a(y,\theta)e^{ir})\Big|_{y=x}$$

where $r(x,y,\theta) = \psi(y,\theta) - \psi(x,\theta) - \langle d_x\psi(x,\theta), y-x \rangle$

and p *is an amplitude for* P.

iii) *in particular*

(2.1.5) $b = p(x,x,d_x\psi(x,\theta))\, a(x,\theta)$ *modulo* S^{m+q-1} .

We shall see that the general term of (2.1.4) is in

$S^{m+q-\frac{|\alpha|}{2}-|\beta|}$, and consequently this asymptotic sum is meaningful.

PROOF: From Proposition 2.5 of Chapter III, it suffices to prove the theorem in a conic neighbourhood of a point $(x_o, \theta_o) \in X \times R^N_o$. By replacing u by a $e^{i\psi}$ in the expression (1.2.1) for P, we obtain

(2.1.6) $b(x,\theta) =$

$$e^{-i\psi(x,\theta)} \iint e^{i(x-y)\xi + i\psi(y,\theta)} p(x,y,\xi) a(y,\theta)\, dy\, d\!\!\!/\xi$$

where by hypothesis we can reduce to the case where the function $y \to p(x,y,\xi)a(y,\theta)$ has support within a compact subset $K \subset X$.

We utilise the equivalence (2.6.1) of Chapter III, and we put $\tau = |\theta|$, $\omega = \frac{\theta}{|\theta|}$. After having made the change of variables $\xi = \tau\eta$, we obtain

(2.1.7) $b(x,\omega,\tau)$

$$= e^{-i\tau\psi(x,\omega)} \left(\frac{\tau}{2\pi}\right)^n \iint e^{i\tau\Phi(x,\omega,y,\eta)} p(x,y,\tau\eta)a(y,\tau\omega)\,dy\,d\eta$$

with

$$\Phi(x,\omega,y,\eta) = (x-y)\eta + \psi(y,\omega) .$$

We define the number $r = \inf\limits_{\substack{(y,\omega)\in \text{conic supp } a \\ y\in K}} |d_y \psi(y,\omega)|$; assumption (2.1.1) shows that $r > 0$. Let $\mu(\eta) \in C_0^\infty(R^n)$ be a function identical to 1 for $|\eta| \le \frac{r}{4}$ and equal to zero for $|\eta| \ge \frac{r}{2}$.

We begin by proving that the integral

$$I(x,\omega,\tau) = e^{-i\tau\psi} \left(\frac{\tau}{2\pi}\right)^n \iint e^{i\tau\Phi}\, p.a.\mu\, dy\, d\eta$$

is in $S^{-\infty}(X \times S_{N-1} \times R^+)$. The first-order operator

$$L(\omega,y,\eta,D_y) = |d_y \psi(y,\omega) - \eta|^{-2} < d_y \psi(y,\omega)-\eta , D_y >$$

satisfies $\frac{1}{\tau} a\mu L(e^{i\tau\Phi}) = a\mu e^{i\tau\Phi}$. By virtue of the choice of μ, there exists for any integer k and any compact subset $H \subset X \times S_{N-1}$ some C such that for $\tau \ge 1$

$$\sup_{\substack{(x,\omega)\in H \\ y\in K}} \left| \left(\frac{{}^t L}{\tau}\right)^k p(x,y,\tau\eta)a(y,\tau\omega)\mu(\eta) \right| \leq C\,\tau^{m+q-k} .$$

We thereby deduce that I is of rapid decrease with respect to τ and the same proof shows that this is also true of its derivatives.

We now plan to apply the stationary phase theorem to the remaining integral $J(x,\omega,\tau) = b(x,\omega,\tau) - I(x,\omega,\tau)$.

We have $d_y \Phi = d_y \psi(y,\omega) - \eta$, $d_\eta \Phi = x-y$; consequently Φ is stationary at the point (y,η) defined by $y = x$ and $\eta = d_y \psi(x,\omega)$. At this point, the Hessian matrix is given by

$$\Phi'' = \begin{pmatrix} \psi''_{yy} , & -\mathrm{Id} \\ -\mathrm{Id} , & 0 \end{pmatrix} ;$$

hence $\operatorname{sign} \Phi'' = 0$, $|\det \Phi''| = 1$. (*)

If we put $y_0 = x_0$, $\eta_0 = d_y \psi(x_0 , \omega_0)$, Theorem 9.9 of Chapter III shows that there exist neighbourhoods, U, Ω, W of x_0, ω_0 , (y_0 , η_0) in which we can utilise a Morse diffeomorphism; after appropriately shrinking U x Ω, we can assume that W is a neighbourhood of the set

$W_U = \{(x , d_y \psi(x,\omega)) \mid x \in U , \omega \in \Omega\}$. If we let $\rho(y,\eta)$ be a $C_0^\infty(W)$ function, identical to 1 on W_0, we can apply

*) *Translator's Note:* "sign" is here an abbreviation for "signature".

the stationary phase theorem to the integral

$$J_1(x,\omega,\tau) = e^{-i\tau\psi}\left(\frac{\tau}{2\pi}\right)^n \iint e^{i\tau\Phi}\, p.a.\rho\,(1-\mu)\,dy\,d\eta$$

observing that the amplitude satisfies

$$p(x,y,\tau\eta)a(y,\tau\omega)\rho(y,\eta)(1-\mu(\eta)) \in S^{m+q}(X\times S_{N-1}\times X\times \mathbf{R}^n\times \mathbf{R}^+)$$

with τ as the frequency variable. We find

$$J_1 \in S^{m+q}\,(U\times\Omega\times\mathbf{R}^+) \qquad \text{and}$$

(2.1.8) $\quad J_1(x,\omega,\tau) \sim$

$$\sum_{k\geq 0} \frac{\tau^{-k}}{k!}\, S_k(x,\omega,y,\eta,D_y,D_\eta)\left[p(x,y,\tau\eta)a(y,\tau\omega)\right]_{\substack{y=x\\ \eta=d_y\psi(x,\omega)}}$$

where S_k is a differential operator of degree $\leq 2k$ which depends only on k and ψ. It should be noted that μ is zero in the neighbourhood of $\eta = d_y\psi(x,\omega)$ for $(x,\omega) \in$ conic supp a and $x \in K$.

Finally, we prove that we have $J_2 = J - J_1 \in S^{-\infty}(U\times\Omega\times\mathbf{R}^+)$. To do this we consider the first-order operator

$$L = D(< d_y\psi(y,\omega) - \eta\ ,\ D_y > + < x-y, D_y >)$$

where we have put $D = (|d_y\psi(y,\omega)-\eta|^2 + |x-y|^2)^{-1}$ so that

$$(1-\rho(y,\eta))(1-\mu(\eta))D(x,\omega,y,\eta) \in S^{-2}(U\times\Omega\times X\times\mathbf{R}^n) \quad \text{for}$$

the frequency variable η. By construction we have

$$\frac{1}{\tau}(1-\mu)(1-\rho)L\,(e^{i\tau\Phi}) = (1-\mu)(1-\rho)\,e^{i\tau\Phi}\ .$$

Leibniz's formula shows that $\left(\frac{{}^tL}{\tau}\right)^k$ $p\,a$ is a sum of terms of the form

$$\frac{1}{\tau^k}\ c(x,\omega,y,\eta)\ D_y^\alpha\ D_\eta^\beta\ p(x,y,\tau\eta)\ .\ \tau^{|\beta|}\ D_y^\gamma a(y,\tau\omega)$$

with $c \in S^{-k}$ and $|\alpha| + |\beta| + |\gamma| \leq k$. Such a term is bounded above by

$$C\ \tau^{-k}(1 + |\eta|)^{-k}(1 + \tau|\eta|)^{m-|\beta|}\ \tau^{|\beta|}\ \tau^q$$

and since $|\eta| \geq \frac{r}{4}$ on the support of $1 - \mu$, this is also bounded above by $C\ \tau^{m+q-k}\ (1 + |\eta|)^{m-k}$

These upper bounds imply that J_2 is of rapid decrease with respect to τ and we proceed in the same way for its derivatives.

Finally, we have shown that $b \sim J_1$, which proves (2.1.3), and it remains to give the asymptotic expansion of J_1. By replacing τ by $\frac{\theta}{|\theta|}$ in (2.1.8), we obtain

(2.1.9) $$b(x,\theta) \sim \sum_{k\geq 0} b_k(x,\theta)$$

where $b_k \in S^{m+q-k}(X \times \mathbf{R}^N)$ is given by

(2.1.10) $$b_k(x,\theta) = \sum_{|\alpha+\beta+\gamma|\leq 2k} c_{k,\alpha,\beta,\gamma}(x,\theta) D_\xi^\alpha\ D_y^\beta p(x,x,d_x\psi(x,\theta))\partial_x^\gamma\ a(x,\theta)$$

and $c_{k,\alpha,\beta,\gamma}(x,\theta)$ is a homogeneous function of degree $|\alpha| - k$ with respect to θ, which depends neither on a nor on p but only on ψ. The degree of the general term of (2.1.10) is bounded above by $m + q - \frac{1}{2}\ |\alpha + \beta + \gamma|$; consequently it tends to $-\infty$, when $|\alpha|$ or $|\beta|$ or $|\gamma|$ tends to $+\infty$, which allows us to apply Remark 4.3 of Chapter III and therefore consider (2.1.9) as an asymptotic sum with respect to (α, β). We find

$$(2.1.11) \qquad b(x,\theta) \sim \sum_{\alpha,\beta} D_\xi^\alpha D_y^\beta p(x,x,d_x\psi(x,\theta)).d_{\alpha,\beta}(x,\theta)$$

where $d_{\alpha,\beta} \in S^{q+\frac{1}{2}|\alpha| - \frac{1}{2}|\beta|}(X \times R^N)$ and depends only on α, β, ψ and a. We take advantage of the fact that $d_{\alpha,\beta}$ is independent of p to determine it by taking p in the particular form

$$p(x,y,\xi) = q(x,\xi)\, g(y)$$

where q is a polynomial with respect to ξ and $g \in C^\infty(X)$. As we observed at the beginning of Section 1, this means that P is the differential operator $Q(x,D) \circ g$. We can thus directly calculate

$$b(x,\theta) = e^{-i\psi(x,\theta)} Q(x,D)\,(g.a\, e^{i\psi})$$

by applying the Leibniz-Hormander formula (1.1.3) of Chapter I, and we obtain

$$(2.1.12) \qquad b(x,\theta) = e^{-i\psi} \sum_\beta \frac{1}{\beta!} D_y^\beta g(x) \,.\, (\partial_\xi^\beta Q(y,D_y)(a\, e^{i\psi}))\big|_{y=x}$$

If we put $\psi(y,\theta) = \psi(x,\theta) + \langle d_y\psi(x,\theta), y-x \rangle + r$, formula (1.1.4) of Chapter I shows that

$$e^{-i\psi} \partial_\xi^\beta Q(y,D_y)(a\, e^{i\psi}) = \sum_\alpha \frac{1}{\alpha!} \partial_\xi^{\alpha+\beta} Q(x,d_x\psi(x,\theta)) D_y^\alpha (a\, e^{ir})$$

and substituting into (2.1.12), we obtain

$$(2.1.13) \qquad b = \sum_{\alpha,\beta} \frac{1}{\alpha!\,\beta!} (D_y^\beta \partial_\xi^{\alpha+\beta} p)(x,x,d_x\psi(x,\theta)).D_y^\alpha(a\, e^{ir})\big|_{y=x}.$$

By identification with (2.1.11), we deduce from this the expression for the terms $d_{\alpha\beta}$, which proves the formula (2.1.4).

As an application of Theorem 2.1, we give below a characterisation of pseudo-differential operators on an open subset $X \subset \mathbb{R}^n$.

PROPOSITION 2.2: *Let* P *be a continuous linear operator from* $C_0^\infty(X)$ *into* $C^\infty(X)$. *Then,* $P \in L^m(X)$ *if and only if we have*

(2.2.1) $e^{-ix\xi} P(a\, e^{ix\xi}) \in S^m(X \times \mathbb{R}^n)$ *for all* $a \in C_0^\infty(X)$.

If we further assume that P *is proper, it suffices to have* (2.2.1) *with* a = 1.

PROOF: It remains to prove that the condition is sufficient. Let (ρ_j) be a partition of unity on X with functions $\rho_j \in C_0^\infty(X)$ and let β_j be functions in $C_0^\infty(X)$, with $\beta_j \in C_0^\infty(X)$ identical to 1 on supp ρ_j; we assume that the supp β_j are also locally finite. For $u \in C_0^\infty(X)$, by Fourier inversion we have:

$$(\rho_j u)(x) = \int e^{ix\xi}\, (\widehat{\rho_j u})(\xi)\, đ\xi = \int \beta_j(x) e^{ix\xi}\, (\widehat{\rho_j u})(\xi)\, đ\xi .$$

Interpreting this last integral as an integral with respect to ξ with values in $C_0^\infty(X)$, the continuity of P allows us to write

$$(2.2.2) \qquad P(\rho_j u)(x) = \int e^{ix\xi}\, p_j(x,\xi)\, \widehat{\rho_j u}(\xi)\, đ\xi$$
$$= \iint e^{i(x-y)\xi}\, p_j(x,\xi)\rho_j(y)u(y)dy\, đ\xi$$

with

$$p_j(x,\xi) = e^{-ix\xi}\, P(\beta_j(x)e^{ix\xi}) \in S^m(X \times \mathbb{R}^n) \quad .$$

Since $u = \sum_j \rho_j u$, we therefore deduce that

$$Pu = \iint e^{i(x-y)\xi}\, p(x,y,\xi)u(y)dy\, đ\xi$$

where the amplitude

$$p(x,y,\xi) = \sum_j p_j(x,\xi)\, \rho_j(y)$$

is clearly an element of $S^m(X \times X \times \mathbb{R}^n)$ since this sum is locally finite with respect to y .

Since P is proper, it defines a continuous operator

$$C^\infty(\mathbb{R}^n) \to C^\infty(X) \xrightarrow{P} C^\infty(X)$$

where the first arrow denotes the restriction to X. If we apply this operator to the integral

$$u(x) = \int e^{ix\xi}\, \hat{u}(\xi)\, đ\xi \quad , \quad u \in \mathcal{S}(\mathbb{R}^n) \ ,$$

considered as an integral with values in $C^\infty(\mathbb{R}^n)$, we obtain

$$(2.2.3) \qquad (Pu)(x) = \int e^{ix\xi}\, p(x,\xi)\hat{u}(\xi)\, đ\xi \quad , \quad u \in \mathcal{S}(\mathbb{R}^n) \ , \ x \in X$$

with

$$(2.2.4) \qquad p(x,\xi) = e^{-ix\xi}\, P(e^{ix\xi}) \in S^m(X \times \mathbb{R}^n) \quad .$$

This leads us to:

PROPOSITION 2.3: *If we let* P *be a proper operator taken from* $L^m(X)$, *then there exists a unique* $p \in S^m(X \times \mathbf{R}^n)$ *such that we have* (2.2.3). *Moreover, if* $p(x,y,\xi)$ *is an amplitude for* P, *we have*

$$p(x,\xi) \sim \sum_{\alpha} \frac{1}{\alpha!} \partial_\xi^\alpha D_y^\alpha p(x,y,\xi)\big|_{y=x} \tag{2.2.5}$$

We say that $p(x,\xi)$ *is the complete symbol of* P *and we write* $P = p(x,D)$.

PROOF: We have just proved the existence of p ; the uniqueness arises out of the density of $\mathcal{S}(\mathbf{R}^n)$. The expansion (2.2.5) follows from application of (2.1.4) to the particular case (2.2.4).

3. SYMBOL OF A PSEUDO-DIFFERENTIAL OPERATOR

3.1 COMPLETE SYMBOL: We shall now extend the notion of a complete symbol to an arbitrary operator $P \in L^m(X)$. By utilising Proposition 1.5, we can write $P = P' + R'$ and we denote by $p'(x,\xi)$ the complete symbol of the proper operator P'. If we consider a different decomposition of this type, $P = P'' + R''$, the difference $p' - p''$ between the corresponding complete symbols is in $S^{-\infty}(X \times \mathbf{R}^n)$ since the operator $P' - P = R'' - R'$

is both proper and in $L^{-\infty}(X)$. If we denote by $\sigma(P)(x,\xi)$ the equivalence class of p' in $S^m / S^{-\infty}$, we have thus defined a linear mapping

$$L^m(X) \ni P \longrightarrow \sigma(P) \in S^m(X \times \mathbb{R}^n) / S^{-\infty}(X \times \mathbb{R}^n) .$$

This is clearly surjective, and Proposition 1.4 shows that its kernel is $L^{-\infty}(X)$; therefore we have an isomorphism

$$(3.1.1) \quad L^m / L^{-\infty} \xrightarrow{\sigma} S^m / S^{-\infty}$$

called the isomorphism of the complete symbol.

REMARK 3.1.2. : The above shows that any operator belonging to $L^m(X)$ can be written, to within a regularising operator, in the form $p(x,D)$ with $p \in S^m(X \times \mathbb{R}^n)$.

3.2. PRINCIPAL SYMBOL. If $P \in L^m(X)$, it is clear that we have the equivalence

$$P \in L^{m-1}(X) \Longleftrightarrow \sigma(P) \in S^{m-1} / S^{-\infty} .$$

Consequently, we deduce from (3.1.1), by passage to the quotient, a simpler isomorphism:

$$(3.2.1) \quad L^m / L^{m-1} \xrightarrow{\sigma_m} S^m / S^{m-1} ,$$

called the isomorphism of the principal symbol of degree m.

If $P = P(x,D)$ is a differential operator, then $\sigma_m(P)$ admits a canonical representative in $S^m(X \times \mathbb{R}^n)$, namely the principal part $P_m(x,\xi)$ of the polynomial $P(x,\xi)$.

Given $P \in L^m(X)$, we shall often commit the notational abuse of identifying $\sigma_m(P)$ with a representative in S^m, and we shall do the same for $\sigma(P)$.

4. ALGEBRA AND SYMBOLIC CALCULUS OF P.D.O.'s

In Chapter I we encountered the definition of the transpose ${}^tP$ and that of the adjoint P^* of an operator P on an open subset of $\mathbb{R}^n$.

For p.d.o.'s we have:

THEOREM 4.1 : *If* $P \in L^m(X)$, *then* ${}^tP$ *and* P^* *are in* $L^m(X)$ *and the complete symbols admit the following expansions:*

$$\sigma({}^tP)\,(x,\xi) \sim \sum_\alpha \frac{1}{\alpha!}\,\partial_\xi^\alpha D_x^\alpha\,\sigma(P)\,(x,-\xi) \tag{4.1.1}$$

$$\sigma(P^*)\,(x,\xi) \sim \sum_\alpha \frac{1}{\alpha!}\,\partial_\xi^\alpha D_x^\alpha\,\overline{\sigma(P)(x,\xi)} \tag{4.1.2}$$

PROOF: Since the transpose or the adjoint of an operator with C^∞ kernel is an operator with C^∞ kernel, we will not restrict the generality by assuming that P is proper.

If we put $p(x,\xi) = \sigma(P)(x,\xi)$, then P can be written

$$Pu(x) = \iint e^{i(x-y)\xi}\,p(x,\xi)u(y)dy\,đ\xi \quad .$$

By noting that we have

$$\langle Pu , v\rangle = \langle u , {}^tPv\rangle \quad , \quad v \in C_0^\infty(X) \quad ,$$ we obtain

$$({}^tP\, v)(y) = \iint e^{i(y-x)\xi}\, p(x, -\xi)\, v(x)\, dx\, d\!\!\!/\xi \quad .$$

We therefore deduce that ${}^tP \in L^m(X)$, and the expansion (4.1.1) then follows from (2.2.5)

The P^* case can be deduced from this since $P^* = {}^t\bar{P}$.

For composition, we have

THEOREM 4.2 : *Suppose we have* $P \in L^m(X)$ *and* $Q \in L^{m'}(X)$; *we assume that at least one of these is a proper operator. Then,*

(4.2.1) $$Q \circ P \in L^{m+m'}(X)$$

and for the complete symbol we have

(4.2.2.) $$\sigma(Q \circ P)(x,\xi) \sim \sum_\alpha \frac{1}{\alpha!}\, \partial_\xi^\alpha\, \sigma(Q)(x,\xi) \,.\, D_x^\alpha\, \sigma(P)(x,\xi) \quad .$$

PROOF: We shall not restrict the generality by assuming that both of these operators are proper. We have

(4.2.3) $$e^{-ix\xi}(Q \circ P)(e^{ix\xi}) = e^{-ix\xi}\, Q\, [e^{ix\xi}\, e^{-ix\xi}\, P(e^{ix\xi})]$$
$$= e^{ix\xi} Q(\sigma(P) e^{ix\xi}) \quad .$$

Consequently (4.2.1) can be deduced from (2.1.3) and from Proposition 2.2. The expansion (4.2.2) follows from application of (2.1.4) to (4.2.3).

REMARK 4.3 : The composition of proper p.d.o.'s induces, via the isomorphism of the complete symbol, an associative law of composition on $S^{+\infty} / S^{-\infty}$. This law is denoted as $\#$ and can be defined directly by

$$(4.3.1) \qquad (q \# p)(x,\xi) \sim \sum_{\alpha} \frac{1}{\alpha!} \partial_{\xi}^{\alpha} q(x,\xi) D_x^{\alpha} p(x,\xi) \quad .$$

Finally, we have invariance of the p.d.o.'s under diffeomorphism.

THEOREM 4.4 : *Let* χ *be a diffeomorphism between two open subsets* X,Y *of* $\mathbb{R}^n$. *Suppose* $P \in L^m(X)$ *and let* $\chi_* P$ *denote the operator transported by* χ *onto* Y. *Then*

$$(4.4.1) \qquad \chi_* P \in L^m(Y)$$

and for the complete symbol we have

$$(4.4.2) \quad \sigma(\chi_* P)(y,\eta) \sim \sum_{\alpha} \frac{1}{\alpha!} \partial_{\xi}^{\alpha} \sigma(P)(x, {}^t\chi'(x).\eta) D_z^{\alpha}(e^{ir})\big|_{z = x}$$

with $r(x,z,\eta) = \langle \chi(z) - \chi(x) - \chi'(x)(z-x) \, , \, \eta \rangle$.

PROOF: We will not reduce the generality by assuming that P is proper; in this case $\chi_* P$ is also proper and we utilise the characterisation in Proposition 2.2. We have

$$(4.4.3) \quad e^{-iy.\eta}(\chi_* P)(e^{iy\eta}) = e^{-i\chi(x).\eta} P(e^{i\chi(x).\eta})\big|_{x = \chi^{-1}(y)} ;$$

consequently, (2.1.3) of Theorem 2.1 shows that the left-hand side of (4.4.3) is in $S^m(Y \times \mathbb{R}^n)$. This left-hand side is also equal to $\sigma(\chi_* P)(y,\eta)$; hence the expansion (4.4.2) by taking in (2.1.4) :

$$a = 1 \quad , \quad p(x,y,\xi) = \sigma(P)(x,\xi) \quad , \quad \phi(x,\eta) = \chi(x).\eta .$$

4.5 REDUCED SYMBOLIC CALCULUS. For principal symbols, it follows from the expression giving the first term of the expansions (4.1.1.), (4.1.2),(4.2.2),(4.4.2), that we have

$$(4.5.1) \qquad \sigma_m({}^tP)(x,\xi) = \sigma_m(P)(x,-\xi)$$

$$(4.5.2) \qquad \sigma_m(P^*)(x,\xi) = \overline{\sigma_m(P)(x,\xi)}$$

$$(4.5.3) \qquad \sigma_{m+m'}(Q \circ P)(x,\xi) = \sigma_{m'}(Q)(x,\xi) \cdot \sigma_m(P)(x,\xi)$$

$$(4.5.4) \qquad \sigma_m(\chi_* P)(y,\eta) = \sigma_m(P)(x, {}^t\chi'(x).\eta) \quad \text{with } y = \chi(x) .$$

As an application of (4.2.2), we deduce (under the conditions of Theorem 3.2) that the commutator

(4.5.5) $[P , Q] = P \circ Q - Q \circ P$ is in $L^{m+m'-1}(X)$ and

$$(4.5.6) \qquad \sigma_{m+m'-1}([P , Q]) = \frac{1}{i} \{\sigma_m(P) , \sigma_{m'}(Q)\} ,$$

where the Poisson bracket { } is defined by

$$\{f,g\}(x,\xi) = \sum_{j=1}^{n} (\partial_{\xi_j} f \cdot \partial_{x_j} g - \partial_{x_j} f \cdot \partial_{\xi_j} g)(x,\xi) .$$

REMARK 4.6 : Case of vector-valued p.d.o.'s.

Let E, F be finite dimensional vector spaces on $\mathbf{C}$. We denote by $L^m(X; E, F)$ the space of p.d.o.'s of the form (1.2.1) when the amplitude $p \in S^m(X \times \mathbf{R}^n ; \mathcal{L}(E , F))$. Then, such operators act continuously from $C_0^\infty(X; E)$ into $C^\infty(X; F)$ and possess all the properties of scalar-valued p.d.o.'s with the exception of (4.5.5) and (4.5.6). For example, Theorem 2.1 follows from Remark 9.11 of Chapter III; it is of course necessary to take $a \in S^q(X \times \mathbb{R}^N ; E)$ and we then find $b \in S^{m+q}(X \times \mathbb{R}^N ; F)$. The principal symbol of $P \in L^m(X ; E , F)$ is clearly an element $\sigma_m(P) \in S^m(X \times \mathbb{R}^n ; \mathcal{L}(E, F)) / S^{m-1}$. For the composition, we assume $Q \in L^{m'}(X ; F, G)$ and we find that $Q \circ P \in L^{m+m'}(X ; E , G)$.

Finally, if we take inner products (resp. Hermitian products) on E and F, the transpose (resp. the adjoint) of $P \in L^m(X ; E , F)$ is an operator ${}^tP$ (resp. P^*) belonging to $L^m(X; F, E)$.

5. P.D.O.'s ON MANIFOLDS

In this section X denotes a manifold of dimension n. Following the definition of differential operators (Definition 7.6 of Chapter I) and from the characterisation of p.d.o.'s described in Section 2, it is natural to state:

DEFINITION 5.1 : *Suppose* $m \in \mathbb{R} \cup \{-\infty\}$; *we denote by* $L^m(X)$ *the set of continuous operators* $P : C_0^\infty(X) \to C^\infty(X)$ *which satisfy the following condition:*

for any real-valued function $\phi(x, \theta) \in C^\infty(X \times \mathbb{R}_0^N)$, *homogeneous of degree* 1 *with respect to* θ, *and for any amplitude* $a \in S^q(X \times \mathbb{R}^N)$ *with compact support in* x *such that* $d_x\phi(x,\theta) \neq 0$ *for* $(x,\theta) \in$ conic supp a, *we have*

$$e^{-i\psi} P(a\, e^{i\psi}) \in S^{m+q}(X \times \mathbb{R}^N_0) \quad . \tag{5.1.1}$$

Firstly, Theorem 2.1 and Proposition 2.2 show that this definition coincides with the former definition when X is an open subset of $\mathbb{R}^n$. If X is a manifold and we have $P \in L^m(X)$, and if we let U be an open subset of X, we define the operator $P|_U$ as being the compose

$$C^\infty_0(U) \hookrightarrow C^\infty_0(X) \longrightarrow C^\infty(X) \longrightarrow C^\infty(U)$$

where the first arrow is the natural injection and last arrow denotes the restriction to U; Definition 5.1 shows that $P|_U \in L^m(U)$.

Definition 5.1 is obviously invariant under diffeomorphism; consequently if $\chi : X \supset U \longrightarrow \tilde{U} \subset \mathbb{R}^n$ is a chart of X, the transported operator $\chi_*(P|_U) \in L^m(\tilde{U})$.

PROPOSITION 5.2 : *We assume that* X *is equipped with a strictly positive density. Then the kernel distribution of a p.d.o.* $P \in L^m(X)$ *is* C^∞ *outside of* Diag $(X \times X)$. *Furthermore, we have the following equalities:*

(5.2.1.) $\{$*Operators with kernels*

$$\in C^\infty(X \times X)\} = L^{-\infty}(X) = \bigcap_m L^m(X) .$$

PROOF: Suppose $(x_0, y_0) \notin \mathrm{Diag}(X \times X)$ and let U, V be open neighbourhoods of x_0, y_0 chosen sufficiently small for them to be also disjoint coordinate patches. Consequently in the open

set $W = U \cup V$ we can define a chart $\chi : W \longrightarrow \tilde{W} \subset \mathbb{R}^n$.
Furthermore, we know (see (8.2.1), Chapter I) that the restriction of P to $W \times W$ is transported by (χ,χ) into a distribution $\tilde{K}$ such that

(5.2.2) $$\tilde{K} = K_{\tilde{P}}\, g$$

where g is a C^∞ function and $\tilde{P} = \chi_* P$.

Since $K_{\tilde{P}}$ is C^∞ outside of the diagonal of $\tilde{W} \times \tilde{W}$, we deduce that the kernel of P is C^∞ in $U \times V$. The same method allows us to prove the inclusion

$$\bigcap_m L^m(X) \subset \{\text{operators with kernel in } C^\infty(X \times X)\} .$$

A proof analogous to that of Proposition 9.2 of Chapter III proves the inclusion: {operators with C^∞ kernels} $\subset L^{-\infty}(X)$. Finally, we obviously have $L^{-\infty}(X) \subset \bigcap_m L^m(X)$, and this concludes the proof of (5.2.1).

The quotient $L^m / L^{-\infty}$ is a sheaf on X; this is

PROPOSITION 5.3 : *Let there be a covering of X by open sets (U_α); we specify operators $P_\alpha \in L^m(U_\alpha)$ such that*

$$P_\alpha|_{U_\alpha \cap U_\beta} \equiv P_\beta|_{U_\alpha \cap U_\beta} .$$

Then, there exists $P \in L^m(X)$ unique modulo $L^{-\infty}(X)$ such that

$P|_{U_\alpha} \equiv P_\alpha$ *(the sign $\equiv$ signifies modulo $L^{-\infty}$).*

PROOF: Taking a finer covering where appropriate, we will not restrict the generality by assuming that the U_α are relatively compact and locally finite. Suppose (ρ_α) is a subordinated partition of unity with functions $\rho_\alpha \in C_0^\infty(U_\alpha)$ and suppose we have functions $\mu_\alpha \in C_0^\infty(U_\alpha)$ with μ_α identical to 1 on supp ρ_α.

We define P by

$$P\,u = \sum_\alpha (\mu_\alpha P_\alpha \rho_\alpha)u \quad , \quad u \in C_0^\infty(X) \quad . \tag{5.3.1}$$

It is clear that P is a continuous operator from $C_0^\infty(X)$ into $C^\infty(X)$ since the sum (5.3.1) is finite for all $u \in C_0^\infty(X)$. Taking a and ψ as in Definition 5.1, we have

$$e^{-i\psi}\; P(a\, e^{i\psi}) = \sum_\alpha \mu_\alpha\, e^{-i\psi} \quad P_\alpha(\rho_\alpha\, a\, e^{i\psi}) \quad , \tag{5.3.2}$$

the right-hand side is a finite sum of symbols in $S^{m+q}(X \times \mathbb{R}^N)$; consequently we clearly have $P \in L^m(X)$.

We now show that $P_{|U_\beta} \equiv P_\beta$. Since supp $\rho_\alpha \cap U_\beta \subset U_\alpha \cap U_\beta$, we have

$$\begin{aligned} (\mu_\alpha P_\alpha \rho_\alpha)_{|U_\beta} &\equiv (\mu_\alpha P_\beta \rho_\alpha)_{|U_\beta} \\ &\equiv P_\beta \cdot \rho_\alpha + R_{\alpha\beta} \end{aligned} \tag{5.3.3}$$

where $R_{\alpha\beta} = (\mu_\alpha - 1)\, P_\beta\, \rho_\alpha$ is an operator with C^∞ kernel in $U_\beta \times U_\beta$.

Substituting (5.3.3) into (5.3.1), we obtain

$$P_{|U_\beta} \equiv \sum_\alpha (P_\beta \rho_\alpha + R_{\alpha\beta})$$

$$\equiv P_\beta \quad .$$

REMARK 5.4 : The above proposition and its proof remain valid if we everywhere replace the equalities modulo $L^{-\infty}$ by equalities modulo $L^{m'}$ with $m' < m$.

This sheaf property allows us to give an equivalent form, which is possibly more natural, for the definition of p.d.o.'s on X.

PROPOSITION 5.5 : *Let* P *be a continuous linear operator* $C_0^\infty(X) \to C^\infty(X)$. *Then* $P \in L^m(X)$ *if and only if for any chart* $\chi : U \to \tilde{U} \subset \mathbb{R}^n$, *the operator* $\chi_*(P_{|U})$ *is in* $L^m(\tilde{U})$.

PROOF: We saw following Definition 5.1 that this was necessary. In order to show that it is also sufficient, it suffices to apply Proposition 5.3 taking a covering of X by open coordinate patches.

Finally, we note that Proposition 1.5 and its proof remain valid when X is a manifold.

6. SYMBOLIC CALCULUS OF P.D.O.'s ON A MANIFOLD

6.1. DEFINITION OF THE PRINCIPAL SYMBOL OF AN OPERATOR $P \in L^m(X)$.

With each chart $\chi : U \to \tilde{U} \subset \mathbb{R}^n$ we associate the symbol $\sigma_m(\chi_* P_{|U}) \in S^m(\tilde{U} \times \mathbb{R}^n) / S^{m-1}$. Formula (4.5.4) shows that the diffeomorphism χ^{-1} allows us to lift it to an element denoted by $\sigma_U(P) \in S^m(T^*U \setminus 0) / S^{m-1}(T^*U \setminus 0)$ independently of the choice of the chart diffeomorphism χ defined on U. For the same reason we have $\sigma_U(P) = \sigma_V(P)$ in $T^*(U \cap V) \setminus 0$, thus the combination property of the symbols (see Proposition 2.10 of Chapter III) proves the existence of a unique element $\sigma \in S^m(T^*X \setminus 0) / S^{m-1}$ such that $\sigma_{|U} = \sigma_U$ for any open subset of charts. This symbol σ is called the principal symbol of degree m of P and is denoted by $\sigma_m(P)$.

The linear mapping

$$P \in L^m(X) \longrightarrow \sigma_m(P) \in S^m(T^*X \setminus 0) / S^{m-1}$$

is surjective; Remark 5.4 allows us to reduce to considering open subsets of $\mathbb{R}^n$ in order to prove this. Furthermore, Proposition 5.5 shows that $\sigma_m(P) \in S^{m-1} / S^{m-1}$ if and only if $P \in L^{m-1}(X)$. Hence we have the isomorphism of the principal symbol of degree m

$$L^m(X) / L^{m-1}(X) \xrightarrow{\sigma_m} S^m(T^*X \setminus 0) / S^{m-1}(T^*X \setminus 0) \tag{6.1.1}$$

For composition we have:

THEOREM 6.2 : *Suppose that* $P \in L^m(X)$, $Q \in L^{m'}(X)$, *where one at least is a proper operator; then* $Q \circ P \in L^{m+m'}(X)$ *and*

(6.2.1) $$\sigma_{m+m'}(Q \circ P) = \sigma_{m'}(Q) \cdot \sigma_m(P) \quad .$$

PROOF: From Definition 5.1, we can immediately show that

$$Q \circ P \in L^{m+m'}(X) \quad .$$

The definition of the principal symbol shows that if it is sufficient to verify (6.2.1) in an open coordinate patch, U. Suppose we have φ , $\psi \in C_0^\infty(U)$ with $\varphi = 1$ in the neighbourhood of supp ψ. We have

$$\psi\, Q\, P = \psi\, Q\, \varphi\, P + \psi\, Q(1-\varphi)P$$

$$\equiv \psi\, Q\, \varphi\, P \quad \text{since} \quad \psi\, Q(1-\varphi) \in L^{-\infty}(X) \quad .$$

In particular, we have

$$\psi(Q\,P)_{|U} \equiv \psi(Q_{|U})\varphi(P_{|U}) \quad ,$$

Formula (4.5.3) applied in the open set $\tilde{U} = \chi(U) \subset \mathbb{R}^n$, shows that

$$\psi(x)\sigma_{m+m'}(Q.P)(x,\xi) = \psi(x)\sigma_{m'}(Q)(x,\xi)\varphi(x)\sigma_m(P)(x,\xi) \quad \text{in} \quad T^*U \setminus 0$$

$$= \psi(x)\sigma_{m'}(Q)(x,\xi)\sigma_m(P)(x,\xi) \ ,$$

which concludes the proof since ψ is arbitrary in $C_0^\infty(U)$.

THEOREM 6.3 : *Assume that* X *is endowed with a strictly*

positive density. *If* $P \in L^m(X)$, *we then have*

$$(6.3.1) \qquad {}^tP \in L^m(X) \quad , \quad \sigma_m({}^tP)(x,\xi) = \sigma_m(P)(x,-\xi)$$

$$(6.3.2) \qquad P^* \in L^m(X) \quad , \quad \sigma_m(P^*)(x,\xi) = \overline{\sigma_m(P)(x,\xi)} \; .$$

PROOF: Letting U be an open subset of X, we clearly have

$$({}^tP)_{|U} = ({}^tP_{|U}) \quad , \quad P^*_{|U} = (P_{|U})^* \; ;$$

consequently (6.3.1) and (6.3.2) follow from Theorem 4.1 and from Proposition 5.5.

PROPOSITION 6.4 : *Any p.d.o.* $P \in L^m(X)$ *extends continuously into an operator from* $\mathcal{E}'(X)$ *into* $\mathcal{D}'(X)$.

PROOF: We have $P = {}^t({}^tP)$; now Theorem 6.3 implies that ${}^tP$ is continuous from $C_o^\infty(X)$ into $C^\infty(X)$; consequently Proposition 6.4 follows from Remark 8.4 of Chapter I.

DEFINITION 6.5 : *A p.d.o.* $P \in L^m(X)$ *is called classical, if for any chart* $U \subset X \xrightarrow{\chi} \tilde{U} \subset R^n$, *the complete symbol of the operator* $\chi_*(P_{|U}) \in L^m(\tilde{U})$ *is classical. We denote by* $L_c^m(X)$ *the subspace of classical p.d.o.'s of degree* m.

Formulas (4.4.2), (4.1.1), (4.1.2) and (4.2.2) show that this class is stable under diffeomorphism, under transposition, under passage to the adjoint, under restriction and under composition, by using the proof of Theorem 6.2.

The principal symbol of an operator $P \in L^m_c$ admits a canonical representative $\sigma_m(P)$ which is a $C^\infty(T^*X \setminus 0)$ function and is homogeneous of degree m with respect to ξ since this is true in any local chart. Moreover, we have

PROPOSITION 6.6 : *Suppose* $P \in L^m_c(X)$. *Then for* $(x_0, \xi_0) \in T^*X \setminus 0)$, *we have*

$$\sigma_m(P)(x_0,\xi_0) = \lim_{\tau \to \infty} \tau^{-m}\left[e^{-i\tau\psi(x)} P(a(x)e^{i\tau\psi(x)})\right]_{x=x_0} \tag{6.6.1}$$

where: $a \in C^\infty_0(X)$ *and has the value* 1 *in the neighbourhood of* x_0 , $\psi \in C^\infty(X)$, $d\psi(x_0) = \xi_0$, $d\psi \neq 0$ *on* supp a.

PROOF: Let U be an open coordinate patch containing x_0, and suppose we have $\alpha, \beta \in C^\infty_0(U)$ with $\alpha = 1$ in the neighbourhood of supp β and $\beta(x_0) = 1$. At the point $x = x_0$, we have

$$e^{-i\tau\psi} P(a\, e^{i\tau\psi}) = e^{-i\tau\psi}\, \beta\, P(\alpha\, a\, e^{i\tau\psi}) + e^{-i\tau\psi}\, \beta P(1-\alpha)(a e^{i\tau\psi}).$$

The second term on the right-hand side is in $S^{-\infty}(\mathbb{R}_+)$ and to calculate the first term we can replace P by $P_{|U}$ and consequently reduce to the case where X is an open subset of $\mathbb{R}^n$. In this case we can replace P by $\sigma_P(x,D)$ and apply Theorem 2.1; we obtain

$$e^{-i\tau\psi}P(a e^{i\tau\psi})(x_0) = \sigma_m(P)(x_0\, ,\, \tau\, d\psi(x_0))a(x_0) \qquad \text{mod } S^{m-1}(\mathbb{R}^+)$$

$$= \tau^m \sigma_m(P)(x_0,\xi_0) \qquad \text{mod } O(\tau^{m-1}) \quad ,$$

which proves (6.6.1).

REMARK 6.7: Case of p.d.o.'s operating on sections of bundles. Let $\mathcal{F}$, $\mathcal{G}$ be vector bundles above X, having as fibre finite-dimensional vector spaces E and F on C. We denote by $\mathcal{L}(\mathcal{F},\mathcal{G})$ the vector bundle above X of the morphisms of $\mathcal{F}$ into $\mathcal{G}$. We say that a continuous operator P : $C_0^\infty(X\,;\mathcal{F}) \to C^\infty(X\,;\mathcal{G})$ is an element of $L^m(X\,;\mathcal{F},\mathcal{G})$ if for any trivialisations f and g of $\mathcal{F}$ and $\mathcal{G}$ above a local chart $U \overset{X}{\to} \tilde{U}$ of X we have (with the notation of (7.8.2) of Chapter I):

$$P_{f,g} \in L^m(\tilde{U}\,;E,F) \quad .$$

We denote by π_X the canonical projection $T^*X\setminus 0 \to X$ and we denote by $\pi_X^*\mathcal{F}$, $\pi_X^*\mathcal{G}$ the bundles on $T^*X\setminus 0$ which are inverse images of $\mathcal{F}$ and $\mathcal{G}$ under π_X. The principal symbol of $P_{f,g}$ is an element of $S^m(\tilde{U}\times R^n\,;\mathcal{L}(F,G))$ which can be lifted into an element $\sigma_U \in S^m(T^*U\setminus 0\,;\mathcal{L}(\pi_u^*\mathcal{F},\pi_u^*\mathcal{G}))$ independent, modulo S^{m-1}, of the choice of f and g. These elements σ_U combine into an element of $S^m(T^*X\setminus 0\,;\mathcal{L}(\pi_X^*\mathcal{F},\pi_X^*\mathcal{G}))/S^{m-1}$ called the principal symbol of P and denoted by $\sigma_m(P)$. The generalisation of the results of Sections 5 and 6 to p.d.o.'s of type $L^m(X\,;\mathcal{F},\mathcal{G})$ presents no new difficulties. Likewise, Definition 6.5 and the properties of classical p.d.o.'s generalise to the spaces $L_c^m(X\,;\mathcal{F},\mathcal{G})$.

7. ELLIPTIC PSEUDO-DIFFERENTIAL OPERATORS

In this section, X denotes a manifold.

To the concept of an elliptic symbol there corresponds the concept of an elliptic operator.

DEFINITION 7.1: *If* $P \in L^m(X)$, *we say that* P *is elliptic of degree* m *if its principal symbol* $\sigma_m(P)$ *is an elliptic symbol of degree* m *in* $S^m(T^*X \setminus 0)$.

EXAMPLE 7.2: If P is a differential operator or more generally a classical p.d.o. of degree m , then it is elliptic if and only if the homogeneous function $\sigma_m(P)$ does not vanish on $T^* X \setminus 0$.

In fact, elliptic operators are those which are invertible in the sense of the following theorem:

THEOREM 7.2: *The operator* $P \in L^m(X)$ *is elliptic of degree* m *if and only if there exists a proper operator* $Q \in L^{-m}(X)$ *such that* P.Q. ≡ Identity. *Such an operator* Q *then satisfies* Q.P. ≡ Identity; *it is unique modulo* $L^{-\infty}$ *and is termed a parametrix of* P.

PROOF: The existence of Q clearly implies the ellipticity of $\sigma_m(P)$ since we have

$$\sigma_m(P) \cdot \underline{\sigma}_m(Q) - 1 \in S^{-1}(T^*X \setminus 0) \quad .$$

In order to prove the existence of Q, we make use of an approximation procedure which is important in its own right.

By hypothesis, we know that there exists $q_o \in S^{-m}(T^*X \setminus 0)$ such that

(7.2.1) $$\sigma_m(P).q_o - 1 \in S^{-1}(T^*X \setminus 0) .$$

Let $Q_o \in L^{-m}(X)$ be proper such that $\sigma_m(Q_o) = q_o$, then (7.2.1) implies

(7.2.2) $$P.Q_o - I = R_o \in L^{-1}(X) \quad .$$

We now prove the following:

LEMMA 7.3 : *We take a sequence of p.d.o.'s* $R_k \in L^{m_k}(X)$ *such that* $m_k \searrow_{k\infty} -\infty$. *Then:*

i) *There exists* $S \in L^{m_o}(X)$ *proper, such that for any integer* h

(7.3.1) $$(S - \sum_{k=0}^{h} R_k) \in L^{m_{h+1}}(X) .$$

Such an operator S *is unique modulo* $L^{-\infty}(X)$ *and we write*

$$S \sim \sum_{k\geq 0} R_k$$

ii) *If* $R \in L^{-1}(X)$ *is proper, and if* $T \in L^{o}(X)$ *is proper such that*

(7.3.2) $$T \sim \sum_{k\geq 0} (-1)^k R^k$$

then we have

(7.3.3) $$(I + R).T \equiv T.(I + R) \equiv Id \quad .$$

PROOF: If $U \xrightarrow{X} \tilde{U}$ is a local chart of X, we can define $S_U \in L^{m_0}(U)$ by considering the asymptotic sum of the complete symbols in $\tilde{U}$ of the operators $X_*(R_k|_U)$ (this being possible in view of Theorem 4.2 of Chapter III). Then, we define S satisfying (7.3.1) by combining the S_U by virtue of Proposition 5.3. The uniqueness modulo $L^{-\infty}$ is obvious from (7.3.1). The verification of (7.3.3) is immediate from the definition of T.

We return now to the proof of the theorem. We take $Q = Q_o . T$ where T is defined by (7.3.2) and R is a proper operator equal to R_o modulo $L^{-\infty}$; then $Q \in L^{-m}(X)$ is proper and

$$P . Q \equiv Id \quad . \tag{7.3.4}$$

The same method will allow us to construct a proper $Q' \in L^{-m}(X)$ such that

$$Q' . P \equiv Id \quad . \tag{7.3.5}$$

Multiplying (7.3.5) on the right by Q, (7.3.4) implies that

$$Q \equiv Q'$$

which proves that Q is a parametrix of P and that a parametrix is unique modulo $L^{-\infty}$.

REMARK 7.4 : Case of p.d.o.'s operating on sections of bundles .

We once more adopt the notation of Remark 6.7. Suppose $P \in L^m(X ; \mathcal{F} , \mathcal{G})$; we say that P is right elliptic if there exists a symbol

$q_0 \in S^{-m}(T^*X \setminus 0 \; ; \; \mathcal{L}(\pi_X^*\mathcal{G}, \pi_X^*\mathcal{F}))$ such that

$$\sigma_m(P) \cdot q_0 - \text{Id} \in S^{-1}(T^*X \setminus 0 \; ; \; \mathcal{L}(\pi_X^*\mathcal{G}, \pi_X^*\mathcal{G})) \; .$$

The proof of Theorem 7.2 shows that P is right elliptic if and only if there exists $Q \in L^{-m}(X \; ; \mathcal{G}, \mathcal{F})$ such that $P.Q \equiv I$; we say that Q is a right parametrix of P.

We have the analogue of this "on the left", and we have the concept of a two-sided parametrix when $\mathcal{F}$ and $\mathcal{G}$ are of the same rank.

For operators of the classical type, we have a simple criterion of ellipticity; this is

PROPOSITION 7.5 : *Supposing* $P \in L_c^m(X \; ; \mathcal{F}, \mathcal{G})$, *then* P *is left elliptic (right elliptic) if and only if for any* $(x,\xi) \in T^*X \setminus 0$ *the linear mapping* $\sigma_m(P)(x,\xi) \in \mathcal{L}(\mathcal{F}_x, \mathcal{G}_x)$ *is injective (surjective). Moreover, if the bundles* $\mathcal{F}$ *and* $\mathcal{G}$ *are of the same rank, any left (right) parametrix of* P *is a two-sided parametrix.*

PROOF: To prove that P admits a left parametrix, we shall now define the adjoint P^* of P. To do this, we endow X with a strictly positive density and we define a Hermitian structure on $\mathcal{F}$ and on $\mathcal{G}$ (we can do this locally by means of a trivialisation and we re-combine by partition of unity). Thus the spaces $C_0^\infty(X,\mathcal{F})$ and $C_0^\infty(X\,;\mathcal{G})$ are each equipped with an inner product denoted respectively as $(\ ,\ )_{\mathcal{F}}$ and $(\ ,\ )_{\mathcal{G}}$. The adjoint P^* of P is defined by the equality

$$(P\,u\,,\,v)_{\mathcal{G}} = (u\,,\,P^*v)_{\mathcal{F}} \qquad u \in C_0^\infty(X,\mathcal{F})\,,\; v \in C_0^\infty(X,\mathcal{G})$$

and we know (see Remark 6.7) that this is an operator belonging to $L_c^m(X\,;\mathcal{G},\,\mathcal{F})$ whose principal symbol satisfies

$$\sigma_m(P^*)(x,\xi) = (\sigma_m(P)(x,\xi))^*$$

for the Hermitian structures on $\mathcal{F}_x$ and $\mathcal{G}_x$. If $\tilde{P}^*$ is a proper operator equivalent to P^*, it thus has the same principal symbol as P^*. The principal symbol $\tilde{P}^*P$ is for all $(x,\xi) \in T^*X\setminus 0$ the injective linear mapping $(\sigma_m(P)(x,\xi))^* \cdot \sigma_m(P)(x,\xi)$ from $\mathcal{F}_x$ into $\mathcal{F}_x$; it is therefore invertible. Consequently the operator $\tilde{P}^*P$ is elliptic of degree 2m. Let $Q \in L^{-2m}(X\,;\,\mathcal{F}\,,\,\mathcal{F})$ be a parametrix of this operator; we have

$$(7.5.1) \qquad Q\,.\,\tilde{P}^*.\,P = I + R \quad \text{with} \quad R \in L^{-\infty}(X\,;\mathcal{F},\mathcal{F})\,.$$

The equality (7.5.1) shows that the operator $Q\tilde{P}^* \in L^{-m}(X\,;\mathcal{G},\,\mathcal{F})$ is a left parametrix of P.

The other assertions of the proposition are immediately obvious.

7.6 CASE OF SYSTEMS WHICH ARE ELLIPTIC IN THE SENSE OF DOUGLIS-NIRENBERG.

We say that a system $P = (P_{jk})_{\substack{j=1,\dots,J\\ k=1,\dots,K}}$ where $P_{jk} \in L_c^{m_{jk}}(X)$ is of the type $(s,t) \in \mathbb{R}^J \times \mathbb{R}^K$ if $m_{jk} = t_k - s_j$ for any $(j,k) \in J \times K$. We designate as the principal symbol of P of type (s,t) the function from $C^\infty(T^*X \setminus 0 ; \mathcal{L}(\mathbb{C}^K, \mathbb{C}^J))$ defined by the matrix

$$\sigma_{s,t}(P) = \left(\sigma_{t_k - s_j}(P_{jk}) \right)_{j,k} .$$

A system P of type (s,t) is said to be left (right) elliptic in the sense of Douglis-Nirenberg, if for any $(x,\xi) \in T^*X \setminus 0$ the matrix $\sigma_{s,t}(P)(x,\xi)$ is injective (surjective). We have:

THEOREM 7.7 : *A system* P *of type* (s,t) *is left (right) elliptic if and only if there exists a system* Q *of type* (t,s) *which is proper and such that* $Q \cdot P \equiv I$ $(P \cdot Q \equiv I)$. *Moreover, when* $K = J$, *one-sided ellipticity is equivalent to the existence of a two-sided parametrix* Q.

PROOF: We equip X with a Riemannian metric so as to be able to define the length $|\xi|$ of a cotangent vector. With $r \in \mathbb{R}$, we denote by Λ_r a proper operator from $L_c^r(X)$ having $|\xi|^r$ for its principal symbol. The operator Λ_r is thus elliptic and we denote by Λ_r^{-1} a parametrix of Λ_r. With the vector $s \in \mathbb{R}^J$, we associate the diagonal system of type $(s,0)$ defined by the matrix Λ_s with $(\Lambda_s)_{jk} = \Lambda_{s_j}$ if $j = k$, and 0 otherwise. For $t \in \mathbb{R}^K$ we similarly define a diagonal system Λ_t^{-1} of type $(0,t)$. Then, the formula

$$\tilde{P} = \Lambda_s \cdot P \cdot \Lambda_t^{-1}$$

places systems P of type (s,t) in bijection with the ordinary systems $\tilde{P} \in L^0(X ; \mathbb{C}^K , \mathbb{C}^J)$. Using the symbols it can be immediately shown that P is left (right) elliptic in the sense (s,t) if and only if $\tilde{P}$ is left (right) elliptic. Consequently, to prove Theorem 7.7 it suffices to use Proposition 7.5 and to observe that if $\tilde{Q}$ is a left (right) parametrix of $\tilde{P}$, then $Q = \Lambda_t^{-1} \cdot \tilde{Q} \cdot \Lambda_s$ is a left (right) parametrix of P.

REMARK 7.8 : The definitions and results concerning systems of type (s,t) generalise immediatcly to the case where $\mathbb{C}^K$ and $\mathbb{C}^J$ are replaced by bundles $\mathcal{F}$ and $\mathcal{G}$ above X which decompose into direct sums of bundles:

$$\mathcal{F} = \bigoplus_{k=1}^{K} \mathcal{F}_k \qquad \mathcal{G} = \bigoplus_{j=1}^{J} \mathcal{G}_j \quad .$$

8. P.D.O.'s AND SOBOLEV SPACES

Let X be an open subset of $\mathbb{R}^n$. To prove the continuity of the p.d.o.'s, we begin by showing that a positive elliptic operator admits a kind of approximate square root.

LEMMA 8.1 : *Let* $C \in L^0(X)$ *be such that its principal symbol* $\sigma_0(C)$ *is real and satisfies for any compact subset* $K \subset X$

$$\varliminf_{|\xi|\infty} \left(\inf_{x \in K} \sigma_0(C)(x,\xi) \right) > 0 \quad . \tag{8.1.1}$$

Then, there exists a proper $B \in L^0(X)$ *such that*

$$C = B^*B + R \quad \textit{with} \quad R \in L^{-\infty}(X) . \tag{8.1.2}$$

PROOF: The assumption (8.1.1) means that for any compact subset $K \subset X$, there exists $c > 0$ such that

$$\sigma_0(C)(x,\xi) \geq c \quad \text{for} \quad x \in K \text{ and } |\xi| \geq \frac{1}{c} \quad .$$

Thus $\sigma_0(C)$ is an elliptic symbol and we can construct a function $\rho \in C^\infty(X \times \mathbb{R}^n)$ of the same type as that utilised during the proof of Proposition 3.2 in Chapter III. Then $b_0 = \rho\sqrt{\sigma_0(C)} \in S^0(X \times \mathbb{R}^n)$ and b_0 is elliptic. If $B_0 \in L^0(X)$ is a proper operator with principal symbol b_0, by construction it satisfies

$$C - B_0^* \cdot B_0 = R_{-1} \in L^{-1}(X) \quad .$$

We can then construct by recurrence proper operators $B_j \in L^{-j}(X)$ such that for any integer $j \geq 0$, we have

$$C - (B_0 + \dots + B_j)^* (B_0 + \dots + B_j) = R_{-j-1} \in L^{-j-1}(X). \tag{8.1.3}$$

Knowing the B_k for $k < j$, the condition (8.1.3) means that B_j must satisfy

$$R_{-j} - B_j^* B_0 - B_0^* B_j = R_{-j-1} \in L^{-j-1}(X) \quad . \tag{8.1.4}$$

Noting that $\sigma_{-j}(R_{-j})$ is real, the condition (8.1.4) will be satisfied if the principal symbol b_j of B_j satisfies

$$2\, b_o \cdot b_j = \sigma_{-j}(R_{-j}) \quad ;$$

this is possible since b_o is elliptic.

It now only remains to take $B \sim \sum_{j\geq 0} B_j$ in order to obtain (8.1.2).

REMARK 8.2 : The same proof allows us to construct an approximate square root of C, i.e. a proper $B \in L^o(X)$ such that $C \equiv B^2$.

REMARK 8.3 : This lemma generalises immediately to the case where $C \in L^m(X)$. We need of course to replace $\sigma_o(C)$ by $|\xi|^{-m}\, \sigma_m(C)(x,\xi)$ in (8.1.1), and we find B in $L^{m/2}(X)$.

As an application of this lemma, we have

THEOREM 8.4 : *Let* $P \in L^o(X)$ *be proper, such that there exists* M *such that for any compact subset* $K \subset X$

$$(8.4.1) \qquad \overline{\lim_{|\xi|\infty}} \left(\sup_{x \in K} |\sigma_o(x,\xi)| \right) < M \quad .$$

Then, there exists a proper and self-adjoint $R \in L^{-\infty}(X)$ *such that*

$$(8.4.2) \qquad \|Pu\|_o^2 \leq M\, \|u\|_o^2 + (Ru\,,u) \quad \text{for} \quad u \in C_o^\infty(X).$$

In particular, any p.d.o. belonging to $L^{o}(X)$ *is continuous from* $L^2_{comp}(X)$ *into* $L^2_{loc}(X)$.

PROOF: The assumption (8.1.1) implies that the operator $C = M^2 - P^*P$ satisfies the assumptions of Lemma 8.1. Consequently, there exists a proper $B \in L^{o}(X)$ and $R \in L^{-\infty}(X)$ such that

$$(8.4.3) \qquad P^*P + B^*B = M^2 + R \quad .$$

The equality (8.4.3) shows that R is self-adjoint and allows us to write

$$(Pu , Pu) + (Bu , Bu) = M^2(u , u) + (Ru , u) \quad \text{for } u \in C^{\infty}(X)$$

where (,) denotes the inner product in $L^2(X)$; this proves (8.4.2).

THEOREM 8.5 : *If* $P \in L^m(X)$, *then for any* $s \in \mathbb{R}$, P *is continuous from* $H^s_{comp}(X)$ *into* $H^{s-m}_{loc}(X)$. *If in addition* P *is proper, it is continuous from* $H^s_{comp}(X)$ *into* $H^{s-m}_{comp}(X)$ *and from* $H^s_{loc}(X)$ *into* $H^{s-m}_{loc}(X)$.

PROOF: Since the theorem is trivial for a regularising operator, we will not restrict the generality by assuming that P is proper. We recall that the norm H^s of $u \in C^{\infty}_{o}(\mathbb{R}^n)$ is defined by

$\|u\|_s = \|\Lambda_s u\|_o$ where Λ_s here denotes the operator whose complete symbol is $(1 + |\xi|^2)^{s/2}$. Let Λ'_s

be a proper operator equivalent to Λ_s and let $\Lambda_s'^{-1}$ be a parametrix of Λ_s' : $\Lambda_s'^{-1} \cdot \Lambda_s' = I + R_s$. For $u \in C_o^\infty(X)$, we obtain

$$\|Pu\|_{s-m} = \|\Lambda_{s-m} Pu\|_o \leq \|\Lambda_{s-m} P \Lambda_s'^{-1} \Lambda_s' u\|_o + \|\Lambda_{s-m} P R_s u\|_o ,$$

but $\Lambda_{s-m} P \Lambda_s'^{-1} \in L^o(X)$, $\Lambda_{s-m} P R_s \in L^{-\infty}(X)$.

Consequently Theorem 8.5 follows from Theorem 8.4.

We now give a result which is known by the name of Gårding's inequality.

THEOREM 8.6 : *If* $P \in L^m(X)$, *we assume that there exists* $a > 0$ *such that for any compact subset* $K \subset X$

$$(8.6.1) \qquad \underline{\lim}_{|\xi|\to\infty} \left[|\xi|^{-m} \left(\inf_{x \in K} \operatorname{Re} \sigma_m(P)(x,\xi) \right) \right] \geq a .$$

Then, for any $\varepsilon > 0$ *and any compact subset* K, *and for all* $s < \frac{m}{2}$, *there exists* C *such that*

$$(8.6.2) \qquad \operatorname{Re}(Pu , u) \geq (a-\varepsilon) \|u\|^2_{\frac{m}{2}} - C\|u\|^2_s , \quad u \in \mathcal{D}_K .$$

PROOF: We define the operator

$$C = \frac{1}{2}(P + P^*) - \Lambda^*_{\frac{m}{2}} (a - \frac{\varepsilon}{2}) \cdot \Lambda_{\frac{m}{2}} ,$$

the assumption (8.6.1) allows us to apply Lemma 8.1 (or more precisely Remark 8.3) to this operator; thus there exists a proper $B \in L^{m/2}(X)$ and $R \in L^{-\infty}(X)$ such that

$$\frac{1}{2}(P + P^*)u = (a - \frac{\varepsilon}{2}) \Lambda^*_{\frac{m}{2}} \Lambda_{\frac{m}{2}} u + B^*Bu + Ru \text{ , with } u \in C_0^\infty(X) .$$

Taking the inner product with u, we obtain

$$\text{(8.6.3)} \qquad \mathrm{Re}(Pu, u) \geq (a - \frac{\varepsilon}{2}) \|u\|^2_{\frac{m}{2}} - |(Ru, u)| .$$

In addition, we always have the upper bounds

$$|(Ru, u)| \leq \|u\|_{\frac{m}{2}} \|Ru\|_{-\frac{m}{2}}$$

$$\leq \frac{\varepsilon}{2} \|u\|^2_{\frac{m}{2}} + C_1 \|Ru\|^2_{-\frac{m}{2}} .$$

Finally, for any K, there exists C_2 such that

$$\|Ru\|^2_{-\frac{m}{2}} \leq C_2 \|u\|^2_s \qquad \text{for} \quad u \in \mathcal{D}_K ,$$

hence we deduce (8.6.2) with $C = C_1 C_2$, by substituting the upper bound for $|(R u, u)|$ into (8.6.3).

REMARK 8.7 : Generalisations to the case of systems.

Let E,F be Hermitian vector spaces of finite dimension. We now give a brief indication of how the results of section 8 generalise to the case where P is a p.d.o. belonging to $L^m(X ; E, F)$. By selecting bases of E and F, we can reduce to the scalar case in order to prove the continuity of P from $H^s_{comp}(X ; E)$ into $H^{s-m}_{loc}(X ; F)$. For Gårding's inequality, we assume that $P \in L^m_c(X ; E , E)$ and we replace the inequality (8.6.1) by the inequality

$$\frac{1}{2}\left[\sigma_m(P)(x,\xi) + \sigma_m(P)^*(x,\xi)\right] \geq a.I \ , \quad x \in K \ , \quad |\xi| = 1 \ ,$$

where the sign ≥ is taken in the sense of the order respecting the Hermitian operators.

Next we note that we do not need the full precision of Lemma 8.1 to prove Gårding's inequality; it suffices to prove the existence of a proper $B \in L^{m/2}_c(X ; E , E)$ such that

$$C = B^* . B + R_1 \ , \quad \text{with } R_1 \in L^{m-1}(X ; E , E) \ .$$

To do this, we take for B a proper operator with homogeneous principal symbol equal to $\sqrt{\sigma_m(C)(x,\xi)}$ (here $\sqrt{\ }$ denotes the positive square root in the positive Hermitian operators). The remainder of the proof of (8.6.2) is practically unchanged.

The continuity of the p.d.o.'s allows us to give a new definition of Sobolev spaces:

PROPOSITION 8.8: *If* $s \in \mathbb{R}$, *we have the equivalence*

$$\text{(8.8.1)} \qquad u \in H^s_{loc}(X) \Longleftrightarrow \begin{cases} u \in \mathcal{D}'(X), \text{ and for all } P \in L^s(X) \\ \text{proper, we have } Pu \in L^2_{loc}(X) \ . \end{cases}$$

Moreover, the semi-norms $\|\varphi P u\|_0$, *for* $\varphi \in C^\infty_0(X)$ *and proper* $P \in L^s(X)$, *also define the topology of* $H^s_{loc}(X)$.

PROOF: If $u \in H^s_{loc}(X)$, Theorem 8.5 implies that $P\,u \in L^2_{loc}(X)$ and that for every $\varphi \in C^\infty(X)$ there exist $\psi \in C^\infty_0(X)$ and C such that $\|\varphi\, P\,u\|_0 \leq C\,\|\psi\, u\|_s$ for every $u \in H^s_{loc}(X)$.

Conversely, suppose $P \in L^s(X)$ is elliptic and proper, and let $Q \in L^{-s}(X)$ be a parametrix of Q. For $u \in \mathcal{D}'(X)$, we have:

$$u = QPu + Ru \quad , \quad \text{where } R \in L^{-\infty}(X) \text{ is proper}.$$

Thus if $Pu \in L^2_{loc}(X)$, we have $u \in H^s_{loc}(X)$ from Theorem 8.5, and this also shows that for $\varphi \in C_0(X)$, $\|\varphi\, u\|_s$ is bounded above by a quantity of the form $\|\psi_1 Pu\|_0 + \|\psi_2 R\,u\|_0$, with $\psi_1, \psi_2 \in C^\infty_0(X)$.

From this we deduce the invariance of Sobolev spaces under diffeomorphism.

COROLLARY 8.9: *Let* χ *be a diffeomorphism between two open subsets* X, Y *of* $\mathbb{R}^n$. *Suppose* $u \in \mathcal{D}'(X)$; *then*

$$(8.9.1) \qquad u \in H^s_{loc}(X) \iff \chi_* u \in H^s_{loc}(Y) \;;$$

moreover, χ_* *defines a homeomorphism between these two Sobolev spaces.*

The proof is immediate in view of Proposition 8.8 and the invariance under diffeomorphism of spaces of proper pseudo-differential operators of given degree.

As regards manifolds, we have

THEOREM 8.10: *If* X *is a* C^∞ *manifold of dimension* n, *then Theorem* 8.5 *generalises word for word to this case.*

PROOF: Let K be a compact subset of X and let $(U_j)_{j=1,\dots,N}$ be a finite covering of K by open coordinate patches U_j of charts $\chi_j : U_j \to \tilde{U}_j$ of X. We put $U_o = X \setminus \bigcup_i^N U_j$. In view of Theorem 2.19 of Chapter II, we have to show that if $u \in H_K^s(X)$ then $Pu \in H_{loc}^{s-m}(U_j)$ and depends continuously on u $(0 \le j \le N)$. Suppose we have $\varphi_j \in C_o^\infty(U_j)$ and $\psi_j \in C_o^\infty(U_j)$ equal to 1 in the neighbourhood of supp φ_j. We have

$$\varphi_j P u = \varphi_j P \psi_j u + R_j u \quad \text{where} \quad R_j = \varphi_j P(1-\psi_j) .$$

Since $\psi_o u = 0$ and $R_j \in L^{-\infty}(X)$, we are led to study $\varphi_j P \psi_j u$ for $1 \le j \le N$. We have

$$\chi_*(\varphi_j P \psi_j u) = \tilde{\varphi}_j \tilde{P}_j \tilde{\psi}_j u , \quad \text{where } \tilde{\psi}_j u = \chi_{j*} \psi_j u , \quad \tilde{\varphi}_j = \chi_{j*} \varphi_j ,$$

$P_j = \chi_{j*}(P|U_j)$. Consequently Theorem 8.5 together with Definition 2.18 of Chapter II, show that $\varphi_j P \psi_j u \in H_{loc}^{s-m}(X)$. The continuity is established in the same manner.

Elliptic p.d.o.'s are hypoelliptic; more precisely we have

THEOREM 8.11: (On elliptic regularity). *Let* X *be a* C^∞ *manifold and let* $P \in L^m(X)$ *be a p.d.o. which is elliptic and proper.*

i) *If* $u \in \mathcal{D}'(X)$, *we have the equivalences:*

(8.11.1) $$Pu \in H^s_{loc}(X) \iff u \in H^{s+m}_{loc}(X)$$

(8.11.2) $$Pu \in C^\infty(X) \iff u \in C^\infty(X) \quad , \quad \textit{and}$$

more generally $$\text{sing supp } Pu = \text{sing supp } u \ .$$

ii) *If we have* s, t $\in \mathbb{R}$, *then for any compact subset* $K \subset X$, *there exists* C *such that*

(8.11.3) $$\|u\|_{s+m} \leq C(\|Pu\|_s + \|u\|_t) \ , \quad u \in C^\infty_K(X) \ .$$

PROOF: Letting $Q \in L^{-m}(X)$ be a parametrix of P, the proof is based on the equality

(8.11.4) $$u = Q \cdot Pu + Ru \ , \quad \text{where } u \in \mathcal{D}'(X) \text{ and } R \in L^{-\infty}(X).$$

Thus (8.11.1), (8.11.2) and (8.11.3) follow from the continuity of the p.d.o.'s. We always have the inclusion (see (1.3.2))

(8.11.5) $$\text{sing supp } Pu \subset \text{sing supp } u \ ,$$

and the inverse inclusion follows from (8.11.4) and from the inclusion (8.11.5) for Q.

REMARK 8.12: Let $\mathcal{F}$ and $\mathcal{G}$ be complex vector bundles of finite dimension over X. Taking $P \in L^m(X ; \mathcal{F} , \mathcal{G})$, we leave to the reader the task of showing that a proof similar to that of Theorem 8.10 allows us to prove that P is continuous from

$H^s_{comp}(X ; \mathcal{F})$ into $H^{s-m}_{loc}(X ; \mathcal{G})$. Furthermore, if P is proper and left elliptic, Theorem 8.11 and its proof generalise immediately. In the case of classical p.d.o.'s, we have a converse to the inequality (8.11.3), as follows:

PROPOSITION 8.13: *If* $P \in L^m_c(X ; \mathcal{F}, \mathcal{G})$, *and* $t < m$, *we assume that for any compact set* $K \subset X$, *there exists* C *such that*

$$(8.13.1) \qquad \|u\|_m \leq C \, (\|Pu\|_0 + \|u\|_t), \quad u \in C^\infty_K(X ; \mathcal{F}) .$$

Then, P *is left elliptic.*

PROOF: From Proposition 7.5, it suffices to prove that for all $(x_0, \xi_0) \in T^*X \setminus 0$, the linear mapping $\sigma_m(P)(x, \xi)$ is injective from $\mathcal{F}_x$ into $\mathcal{G}_x$. We begin by reducing to the case of a system of p.d.o.'s in $\mathbb{R}^n$. Let U be a neighbourhood of x_0 which is an open coordinate patch of X over which the bundles $\mathcal{F}$ and $\mathcal{G}$ are isomorphic to $U \times \mathbb{C}^M$ and $U \times \mathbb{C}^N$. Suppose we have $\varphi \in C^\infty_0(U)$ identical to 1 in the neighbourhood of x_0 and $\psi \in C^\infty_0(U)$ identical to 1 in the neighbourhood of supp φ. If we fix K = supp φ, and if $u \in C^\infty_K(X)$ we have

$$Pu = P\varphi u = \psi P \varphi u + (1-\psi) P \varphi u .$$

Since $(1-\psi)P\varphi$ is regularising, we can replace P by $\psi P \varphi$ in the inequality (8.13.1) and this allows us to work within the open set U. Consequently, we will not reduce the generality by taking

$$X = \mathbb{R}^n, \quad \mathcal{F} = X \times \mathbb{C}^M, \quad \mathcal{G} = X \times \mathbb{C}^N .$$

Suppose we have $v \in C_K^\infty(\mathbb{R}^n ; \mathbb{C}^M)$ and let $u_\tau(x) = v(x)e^{i\tau x.\xi_0}$ for $\tau > 0$. We study the behaviour, as $\tau \to +\infty$, of both sides of (8.13.1) when we take $u = u_\tau$.

If we let Λ_m be the p.d.o. with complete symbol $(1 + |\xi|^2)^{m/2}$, we have

$$\| u_\tau \|_m = \| \Lambda_m u_\tau \|_0$$

$$= \| e^{-i\tau x\xi_0} \Lambda_m u_\tau \|_0 .$$

Proposition 6.6 shows that $e^{-i\tau x\xi_0}(\Lambda_m u_\tau)(x) \underset{\tau\to+\infty}{\sim} \tau^m |\xi_0|^m v(x)$, consequently we obtain

$$\lim_{\tau\to+\infty} \left(\tau^{-m} \| u_\tau \|_m \right) = |\xi_0|^m \left(\int_{\mathbb{R}^n} |v(x)|^2_{\mathbb{C}^M} dx \right)^{1/2} .$$

Likewise, we obtain

$$\lim_{\tau\to+\infty} \tau^{-m} \left(\| Pu_\tau \|_0 + \| u_\tau \|_t \right) = \left(\int_{\mathbb{R}^n} |\sigma_m(P)(x,\xi_0) . v(x)|^2_{\mathbb{C}^N} dx \right)^{1/2}$$

By combining (8.13.1) and these two latter inequalities, we obtain

$$|\xi_0|^{2m} \int |v(x)|^2 dx \leq C^2 \int |\sigma_m(P)(x,\xi_0) v(x)|^2 dx$$

and as v is arbitrary, this implies that

$$|v(x)| \leq C \, |\sigma_m(P)(x,\xi_0) . v(x)| . |\xi_0|^{-m}$$

and consequently $\sigma_m(P)(x_0 , \xi_0)$ is injective.

We now assume that X is a compact manifold endowed with a

strictly positive density and we let $\mathcal{F}$ be a Hermitian vector bundle of finite rank over X. If $P \in L^m(X ; \mathcal{F}, \mathcal{F})$, it defines a continuous operator $\mathcal{P}$ from $C^\infty(X ; \mathcal{F})$ into $C^\infty(X ; \mathcal{F})$; we can also consider it as an unbounded operator $\mathbb{P}$ in $L^2(X ; \mathcal{F})$ of dense domain $D = H^m(X ; \mathcal{F})$.

Under these conditions we have

THEOREM 8.14: *Suppose that* P *is elliptic. Then,*

i) *The adjoint of the unbounded operator* ($\mathbb{P}$, D) *is the unbounded operator* ($\mathbb{P}^*$, D^*) *associated with the elliptic p.d.o.* $P^* \in L^m(X ; \mathcal{F})$ *with* $D^* = D = H^m(X ; \mathcal{F})$.

ii) *The operator* $\mathbb{P}$ *is of Fredholm type, i.e.:* $\dim \operatorname{Ker} \mathbb{P} < +\infty$, $\operatorname{Im} \mathbb{P}$ *is a closed set of finite codimension. Moreover,*

$$\operatorname{Ker} \mathbb{P} = \operatorname{Ker} \mathcal{P} \subset C^\infty(X ; \mathcal{F})$$

$$L^2(X , \mathcal{F}) = \operatorname{Ker} \mathcal{P} \oplus \operatorname{Im} \mathbb{P}^* \ . \tag{8.14.1}$$

iii) $\mathbb{P}$ *is self-adjoint if and only if* P = P*, *and in this case we have*

$$L^2(X , \mathcal{F}) = \operatorname{Ker} \mathcal{P} \oplus \operatorname{Im} \mathbb{P} \tag{8.14.2}$$

$$C^\infty(X ; \mathcal{F}) = \operatorname{Ker} \mathcal{P} \oplus \mathcal{P}(C^\infty(X ; \mathcal{F})) \ . \tag{8.14.3}$$

PROOF OF i): Since $C^\infty(X ; \mathcal{F})$ is dense in $D = H^m(X , \mathcal{F})$

the domain D^* of $\mathbb{P}^*$ is defined by

$$(8.14.2)' \quad D^* = \left\{ v \in L^2(X, \mathcal{F}) \;\middle|\; \begin{array}{l} \text{The mapping} \\ u \in C^\infty(X,\mathcal{F}) \to (\mathbb{P}u, v) \in \mathbb{C} \\ \text{is continuous for the} \\ \text{topology of } L^2(X,\mathcal{F}) \end{array} \right\}.$$

We have $\mathbb{P} = P$ on $C^\infty(X)$; since P^* is a continuous operator from $\mathcal{D}'(X, \mathcal{F})$ into itself, we obtain

$$(8.14.3)' \quad (\mathbb{P}u , v) = (Pu , v) = (u , P^*v) \quad \text{for } u \in C^\infty(X, \mathcal{F}) ,$$

(the last bracket is taken in the sense of the pairing between $C^\infty(X, \mathcal{F})$ and $\mathcal{D}'(X, \mathcal{F})$). Substituting into (8.14.2)', we obtain

$$D^* = \left\{ v \in L^2(X, \mathcal{F}) \;\middle|\; P^*v \in L^2(X, \mathcal{F}) \right\} .$$

The elliptic regularity property (see Remark 8.12) implies that $D = D^* = H^m(X; \mathcal{F})$. The equality (8.14.3)' shows that $\mathbb{P}^*v = P^*v$ for $v \in D$, since $\mathbb{P}^*$ is defined by the equality

$$(\mathbb{P}u, v) = (u , \mathbb{P}^*v) \quad , \quad u , v \in D \quad .$$

PROOF OF ii): The elliptic regularity result shows also that $\text{Ker}\ \mathbb{P} = \text{Ker}\ P \subset C^\infty(X; \mathcal{F})$. If we let Q be a left parametrix of P, we have

$$(8.14.4) \qquad Q.P = I + R_1 , \quad \text{with } R_1 \in L^{-\infty}(X; \mathcal{F}) \; .$$

This equality shows that $I + R_1 = 0$ on $\text{Ker}\ \mathbb{P}$, and since R_1 is a compact operator from $C^\infty(X; \mathcal{F})$ (see Theorem 8.5 of Chapter I) this implies that $\dim(\operatorname{Ker} \mathbb{P}) < +\infty$.

If we let Q' be a right parametrix of P, we have

$$(8.14.5) \qquad P.Q' = I + R_2 \quad , \text{ with } \quad R_2 \in L^{-\infty}(X\,;\mathcal{F}) \quad .$$

This equality shows that $\operatorname{Im} \mathbb{P} \supset \operatorname{Im}(PQ') = \operatorname{Im}(I + R_2)$ and since R_2 is a compact operator from $C^\infty(X\,;\mathcal{F})$, we know that $\operatorname{Im}(I + R_2)$ is a closed subset of finite codimension; the same is therefore also true for Im $\mathbb{P}$. We note that if we multiply (8.14.4) on the right by Q', we thereby deduce that Q is equivalent to Q' and it is thus a both-sided parametrix.

It can immediately be shown that $(\operatorname{Im} \mathbb{P}^*)^\perp = \operatorname{Ker} \mathbb{P}$; furthermore the above shows that $\mathbb{P}^*$ is also of Fredholm type; consequently Im $\mathbb{P}^*$ is closed and equal to $(\operatorname{Ker} \mathbb{P})^\perp$, hence (8.14.1).

PROOF OF iii): Following i) and ii), it only remains for us to prove (8.14.3). If we have $u \in C^\infty(X\,;\mathcal{F})$, the decomposition (8.14.2) allows us to write

$$u = u_1 + u_2 \quad , \text{ with } \quad u_1 \in \operatorname{Ker} \mathbb{P} \quad , \; u_2 \in \operatorname{Im} \mathbb{P} \quad .$$

However, $u_2 = u - u_1 \in C^\infty(X\,;\mathcal{F})$; consequently the elliptic regularity result implies that u_2 is the image under $\mathbb{P}$ of an element of $C^\infty(X,\mathcal{F})$ and this proves (8.14.3).

REMARK 8.15: The index of a Fredholm operator $\mathbb{P}$ is the integer defined by

$$\text{index } (\mathbb{P}) = \dim \operatorname{Ker} \mathbb{P} - \operatorname{codim} \operatorname{Im} \mathbb{P} \quad .$$

With the assumptions of Theorem 8.14, we have

$(\operatorname{Im} \mathbb{P})^{\perp} = \operatorname{Ker} P^{*}$; consequently in this case the index can also be expressed by

$$\text{Index } (\mathbb{P}) = \dim \operatorname{Ker} P - \dim \operatorname{Ker} P^{*} = \text{index } (\mathbb{P}) \quad .$$

If we assume furthermore that P is a classical p.d.o., the theorem of ATIYAH-SINGER [1] shows that the index of P can be expressed using topological invariants associated with X, $\mathcal{F}$ and $\sigma_m(P)$, (see also ATIYAH-BOTT-PATODI [1]).

We conclude this section with the following:

REMARK 8.16: Suppose $\rho \in]\frac{1}{2}, 1]$; the space $L^m_\rho(X)$ of p.d.o.'s of type ρ and of degree m can be obtained by replacing the spaces of ordinary symbols by symbols of type ρ. The extension of the results of Chapter III to symbols of type ρ allows us to generalise, with some minor modifications, the definitions and results of this chapter to p.d.o.'s of type ρ.

For the case $\rho = \frac{1}{2}$, we refer to CALDERON-VAILLANCOURT [1], MELIN-SJOSTRAND [1], and for other generalisations we refer to BEALS-FEFFERMAN [1], (see also Exercise 11.7).

9. ELLIPTIC COMPLEXES AND HODGE THEORY

Let X be a manifold and let $E_o,\dots,E_\ell$ be complex vector bundles of finite dimension over X.

In differential geometry we frequently encounter the following situation:

DEFINITION 9.1: *We describe as a complex of differential operators* $P = (P_k)$ *the collection of differential operators of the same degree,* m *say,*

$$P_k : C^\infty(X ; E_k) \longrightarrow C^\infty(X , E_{k+1}) , \qquad k = 0,\dots,\ell-1$$

such that

$$P_{k+1} \circ P_k = 0 . \tag{9.1.1}$$

If we denote by $E_k(x)$ the fibre of E_k above $x \in X$, for $(x,\xi) \in T^*X\setminus 0$, the principal symbol of degree m of P_k is a linear mapping $\sigma_m(P_k)(x,\xi) \in \mathcal{L}(E_k(x) , E_{k+1}(x))$. We then define a sequence of linear mappings

$$0 \to E_o(x) \xrightarrow{\sigma_m(P_o)(x,\xi)} E_1(x) \longrightarrow \dots E_{\ell-1}(x) \xrightarrow{\sigma_m(P_{\ell-1})(x,\xi)} E_\ell(x) \longrightarrow 0 \tag{9.1.2}$$

which also constitute a complex by virtue of (9.1.1).

DEFINITION 9.2: *The complex* (P_k) *is said to be elliptic if for all* $(x,\xi) \in T^*X\setminus 0$, *the sequence* (9.1.2) *is exact.*

REMARK 9.3: A complex reduced to a single operator $P = P_o$

is elliptic if and only if the sequence

$$0 \longrightarrow E_0(x) \xrightarrow{\sigma_m(P)(x,\xi)} E_1(x) \longrightarrow 0$$

is exact, in other words if $\sigma_m(P)(x,\xi)$ is an isomorphism. We are clearly encountering once more the concept of an elliptic operator.

EXAMPLE 9.4: (De Rham complex) We put $E_k = \Lambda^k(T^*X) \otimes \mathbb{C}$ $k = 0,\ldots,n = \dim X$. The C^∞ sections of this bundle constitute the space $\Omega^k(X)$ of complex-valued differential k-forms. The exterior differential of degree k defines a linear mapping $d_k : \Omega^k(X) \longrightarrow \Omega^{k+1}(X)$ and we have $d_{k+1} \circ d_k = 0$.

We saw in Example 7.9 of Chapter I that d_k is a differential operator of degree 1 whose principal symbol at a point $(x,\xi) \in T^*X \setminus 0$ is the linear mapping

$$\omega \in (\Lambda^k T_x^*X) \otimes \mathbb{C} \longrightarrow \sigma_1(d_k)(x,\xi)\omega = i\xi \wedge \omega \in (\Lambda^{k+1} T_x^* X) \otimes \mathbb{C}.$$

We now show that in this case the sequence (9.1.2) is exact. Let $(\theta_1,\ldots\ldots,\theta_n)$ be a basis of T_x^*X such that $\theta_1 = \xi$. If $\omega \in (\Lambda^k T_x^*X) \otimes \mathbb{C}$, it decomposes in the form

$$\omega = \sum_{\alpha_1<\ldots<\alpha_k} \omega_\alpha \, \theta_{\alpha_1} \wedge \ldots \wedge \theta_{\alpha_k} \quad \text{with} \quad \omega_\alpha \in \mathbb{C} .$$

The exactness to degree k arises from the following equivalences:

$$\omega \in \operatorname{Ker} \sigma(d_{k+1})(x,\xi) \iff \xi \wedge \omega = 0$$

$$\iff \omega_\alpha = 0 \quad \text{when} \quad \alpha_1 \neq 1$$

$$\iff \omega = \sum_{1<\alpha_2<\dots<\alpha_k} \omega_\alpha \, \xi \wedge \theta_{\alpha_2} \wedge \dots \wedge \theta_{\alpha_k}$$

$$\iff \omega \in \operatorname{Im} \sigma(d_k)(x,\xi) \; .$$

We can always assume that X is endowed with a strictly positive density μ and that each bundle E_k is endowed with a Hermitian structure. Under these conditions, we define a prehilbert structure $(\ ,\ )_k$ on each $C^\infty(X \,;\, E_k)$ space and the adjoint P_k^* of P_k is a differential operator of degree m

$$P_k^* : C^\infty(X \,;\, E_{k+1}) \longrightarrow C^\infty(X \,;\, E_k) \quad k = 0,\dots,\ell \; .$$

This allows us to define self-adjoint differential operators of degree 2m

$$\Delta_k = P_k^* \circ P_k + P_{k-1} \circ P_{k-1}^* \; : \; C^\infty(X \,;\, E_k) \to C^\infty(X \,;\, E_k)$$
$$k = 0,\dots,\ell \; .$$

PROPOSITION 9.5: *If the complex* (P_k) *is elliptic, then the operators* Δ_k , $k = 0,\dots,\ell$, *are elliptic.*

PROOF: Suppose $(x,\xi) \in T^*X \setminus 0$. It suffices to prove that the mapping

$$\delta_k = \sigma_{2m}(\Delta_k)(x,\xi) \in \mathcal{L}(E_k(x), E_k(x))$$

is injective. We put $\sigma_m(P_k)(x,\xi) = a_k$; then the formulas of the symbolic calculus show that

$$\delta_k = a_k^* \circ a_k + a_{k-1} \circ a_{k-1}^* .$$

If we let λ be a vector from $E_k(x)$ such that $\delta_k\lambda = 0$, then, utilising the metric of $E_k(x)$, we obtain

$$0 = (\delta_k\lambda , \lambda) = (a_k\lambda , a_k\lambda) + (a_{k-1}^*\lambda , a_{k-1}^*\lambda) ,$$

consequently $a_k\lambda = 0$ and $a_{k=1}^*\lambda = 0$. However, since the sequence (9.1.2) is exact, we thereby deduce that $\lambda \in \operatorname{Im} a_{k-1}$ and this implies that $\lambda = 0$.

PROPOSITION 9.6: *If* (P_k) *is an elliptic complex, then there exist proper pseudo-differential operators* Q_k *such that*

$$Q_k \in L^{-m}(X ; E_{k+1}, E_k) \qquad k = 0,\dots,\ell \quad .$$

(9.6.1) $P_{k-1} \circ Q_{k-1} + Q_k \circ P_k = I + R_k$ where R_k is regularising.

We say that the (Q_k) *constitute a parametrix of the complex* (P_k).

PROOF: Let $T_k \in L^{-2m}(X ; E_k , E_k)$ be a parametrix of the elliptic operator Δ_k. We can convince ourselves that

(9.6.2) $$T_{k+1}\, P_k \equiv P_k\, T_k \quad \text{(modulo a regularising operator)}$$

by multiplying on the left by T_{k+1} and on the right by T_k in the immediate equality

$$P_k \circ \Delta_k = \Delta_{k+1} \circ P_k \quad .$$

We put $Q_k = T_k \circ P_k^* \in L^{-m}(X\,;\,E_{k+1}\,,\,E_k)$, which gives

$$P_{k-1}\,Q_{k-1} + Q_k\,P_k = P_{k-1}\,T_{k-1}\,P_{k-1}^* + T_k\,P_k^*\,P_k$$

which, bearing in mind (9.6.2), is equivalent to

$$T_k(P_{k-1}P_{k-1}^* + P_k^*P_k)$$

$$\equiv \mathrm{Id}\ .$$

We have the generalised Hodge theorem:

THEOREM 9.7: *Suppose* $P = (P_k)_{k=0,\dots,\ell}$ *is an elliptic complex on a compact manifold* X. *Then:*

i) *The space* Im P_{k-1} *is a closed subspace, of finite codimension, of* Ker P_k *equipped with the topology induced by* $C^\infty(X, E_k)$. *The quotient space* Ker P_k/Im P_{k-1} *is thus of finite dimension; it is denoted by* $H^k(P)$ *and is called the* k^{th} *cohomology space of the complex* (P_k).

ii) *We put* $\mathcal{H}^k(P) = \mathrm{Ker}\,\Delta_k$, *and we obtain*

$$\mathcal{H}^k(P) = \mathrm{Ker}\,P_k \cap \mathrm{Ker}\,P_{k-1}^* \quad . \tag{9.7.1}$$

Moreover, $\dim \mathcal{H}^k(P) < +\infty$ *and we have the orthogonal direct sum decomposition (for the structure* $(\ ,\ )_k$ *)*

(9.7.2) $$C^\infty(X\,,\,E_k) = \mathcal{H}^k(P) \oplus \operatorname{Im} P_{k-1} \oplus \operatorname{Im} P_k^* \quad .$$

iii) *The canonical mapping* j *of passage to the quotient under* $\operatorname{Im} P_{k-1}$ *in the equality* (9.7.1) *induces a linear isomorphism*

(9.7.3) $$\mathcal{H}^k(P) \xrightarrow[\sim]{\ j\ } H^k(P) \ .$$

PROOF OF i): The equality (9.6.1) induces on the subspace $\operatorname{Ker} P_k$ the equality $P_{k-1}\cdot Q_{k-1} = I + R_k$; consequently the proof of i) is analogous to that of ii) in Theorem 8.14.

PROOF OF ii): The inclusion $\operatorname{Ker} P_k \cap \operatorname{Ker} P_{k-1}^* \subset \mathcal{H}^k(P)$ follows from the definition of Δ_k. To demonstrate the inverse inclusion, we take $\varphi \in \operatorname{Ker} \Delta_k$, utilising the scalar product $(\ ,\ )_k$, and we obtain

$$0 = (\Delta_k \varphi\,,\,\varphi)_k = (P_k^* P_k\,\varphi\,,\varphi) + (P_{k-1}P_{k-1}^*\varphi\,,\varphi)$$

$$= (P_k\varphi\ ,\ P_k\varphi) + (P_{k-1}^*\varphi\ ,\ P_{k-1}^*\varphi)\quad ,$$

thus $\varphi \in \operatorname{Ker} P_k \cap \operatorname{Ker} P_{k-1}^*$.

The application of Theorem 8.14 iii) to the self-adjoint operator Δ_k shows that $\dim \mathcal{H}^k(P) < +\infty$ and gives the orthogonal decomposition

(9.7.4) $$C^\infty(X\,,\,E_k) = \mathcal{H}^k(P) \oplus \operatorname{Im} \Delta_k \quad .$$

The assumption (9.1.1) shows that

(9.7.5) $$(P_{k-1}\varphi , P_k^*\psi) = 0 \quad , \quad \varphi , \psi \in C^\infty(X , E_k) \;;$$

consequently, we have

$$\operatorname{Im} \Delta_k = \operatorname{Im} P_{k-1}P_{k-1}^* \oplus \operatorname{Im} P_k^*P_k$$

so that

(9.7.6) $$\operatorname{Im} \Delta_k \subset \operatorname{Im} P_{k-1} \oplus \operatorname{Im} P_k^* \quad .$$

Equalities (9.7.1) and (9.7.5) show that the subspaces $\mathcal{H}^k(P)$, $\operatorname{Im} P_{k-1}$, $\operatorname{Im} P_k^*$, taken two at a time, are orthogonal; consequently (9.7.2) follows from (9.7.4) and (9.7.6).

PROOF OF iii): Let us first show that j is injective; if we have $\varphi \in \operatorname{Ker} \Delta_k$ such that $j(\varphi) = 0$, we thus have both $\varphi \in \operatorname{Ker} P_{k-1}^*$ and $\varphi \in \operatorname{Im} P_{k-1}$ which implies that $\varphi = 0$. We now show that j is surjective; suppose $\dot{\psi} \in H^k(P)$; this is the class of an element $\psi \in \operatorname{Ker} P_k$. The decomposition (9.7.2) allows us to write

(9.7.7) $$\psi = \psi_0 + P_{k-1}\,\psi_1 + P_k^*\psi_2 \quad , \quad \text{with} \quad \psi_0 \in \mathcal{H}^k(P)$$

$$\psi_1 \in C^\infty(X , E_{k-1}) \; , \; \psi_2 \in C^\infty(X , E_{k+1}) \quad .$$

Applying P_k to (9.7.7), we obtain

$$0 = P_k\psi_0 + P_kP_{k-1}\psi_1 + P_kP_k^*\psi_2$$

and having regard to (9.7.7) and (9.1.1), we are left with

$$0 = P_kP_k^*\psi_2 \quad , \quad \text{so that} \quad P_k^*\psi_2 = 0$$

Consequently, $\psi = \psi_o + P_{k-1}\psi_1$, and this proves that $j(\psi) = j(\psi_o)$.

EXAMPLE 9.4: (continued). In this particular case, the space $H^k(P)$ is called the kth De Rham cohomology group of the manifold X. When X is compact, Theorem 9.7 shows in particular that all these cohomology groups are of finite dimension and that each De Rham cohomology class can be represented uniquely using a harmonic form (Hodge theorem).

For more details concerning the material in this section, we refer the reader to the original article by ATIYAH-BOTT [1] and to the monograph of GILKEY [1].

10. FRIEDRICH'S LEMMA AND GENERALISATIONS

We now concern ourselves with studying the behaviour of the commutator of a p.d.o. $P(x,D)$ with a regularisation $\rho_\varepsilon *$; this study is based on the following theorem which gives a sort of limited expansion with respect to ε of $[P, \rho_\varepsilon *]$ when $\varepsilon \to 0$.

THEOREM 10.1: *Suppose we have* $P \in L^m(\mathbb{R}^n)$ *of the form* $(Pv)(x) = \int e^{ix\xi} p(x,\xi)\hat{u}(\xi) đ\xi$, *where the symbol* p *is independent of* x *for* x *outside a compact subset* K. *We denote by* P_α *the p.d.o. associated with the symbol* $D_x^\alpha p(x,\xi)$ *and we denote by* ρ_ε^α

the function $\epsilon^{-n}\rho^{\alpha}(\frac{x}{\epsilon})$ *with* $\rho^{\alpha}(x) = x^{\alpha}\rho(x)$ *and* $\rho \in C_0^{\infty}(\mathbb{R}^n)$. *If we have* $s \in \mathbb{R}$ *and* $\sigma \geq 0$, *then for all* N, *there exists* C *such that*

$$(10.1.1) \quad P(\rho_\epsilon * u) - \rho_\epsilon * (Pu) = \sum_{0<|\alpha|<N} \frac{\epsilon^{|\alpha|}}{\alpha!}(P_\alpha u) * \rho_\epsilon^\alpha + R_N(\epsilon)\, u$$

with

$$(10.1.2) \quad \|R_N(\epsilon).u\|_{s+\sigma} \leq C\, \epsilon^{N-\sigma}\|u\|_{s+m}$$

for $u \in H^{s+m}$ and $\epsilon > 0$.

PROOF: If P is an operator with constant coefficients the commutator $[P, \rho_\varepsilon *]$ is identically zero, and we can reduce to considering the case where the symbol p is zero for x outside of K. Using Fourier transformation, we can write

$$(Pv)(x) = \iint \widehat{p(\eta-\xi}, \xi)\, e^{ix\eta}\, \hat{v}(\xi)\, đ\xi\, đ\eta \quad , \quad v \in \mathcal{S} \;;$$

hence, for $u \in \mathcal{S}$, we obtain

$$(10.1.3) \quad P(\rho_\epsilon * u)(x) = \iint \widehat{p(\eta-\xi}, \xi)\, e^{ix\eta}\, \hat{\rho}(\epsilon\xi)\, \hat{u}(\xi)\, đ\xi\, đ\eta \,.$$

In similar fashion, we obtain

$$(10.1.4) \quad (\rho_\epsilon * Pu)(x) = \iint \widehat{p(\eta-\xi}, \xi) e^{ix\eta}\, \hat{\rho}(\epsilon\eta)\, đ\xi\, đ\eta \,.$$

We utilise the Taylor formula up to order N to expand with respect to ε the difference

$$(10.1.5) \quad \hat{\rho}(\epsilon\xi) - \hat{\rho}(\epsilon\eta) = \sum_{0<|\alpha|<N} (\xi-\eta)^{\alpha} \frac{\epsilon^{|\alpha|}}{\alpha!} \partial_\xi^\alpha \hat{\rho}(\epsilon\eta) + r_N(\xi,\eta,\epsilon) \,.$$

By substituting (10.1.3) and (10.1.4) into the left-hand side of (10.1.1) and taking account of (10.1.5), we obtain

$$(R_N(\epsilon)u)(x) = \iint \widehat{p}(\eta-\xi, \xi) e^{ix\eta} \, r_N(\xi,\eta,\epsilon)\hat{u}(\xi) \, d\!\!\!/\xi \, d\!\!\!/\eta \; .$$

The integral expression for the remainder r_N shows immediately that, for any $k \geq 0$, there exists C such that

(10.1.6)
$$\begin{cases} |r_N(\xi,\eta,\epsilon)| \leq C \quad \epsilon^N |\xi-\eta|^N (1+\epsilon|\xi|)^{-k} & \text{if} \quad |\xi-\eta| \leq \dfrac{|\xi|}{2} \\ \quad \text{and} & \\ |r_N(\xi,\eta,\epsilon)| \leq C \quad \epsilon^N |\xi-\eta|^N & \text{for all} \quad \xi,\eta \; . \end{cases}$$

In addition, the assumption concerning the support of $p(x,\xi)$ implies that for all $h \geq 0$, there exists C such that

$$(10.1.7) \quad |\widehat{p}(\eta-\xi, \xi)| \leq C\,(1+|\xi|)^m (1+|\eta-\xi|)^{-h} \; .$$

To majorise the norm $\|R_N(\epsilon)u\|_{s+\sigma}$, we specify $v \in \mathcal{S}$ and we consider the inner product

$$< R_N(\epsilon)u, v > = \iint \widehat{p}(\eta-\xi, \xi) \; r_N(\xi,\eta,\epsilon)\hat{u}(\xi)\hat{v}(-\eta) \, d\!\!\!/\xi \, d\!\!\!/\eta \; .$$

We put
$$U(\xi) = (1+|\xi|^2)^{(s+m)/2} |\hat{u}(\xi)|$$

$$\text{and} \quad V(\eta) = (1+|\eta|^2)^{-(s+\sigma)/2} |(\hat{v}(-\eta)| \; .$$

Taking account of (10.1.7) and of Peetre's inequality, there follows

$$(10.1.8) \quad |< R_N(\epsilon)u,v >| \leq$$

$$C \iint |r_N(\xi,\eta,\epsilon)|(1 + |\xi|)^{\sigma}(1 + |\xi-\eta|)^{|s|+|\sigma|-h} u(\xi)v(\eta) d\xi\, d\eta .$$

Noting that for $|\xi-\eta| \leq \frac{|\xi|}{2}$ we have, by virtue of (10.1.6),

$$|r_N(\xi,\eta,\epsilon)| \leq C\ \epsilon^N |\xi-\eta|^N (\epsilon+\epsilon|\xi|)^{-\sigma}$$

and that for $|\xi-\eta| \geq \frac{|\xi|}{2}$ we have $(1 + |\xi|) \leq 2(1 + |\xi-\eta|)$,

we thereby deduce that there exists C such that the integrand of the right-hand side of (10.1.8) is bounded above by

$$C \quad \epsilon^{N-\sigma} (1 + |\xi-\eta|)^{|s|+|\sigma|+N-h} \quad u(\xi)\ v(\eta) \quad .$$

Consequently, by taking h sufficiently large, we thus deduce

$$|< R_N(\epsilon)u,v >| \leq C \quad \epsilon^{N-\sigma} \|u\|_0 \quad \|v\|_0$$

$$\leq C \quad \epsilon^{N-\sigma} \|u\|_{s+m} \|v\|_{-s-\sigma}$$

which proves (10.1.2).

As a particular case, we have Friedrich's Lemma

COROLLARY 10.2: *We adopt the same assumptions as in the preceding theorem, together with the further condition* $\hat{\rho}(0) = 1$. *Then for all* $s \in \mathbb{R}$, *there exists* C *such that*

$$(10.2.1) \qquad \|[P,\ \rho_\epsilon *]u\|_s \leq C\|u\|_{s+m-1}$$

for $u \in H^{s+m-1}$ and $\epsilon > 0$.

Moreover, $[P\ ,\ \rho_\epsilon *]u \xrightarrow[\epsilon \to 0]{} 0$ *in* H^0 *when* $u \in H^{s+m-1}$.

PROOF: The inequality (10.2.1) is obtained by setting

$N = \sigma = 1$ and by replacing s by $s-1$ in the inequality (10.1.2). If $u \in \mathcal{S}$, we know that $\rho_\epsilon * u \xrightarrow[\epsilon \to 0]{} u$ in $\mathcal{S}$ and since $\mathcal{S}$ is dense in H^{s+m-1} the inequality (10.2.1) allows us to extend the conclusion to all $u \in H^{s+m-1}$.

We now give the proof of Theorem 6.5 of Chapter II within a more general setting; this is as follows:

COROLLARY 10.3: *Let* P *be a* p.d.o. *as in Theorem* 10.1; *let there exist* $s \in \mathbb{R}$, *and* ρ *satisfying the supplementary condition* (6.3.3) *of Chapter* II, *with* $h > s+1$. *Then, there exists* C *such that*

$$(10.3.1) \quad \int_0^1 \|[P, \rho_\epsilon *]u\|_0^2 \, \epsilon^{-2s-2} \left(1 + \delta^2/\epsilon^2\right)^{-1} \frac{d\epsilon}{\epsilon} \leq C \|u\|^2_{s+m-1,\delta}$$

for $u \in H^{s+m-1}$ *and* $0 < \delta \leq 1$.

PROOF: We replace $[P, \rho_\epsilon *]u$ by its expression (10.1.1) and we majorise each term. For $0 < |\alpha| < N$, we utilise Theorem 6.4 of Chapter II

$$\int_0^1 \|P_\alpha u * \rho_\epsilon\|_0^2 \quad \epsilon^{-2s-2+2|\alpha|} \left(1 + \delta^2/_{\epsilon}-1\right) \frac{d\epsilon}{\epsilon} \leq C\|P_\alpha u\|^2_{s-|\alpha|,\delta}$$

and since P_α is an operator of degree m, Proposition 6.3 of Chapter II (or rather its immediate generalisation to p.d.o.'s) shows that

$$\|P_\alpha u\|_{s-|\alpha|,\delta} \leq C\|u\|_{s+m-1,\delta}$$

since $|\alpha| \geq 1$.

To majorise the remainder, we utilise (10.1.2) with s replaced by $s-1$ and $\alpha = 1-s$; we obtain

$$\int_0^1 \|R_N(\epsilon)u\|_0^2 \, \epsilon^{-2s-2} \left(1 + \frac{\delta^2}{\epsilon^2}\right)^{-1} \frac{d\epsilon}{\epsilon}$$

$$\leq C\|u\|_{s+m-1}^2 \int_0^1 \epsilon^{2N-4} \left(1 + \frac{\delta^2}{\epsilon^2}\right)^{-1} \frac{d\epsilon}{\epsilon}$$

and taking $N \geq 2$, we have an upper bound given by $C'\|u\|_{s+m-1}^2$.

REMARK 10.4: All the above results are still valid when the symbol of P depends continuously on one parameter varying within a compact subset; in fact, it is sufficient to note that we can choose the constant C in (10.1.7) independent of this parameter.

As an application of Corollary 10.2, we shall prove the identity between a weak solution (i.e. in the distributional sense) and a strong solution; more precisely we have

COROLLARY 10.5: *Let* $P(x,D)$ *be a p.d.o. satisfying the assumptions of Theorem* 10.1, *and of degree* m. *We assume that* $u \in H^{m-1}(R^n)$ *satisfies, in the distributional sense,* $Pu = f$ *with* $f \in L^2(R^n)$. *Then, there exists a sequence*

$u_j \in C^\infty(\mathbb{R}^n) \cap H^{m-1}$ *such that* $u_j \xrightarrow[j\infty]{H^{m-1}} u$ *and* $Pu_j \xrightarrow{L^2} f$; *we say that* u *is a strong solution.*

PROOF: It suffices to put $u_j = u * \rho_{1/j}$. We know that $u_j \xrightarrow{H^{m-1}} u$ and Corollary 10.2 shows that

$$P u_j = \rho_{1/j} * Pu + [P , \rho_{1/j} *] u \xrightarrow[j \infty]{L^2} P u .$$

We conclude with the following:

REMARK 10.6: All the above discussion generalises word for word to the case where P is a matrix-valued p.d.o.

11. ADDITIONAL NOTES AND EXERCISES

11.1 P.D.O.'s WITH UNIFORM SYMBOLS : For certain global problems in $\mathbb{R}^n$ (see Chapter VI) it is convenient to make use of p.d.o.'s whose symbols are in $BS^{+\infty}$ (see Section 10.1 of Chapter I). More specifically, we denote by BL^m the subspace of $L^m(\mathbb{R}^n)$ composed of the p.d.o.'s of the form (1.2.1) where the amplitude p is in $BS^m(\mathbb{R}^{2n} \times \mathbb{R}^n)$.

a) If we have $P \in \bigcap_m BL^m$, this is therefore a regularising operator with kernel $K \in C^\infty(\mathbb{R}^n \times \mathbb{R}^n)$.

Prove that for any multi-indices α, β, the function

$$(11.1.1) \qquad (x-y)^{\beta}\,\partial^{\alpha}_{x,y}\,K(x,y) \text{ is bounded in } \mathbb{R}^n \times \mathbb{R}^n .$$

[Use the expression (1.4.1) for $p \in BS^m$ with $m < -n - |\beta|$]

b) Deduce from this the equalities

$$BL^{-\infty} = \bigcap_m BL^m = \left\{ \begin{array}{l} \text{set of operators with kernel } C^{\infty}(\mathbb{R}^n \times \mathbb{R}^n) \\ \text{satisfying (11.1.1)} . \end{array} \right\}$$

c) Prove that an operator belonging to $BL^{-\infty}$ is continuous from $\mathcal{S}$ into $\mathcal{S}$, from $\mathcal{S}'$ into $\mathcal{S}'$ and from $H^{-\infty}$ into $H^{+\infty}$. [Utilise the expression for P, in terms of its kernel].

d) A continuous linear operator with kernel distribution K is said to be B-proper if there exists C such that

$$\text{supp. } K \subset \{(x,y) \in \mathbb{R}^{2n} \mid |x-y| \le C\} .$$

With $P \in BL^m$, prove that there exists a B proper $P' \subset BL^m$ and $R \in BL^{-\infty}$ such that $P = P' + R$. [Multiply the kernel of P by a function $\chi(x-y)$ where $\chi \in C_0^{\infty}(\mathbb{R}^n)$ is equal to 1 for $|x| \le$ constant].

e) Given that P is a B-proper operator, prove that

$$(11.1.2) \qquad \sigma_P(x\,,\,y) = e^{-ix\xi}\,P(e^{ix\xi}) \in BS^m$$

if and only if $P \in BL^m$. In this case, we have

$$(11.1.3) \qquad Pu(x) = \int e^{ix\xi}\,\sigma_P(x\,,\,\xi)\,\hat{u}(\xi)\,đ\xi \quad , \quad u \in \mathcal{S}$$

[Put $b(x, z, \xi) = p(x, x + z, \xi)$, where p is an amplitude for P, and prove that for all α, β and $k \in \mathbb{N}$, there exists C such that

$$(11.1.4) \qquad |\partial_x^\alpha \partial_\xi^\beta b(x, \hat{\zeta}, \xi)| \leq C(1 + |\xi|)^{m-|\beta|} (1 + |\zeta|)^{-k} ;$$

thus deduce an upper bound for

$$\partial_x^\alpha \partial_\xi^\beta \sigma_P(x, \xi) = \int_{R^n} \partial_x^\alpha \partial_\xi^\beta b(x, \hat{\zeta}, \xi + \zeta) \, d\!\!/\xi$$

by splitting this latter integral into two parts corresponding to $|\zeta| \leq \frac{|\xi|}{2}$ and $|\zeta| \geq \frac{|\xi|}{2}$] .

f) By utilising the decomposition in d) together with the expression (11.1.3) and c), deduce that an operator $P \in BL^m$ acts continuously from $\mathcal{S}$ into $\mathcal{S}$ and from $\mathcal{S}'$ into $\mathcal{S}'$, and that (11.1.2), (11.1.3) remain valid; σ_P is called the complete symbol of P.

g) Let $P \in BL^m$ be a B-proper operator. Apply the Taylor formula with integral remainder to the function $b(x, \hat{\zeta}, \xi + \zeta)$ between ξ and $\xi + \zeta$ to show that for any $k \geq 0$, we have

$$\sigma_P(x, \xi) - \sum_{|\alpha|< k} \frac{1}{\alpha!} \partial_\xi^\alpha D_z^\alpha b(x, z, \xi)\Big|_{z=0}$$

$$= \sum_{|\alpha| = k} \iint_0^1 \frac{(1-t)^k}{\alpha!} \partial_\xi^\alpha b(x, \zeta, \xi + t\zeta) \, \zeta^\alpha \, d\!\!/\zeta \, dt .$$

Thus deduce, by utilising (11.1.4) and Peetre's inequality

$$(1 + |\xi + t\zeta|)^s \leq C_s (1 + |\xi|)^s (1 + t|\zeta|)^{|s|} ,$$

that we have

$$\text{(11.1.5)} \quad \sigma_P(x,\xi) \underset{B}{\sim} \sum_{\alpha \geq 0} \frac{1}{\alpha!} \partial_\xi^\alpha D_y^\alpha p(x, y, \xi) \big|_{y=x}$$

Generalise this result to the case $P \in BL^m$.

h) Suppose $P \in BL^m$. The principal symbol of degree m of P is the class of σ_P in BS^m / BS^{m-1}; we denote this by $\sigma_m(P)$; show that we have the linear isomorphism

$$BL^m / BL^{m-1} \xrightarrow{\sigma_m} BS^m / BS^{m-1} .$$

i) If we have $P \in BL^m$, then ${}^tP$, P^* are in BL^m and

$$\text{(11.1.6)} \quad \sigma_{{}^tP}(x, \xi) \underset{B}{\sim} \sum_{\alpha \geq 0} \frac{1}{\alpha!} \partial_x^\alpha \partial_\xi^\alpha \sigma_P(x, -\xi)$$

$$\text{(11.1.7)} \quad \sigma_{P^*}(x, \xi) \underset{B}{\sim} \sum_{\alpha \geq 0} \frac{1}{\alpha!} \partial_x^\alpha \partial_\xi^\alpha \overline{\sigma_P(x, \xi)}$$

[If $p(x, y, \xi)$ is an amplitude for P, then $p(y, x, -\xi)$ is an amplitude for ${}^tp$ and we obtain (11.1.6) by using (11.1.5). Similarly, for P^*.]

j) If $P \in BL^m$, prove that there exists $b(y, \xi) \in BL^m$ with

$$b(y, \xi) \underset{B}{\sim} \sum \frac{1}{\alpha!} \partial_\xi^\alpha D_y^\alpha \sigma_P(y, \xi)$$

such that

$$\text{(11.1.8)} \quad \widehat{(Pu)}(\xi) = \int e^{-iy\xi} b(y, \xi) u(y)\, dy , \quad u \in \mathcal{S}.$$

[We express ${}^tP$ by means of an expression of the type (11.1.3), i.e. we write ${}^tP = \sigma_{{}^tP}(x, D)$. Then, observing that $P = {}^t({}^tP)$, we thereby deduce that P is defined by the amplitude

$$p(x , y , \xi) = \sigma_{{}^tP}(y , -\xi) = b(y , \xi) \;] .$$

k) Given $Q \in BL^{m'}$ and $P \in BL^m$, prove that

$Q.P \in BL^{m+m'}$ and that we have

$$\sigma_{Q.P}(x , \xi) \underset{B}{\sim} \sum_{\alpha \geq 0} \frac{1}{\alpha!} \partial_\xi^\alpha q(x , \xi) D_x^\alpha \sigma_P(x , \xi) \tag{11.1.9}$$

[Express P by means of (11.1.8) and write Q in the form (11.1.3).] Thus deduce that $[P , Q] \in BL^{m+m'-1}$].

ℓ) We say that $P \in BL^m$ is elliptic if $\sigma_m(P)$ is elliptic in BS^m. Show that P is elliptic if and only if there exists $Q \in BL^{-m}$ such that

$$PQ = I+R \;, \quad QP = I+R' \quad \text{with} \quad R , R' \in BL^{-\infty} .$$

m) Let $C \in BL^0$; we assume that there exists $c > 0$ such that

$$\sigma_0(C)(x , \xi) \geq c \qquad \text{for} \qquad |\xi| \geq 1/c . \tag{11.1.10}$$

Prove that there exists $B \in BL^0$ such that $c = B^*B+R$ with $R \in BL^{-\infty}$. [Use the proof of Lemma 8.1 with the simplification which follows from the fact that the inequality (11.1.10) is uniform with respect to x.]

If $P \in BL^m$, deduce from this that P operates continuously from H^s into H^{s-m}. [Proceed as in Theorem 8.5, observing that

$\Lambda_s \in BL^s$].

n) Let $P \in BL^m$; we assume that there exists $a > 0$ such that

$$\text{(11.1.11)} \qquad \operatorname{Re} P_m(x, \xi) \geq a|\xi|^m \quad \text{for} \quad |\xi| \geq \frac{1}{a} \quad .$$

Prove that for any $\varepsilon > 0$ and any $s < \frac{m}{2}$, there exists C such that we have Gårding's inequality:

$$\text{(11.1.12)} \qquad \operatorname{Re}(Pu, u) \geq (a - \epsilon)\|u\|^2_{\frac{m}{2}} - C\|u\|^2_s \quad , \quad u \in \mathcal{S}.$$

o) Generalise this exercise to the case of matrix-valued operators.

Note: the method of proof indicated in this exercise also provides another means of studying p.d.o.'s belonging to $L^m(X)$. (see HORMANDER [4] for a presentation of this.) This method is also simpler as long as we are not concerned with questions of change of variable, and it allows a number of important generalisations (see BEALS [1]).

(*) 11.2 P.D.O.'s WITH UNIFORM SYMBOLS AND DEPENDENT ON ONE PARAMETER

In Chapter VII we shall make use of p.d.o.'s which depend on a parameter varying within an interval of the form $[\gamma_o, +\infty[$.

We denote by $PS^m(\mathbb{R}^n \times \mathbb{R}^N)$ the set of functions

$p(x, \xi, \gamma) \in C^\infty(\mathbb{R}^n \times \mathbb{R}^N \times [\gamma_o, +\infty[)$ for a certain $\gamma_o > 0$ and such that for all (α, β) there exists C such that

$$|\partial_x^\alpha \partial_\xi^\beta p(x, \xi, \gamma)| \leq C(|\xi| + \gamma)^{m-|\beta|} \text{ for all } x, \xi \text{ and } \gamma \geq \gamma_o .$$

We denote by PL^m the space of p.d.o.'s $P^{(\gamma)}$ of the form

$$(P^{(\gamma)}u)(x) = \int e^{i(x-y).\xi} p(x, y, \xi, \gamma) u(y)dy \, đ\xi$$

where the amplitude $p \in PS^m(\mathbb{R}^{2n} \times \mathbb{R}^n)$.

The results and proofs concerning the p.d.o.'s belonging to BL^m extend almost word for word to p.d.o.'s belonging to PL^m. We shall indicate only the main modifications.

In a) it is necessary to replace $(x - y)^\beta$ by $\gamma^k(x - y)^\beta$, where k is an arbitrary positive integer.

In condition (11.1.2), it is necessary to replace BS^m by PS^m and to change $(1 + |\xi|)$ into $(\gamma + |\xi|)$ in (11.1.4).

In g), Peetre's inequality must be used in the form

$$(\gamma + |\xi + t\zeta|)^s \leq C_s(\gamma + |\xi|)^s(1 + t|\zeta|)^{|s|}, \quad \gamma \geq 1 .$$

All the asymptotic expansions of symbols are still valid as long as we consider them in the graded algebra $PS^{+\infty} = \bigcup_m PS^m$.

In particular if $P^{(\gamma)} \in PL^m$ and $Q^{(\gamma)} \in PL^{m'}$ then

$$[P^{(\gamma)}, Q^{(\gamma)}] \in PL^{m+m'-1} .$$

In ℓ) we say that $P^{(\gamma)}$ is elliptic if $\sigma_m(P)$ is elliptic in PS^m. For example $(\gamma^2 + |\xi|^2)^{s/2}$ is elliptic of degree s.

Let $\gamma \in \mathbb{R} \setminus 0$; we denote by H^s_γ the space H^s endowed with the equivalent norm

$$|||\, u \,|||^2_{s;\gamma} = \int (\gamma^2 + |\xi|^2)^s \, |\hat{u}(\xi)|^2 \, d\!\!\!/\xi .$$

Supposing $P^{(\gamma)} \in PL^m$ and $s \in R$, prove that there exists C independent of γ such that

$$|||\, P^{(\gamma)}u \,|||_{s-m;\gamma} \le C \, |||\, u \,|||_{s;\gamma} , \quad u \in \mathcal{S} .$$

In condition (11.1.11) we need to replace $|\xi|^m$ by $(\gamma + |\xi|)^m$ and we then obtain Gårding's inequality (11.1.12) with norms

$$|||\ .\ |||_{.;\gamma} .$$

Finally, let $p(x, \xi, \gamma)$ be a C^∞ function for $x \in \mathbb{R}^n$, $\xi \in \mathbb{R}^N$, $\gamma \geq 0$, $(\xi, \gamma) \neq 0$. We assume that p is homogeneous of degree m with respect to (ξ, γ) and that all its derivatives are bounded for $|\xi| + \gamma = 1$.

Show that $p \in BS^m(\mathbb{R}^n \times \mathbb{R}^N)$ and thus deduce the concept of an operator "with homogeneous principal symbol" together with the concept of a "classical" operator in $PL^{+\infty}$.

11.3 SECOND PROOF OF THE ESTIMATE (7.6.1) OF CHAPTER 2.

Let $P = p(x,D)$ be a differential operator of principal type of degree m in an open subset Ω of $\mathbb{R}^n$. Show that the differential operator $Q : C^\infty(\Omega) \to C^\infty(\Omega, \mathbb{C}^n)$ defined by $Q\varphi = \left(\frac{\partial p}{\partial \xi_j}(x,D)\varphi\right)_{j=1,\dots,n}$ is left elliptic of degree $m - 1$.

By utilising the elliptic inequality (see (8.11.3) and Remark 8.12) and the Poincaré inequality (see Exercise 7.1 of Chapter II), deduce the inequality (7.6.1) of Chapter II.

11.4 CONVOLUTION WITH CERTAIN HOMOGENEOUS DISTRIBUTIONS.

Let $T \in \mathcal{D}'(\mathbb{R}^n)$ be a distribution which is homogeneous of degree m (see Section 10.2 of Chapter I).
Assume further that T is C^∞ in $\mathbb{R}^n \setminus 0$. Show that the operator of convolution with T is in $L^{-m-n}(\mathbb{R}^n)$.

11.5 ENDOMORPHISMS OF COMPLEXES AND TRACES

Let (E_j, P_j) be an elliptic complex above a compact manifold X endowed with a density dx. We take an endomorphism (T_j) of this complex, i.e. a family of endomorphisms T_j of $C^\infty(X;E_j)$, which for all j satisfy:

$$P_j \circ T_{j+1} = T_{j+1} \circ P_j \quad .$$

a) Show that each T_j induces an endomorphism, denoted by

$H(T_j)$, of the space $H^j(P)$.

b) Assume that the T_j are of finite rank; prove that we have the abstract Lefchetz formula

$$(11.5.1) \qquad \sum_j (-1)^j \text{ Trace } T_j = \sum_j (-1)^j \text{ Trace } H(T_j) \ .$$

[Utilise the exact sequences $0 \to \text{Ker } P_j \to C^\infty(X ; E_j) \to \text{Im } P_j \to 0$ and $0 \to \text{Im } P_{j-1} \to \text{Ker } P_j \to H^j(P.) \to 0$ to reduce to situations of the type $0 \to A \to A \oplus C \to C \to 0$ for which we have

$$\text{Trace } (T_{|A\oplus C}) = \text{Trace } (T_{|A} + \text{Trace } (T_{|C}] \ .$$

c) We define the kernel distribution

$K_j \in \mathcal{D}'(X \times X ; E_j \boxtimes E_j^*)$ of T_j by

$< T_j v , u > = < K_j , v \otimes u >$ for $u \in C^\infty(X , E_j)$ and $v \in C^\infty(X , E_j^*)$. We recall that the sign $\boxtimes$ denotes the outer tensor product.

We assume that $K_j \in C^\infty(X \times X ; E_j \boxtimes E_j^*)$; prove that we again have the inequality (11.5.1) when we put

$$\text{Trace } T_j = \int_X (\text{trace } K_j(x,x)) \, dx$$

[approximate the T_j by operators of finite rank in the sense of the convergence of the kernels in $C^\infty(X \times X ; E_j \boxtimes E_j^*)$] .

11.6 COTLAR-KNAPP-STEIN LEMMA

Let X be a measured set whose measure is denoted by $d\xi$. We take a measurable mapping $\xi \in X \to A_\xi \in \mathcal{L}(\mathcal{H})$ where $\mathcal{H}$ is a Hilbert space. We suppose that there exists a function $h: X \times X \to \mathbb{R}$ which is the kernel of an operator H, bounded in $L^2(X)$ and such that

$$\| A_\xi A_\eta^* \| \le h^2(\xi, \eta) \;,\quad \| A_\xi^* A_\eta \| \le h^2(\xi, \eta) \text{ for } (\xi, \eta) \in X \times X.$$

Then, we propose to show that the integral $\int A_\xi d\xi$ converges weakly and defines a bounded operator A such that $\| A \| \le C$ with $C = \| H \|$.

a) Show that it is sufficient to establish this result when X is of finite measure m and when $\| A_\xi \|$ is uniformly bounded by a constant M; we shall assume this in the following text.

b) Show that for all $n \in \mathbb{N}$, we have $\| A \|^{2n} = \|(A^* A)^n \|$

$$\le \int_{X^{2n}} \| A_{\xi_1}^* A_{\xi_2} , \dots , A_{\xi_{2n-1}}^* A_{\xi_{2n}} \| d\xi_1 \dots d\xi_{2n} \;.$$

[Use the equality $\| A \|^2 = \| A^* A \|$] . Thus deduce that

$$(11.6.1) \quad \| A \|^{2n} \le \int_{X^{2n}} \| A_{\xi_1}^* \|^{\frac{1}{2}} \Big(\prod_{j=1}^{2n-1} \| A_{\xi_j}^* A_{\xi_{j+1}} \|^{\frac{1}{2}} \Big) \| A_{\xi_{2n}} \|^{\frac{1}{2}} d\xi_1 \dots d\xi_{2n}$$

[Use an inequality of the type

$$\| B_1 , \dots , B_{2n} \|^2 \le \| B_1 \| \prod_{j=1}^{2n-1} \| B_j B_{j+1} \| \| B_{2n} \|]$$.

c) Letting $H_{2n-1}(\xi_1 , \xi_{2n})$ be the kernel of the operator H^{2n-1}, use (11.6.1) to show that we have

$$\| A \|^{2n} \leq M \int_{X \times X} H_{2n-1}(\xi_1 , \xi_{2n}) d\xi_1 \, d\xi_{2n} \quad .$$

Thus deduce that $\| A \|^{2n} \leq M \, C^{2n-1} m$; then by letting n tend to + ∞ show that $\| A \| \leq C$.

11.7 CALDERON-VAILLANCOURT THEOREM.

We take a real number $\rho \in [0, 1[$ and an amplitude $a(x, y, \xi) \in C^\infty(\mathbb{R}^n \times \mathbb{R}^n \times \mathbb{R}^n)$ which satisfies for any multi-index α, β

$$\partial_\xi^\alpha \, \partial_{x,y}^\beta \, a = O(\lambda_\xi^{|\beta|-|\alpha|})$$

where we have put $\lambda_\xi = (1 + |\xi|)^\rho$. Let A be the operator defined by

$$\text{(11.7.1)} \quad Au(x) = \iint e^{i(x-y)\xi} \, a(x , y , \xi) \, u(y) dy \, d\!\!/\xi \quad \text{for } u \in C_0^\infty(\mathbb{R}^n) \; .$$

We now have to show that A is continuous in $L^2(\mathbb{R}^n)$; we shall follow the approach of a method described in BOUTET DE MONVEL [3].

a) Give a meaning to the oscillatory integral (11.7.1) by using an integration by parts with the aid of the differential operator

$$L_y = \frac{\chi(\xi)}{|\xi|^2} \sum_{j=1}^{n} \xi_j D_j - (1 - \chi)$$

where $\chi \in C^\infty(\mathbb{R}^n)$ is zero in the neighbourhood of 0 and equal to 1 for $|\xi|$ large.

b) Consider the differential operator

$$L_\xi = (1 + \lambda_\xi^{2N} |x - y|^{2N})^{-1}(1 + \lambda_\xi^{2N} (-\Delta_\xi)^N) \quad .$$

Prove that $L_\xi \cdot e^{i(x-y)\xi} = e^{i(x-y)\xi}$ and that $b = {}^tLa$ satisfies

$$\partial_\xi^\alpha \partial_{x,y}^\beta b = O\left[\lambda_\xi^{|\beta|}(1 + \lambda_\xi |x - y|)^{-2N}\right]$$

for any multi-indices α, β.

c) For $\xi \in \mathbb{R}^n$ we consider the operator A_ξ defined by

$$(A_\xi u)(x) = \int e^{i(x-y)\xi} a(x, y, \xi)\, u(y)dy \quad .$$

By observing that $A_\xi u(x) = \int e^{i(x-y)\xi}\, b(x,, y, \xi)\, u(y)\, dy$ prove that the kernel of A_ξ is bounded above in modulus by $C(1 + |x - y|)^{-2N}$. Thus deduce by choosing $N > \frac{n}{2}$, that the operators A_ξ are uniformly bounded in $L^2(\mathbb{R}^n)$.

d) Prove that the kernel $c_{\xi,\eta}$ of $A_\xi A_\eta^*$ is given by

$$c_{\xi,\eta}(x,y) = e^{i(x.\xi - y.\eta)} \int e^{iz(\eta-\xi)} b(x, z, \xi)\, \bar{b}(y, z, \eta)\, dz.$$

Let M_z be the differential operator defined by

$$M_z = (1 + (\lambda_\xi + \lambda_\eta)^{-2}|\xi - \eta|^2)^{-1}\left[1 - (\lambda_\xi + \xi_\eta)^{-2}\Delta_z\right] \quad .$$

Show that for any $k \in \mathbb{N}$ we have

$$({}^tM)^k(b(x, z, \xi)\bar{b}(y, z, \eta)) \le$$

$$C_k\left[1 + (\lambda_\xi + \lambda_\eta)^{-1}|\xi - \eta|\right]^{-2k}\left[1 + \lambda_\xi |x - z|\right]^{-2N}$$
$$\left[1 + \lambda_\eta |y - z|\right]^{-2N}.$$

Thus deduce that $\|A_\xi A_\eta^*\| \leq C_k' h^2(\xi, \eta)$ with

$$h(\xi, \eta) = [1 + (\lambda_\xi + \lambda_\eta)^{-1} |\xi - \eta|]^{-2k} (\lambda_\xi \lambda_\eta)^{-n}$$

and deduce an analogous upper bound for

$$\| A_\xi^* A_\eta \| .$$

e) Show that the integral $\int h(\xi, \eta) d\eta$ is a bounded function of ξ. [Split this integral into three parts corresponding to the regions $|\eta| \leq \frac{1}{2}(1 + |\xi|)$,

$|\eta| \geq 2(1 + |\xi|)$, and $\frac{1}{2}(1 + |\xi|) \leq |\eta| \leq 2(1 + |\xi|)$].

Thus deduce that the operator H with kernel h is bounded in L^2. In accordance with the Cotlar-Knapp-Stein Lemma, we consider the operator $\mathcal{A}$, bounded in L^2, defined by $\mathcal{A} = \int_{R^n} A_\xi d\xi$.

Show that $\mathcal{A}$ extends the operator A.

CHAPTER 5

ELLIPTIC BOUNDARY-VALUE PROBLEMS

1. INTRODUCTION

Let Ω be a regular open subset of $\mathbb{R}^n$, with boundary $\partial\Omega$; we take, in a neighbourhood of $\partial\Omega$ in $\mathbb{R}^n$, a field of vectors ν transversal to $\partial\Omega$, and we define for $j \in \mathbb{N}$ the trace operator γ_j of order j on $\partial\Omega$ by $\gamma_j u = \left[(\frac{1}{i}\nu)^j u\right]_{|\partial\Omega}$. Suppose we have $P(x, D_x)$, an elliptic differential operator in $\mathbb{R}^n$ of degree m, and consider proper pseudo-differential operators $B_{j,k} \in L^{d_j-k}(\partial\Omega)$ for $j = 0,\ldots, \mu-1$ and $k = 0,\ldots, m-1$ (with μ a positive integer and d_j given reals). We propose to study the existence, uniqueness and regularity of solutions u of the boundary value problem

$$(1.1.)\quad \begin{cases} P(x, D_x)u = f & \text{in} \quad \Omega \\ B(x', D_{x'})\,\gamma u = d & \text{in} \quad \partial\Omega \end{cases}$$

where B is the matrix $(B_{j,k})_{\substack{j = 0,..,\mu-1 \\ k = 0,..,m-1}}$ and where γu is the column matrix $(\gamma_j u)_{j = 0,\ldots,m-1}$.

For example, the Dirichlet problem corresponds to

$$\begin{cases} \Delta u = f & \text{in} \quad \Omega \\ \gamma_0 u = g & \text{in} \quad \partial\Omega \end{cases}$$

and the Neumann problem corresponds to

$$(1.2.) \qquad \begin{cases} \Delta u = f & \text{in} \quad \Omega \\ \gamma_1 u = g & \text{in} \quad \partial\Omega \quad , \quad \nu \text{ being} \end{cases}$$

the inward unit normal to $\partial\Omega$.

There are broadly two types of method for studying (1.1). One of these consists of majorising a norm of u by suitable norms of the data f, g, and then concluding by employing functional analysis arguments. The other type of method consists of reducing (1.1) to a problem on the boundary $\partial\Omega$, then studying this problem by means of the theory of systems of elliptic pseudo-differential equations on a manifold without boundary. Descriptions of the first type of method are given, for example, in AGMON - DOUGLIS - NIRENBERG [1], LIONS - MAGENES [1]. Here we shall describe the second type of method, following the work of HORMANDER [2].

For clarity, we shall start with a formal description of the case in which P has constant coefficients and Ω is bounded. If $\varphi \in C^\infty(\overline{\Omega})$, we denote by φ^0 its extension by 0 outside Ω. Then, for $u \in C^\infty(\overline{\Omega})$, $P(u^0) - (Pu)^0$ is a distribution carried by $\partial\Omega$, which depends only on P and on γu; we put

$$(1.3.)\qquad P(u^{\circ}) = (Pu)^{\circ} + \tilde{P}\,\gamma u \quad .$$

In fact, $\tilde{P}\,\gamma u$ is a multiple layer of order m-1: after transport by a local chart of $\mathbb{R}^n$ in a neighbourhood of a point of $\partial\Omega$ which transforms Ω into $\{ x = (x', x_n) \in \mathbb{R}^n \mid x_n > 0 \}$, $\partial\Omega$ into $\{x_n = 0\}$ and ν into $\frac{\partial}{\partial x_n}$ (in what follows, such a local chart will be called a local chart of $(\bar{\Omega}, \nu)$), P is written

$$(1.4.)\qquad P(x, D_x) = \sum_{j=0}^{m} P_j(x, D_{x'})\, D_{x_n}^j \quad (\text{degree } P_j = m-j)$$

and we have the jump formula (cf (1.4.1), Chapter 1):

$$(1.5.)\qquad \tilde{P}\,\gamma u = \frac{1}{i} \sum_{\ell+k+1 \leq m} P_{\ell+k+1}(x, D_{x'})\, \gamma_\ell u \otimes D^k \delta(x_n).$$

All the above clearly holds for an arbitrary regular open set Ω and for an arbitrary differential operator P.

Carrying out a left convolution of both sides of (1.3) with a fundamental solution E of P = P(D) we obtain, putting Pu = f :

$$(1.6.)\qquad u = (E * f^{\circ})_{|\Omega} + (E * \tilde{P}\,\gamma u)_{|\Omega} \quad .$$

This formula expresses u as the sum of a volume potential and a surface potential; we shall show that these two potentials are in $C^\infty(\bar{\Omega})$, so that taking traces in (1.6) :

$$\text{(1.7.)} \qquad \gamma u = \gamma\left[(E * \tilde{P}\,\gamma u)_{|\Omega}\right] + \gamma\left[(E * f^{\circ})_{|\Omega}\right].$$

In particular, if $u \in C^{\infty}(\bar{\Omega})$ satisfies $Pu = 0$, we obtain $\gamma u = C(\gamma u)$, where $C : C^{\infty}(\partial\Omega, \mathbb{C}^m) \longrightarrow C^{\infty}(\partial\Omega, \mathbb{C}^m)$ is defined by $C\,v = \gamma\left[(E * \tilde{P}v)_{|\Omega}\right]$. It can easily be shown that C is a projector. From (1.7), if $u \in C^{\infty}(\bar{\Omega})$ satisfies (1.1), then $\gamma u \in C^{\infty}(\partial\Omega, \mathbb{C}^m)$ satisfies

$$\begin{cases} (I - C)\,\gamma u = \gamma\left[(E * f^{\circ})_{|\Omega}\right] \\ B\,\gamma u = g \end{cases}$$

and u is given by (1.6).

Conversely, if $v \in C^{\infty}(\partial\Omega, \mathbb{C}^m)$ satisfies

$$\begin{cases} (I - C)v = \gamma\left[(E * f^{\circ})_{|\Omega}\right] \\ B\,v = g \end{cases}$$

and if we put $u = (E * f^{\circ})_{|\Omega} + (E * \tilde{P}v)_{|\Omega}$, then $u \in C^{\infty}(\bar{\Omega})$ and satisfies (1.1).

We thus see that (1.1) reduces to investigating the system of boundary operators $\begin{pmatrix} I - C \\ B \end{pmatrix}$.

In what follows, we show that the potentials

have the stated regularity at the boundary, and then go on to investigate the operator C (called the Calderón projector). Finally, we apply the results obtained to the investigation of (1.1) under certain ellipticity assumptions.

2. REGULARITY OF THE POTENTIALS AT THE BOUNDARY

We use the notation defined at the start of the Introduction. Let $u \in \mathcal{D}'(\Omega)$. We say that u has sectional traces up to order k ($k \in \mathbb{N}$) on $\partial\Omega$ if, for any local chart $\chi : U \longrightarrow V \subset \mathbb{R}^n$ of $(\bar{\Omega}, \nu)$ and for any $\varphi \in C_0^\infty(U)$, φu is transported into a function of class C^k in x_n (for $x_n \geq 0$ sufficiently small) with values in $\mathcal{D}'(R^{n-1}_{x'})$. We then naturally define the traces $\gamma_0 u = u_{|\partial\Omega}$, $\gamma_j u = \gamma_0 \left[(\frac{1}{i}\, \nu)^j u\right]$ $(0 \leq j \leq k)$ as distributions on $\partial\Omega$. For example, if $u \in H^s_{loc}(\bar{\Omega})$, then u has sectional traces on $\partial\Omega$ up to order k for $k < s - \frac{1}{2}$, and $\gamma_j u \in H^{s-j-1/2}_{loc}(\partial\Omega)$ for $0 \leq j \leq k$ (see Chapter II).

If $u \in \mathcal{D}'(\Omega)$ has a sectional trace of order 0 on $\partial\Omega$, we naturally define its extension $u°$ by 0 outside Ω. Let $v \in \mathcal{D}'(\partial\Omega)$ and $j \in \mathbb{N}$; by means of local charts of $(\bar{\Omega}, \nu)$, we define the layer $v \otimes D^j \delta \in \mathcal{D}'_{\partial\Omega}(R^n)$. If $u \in \mathcal{D}'(\Omega)$ has sectional traces up to order m, we again have the jump formula (1.3), (1.5).

Now let $A \in L^{\mu}_{c}(\mathbb{R}^n)$ be a classical pseudo-differential operator. We introduce the conditions

(2.1.) *A is proper, and each term of the complete symbol* $\sum_{j\geq 0} a_j(x, \xi)$ *of A is a rational fraction in* ξ.

Using the formulas of the symbolic calculus for p.d.o.'s, it can be shown that the conditions (2.1) are preserved under diffeomorphism, transposition, composition, and passage to the parametrix in the elliptic case. In particular, they are thus satisfied when A is the proper parametrix of an elliptic differential operator.

THEOREM 2.2 : *Suppose we have* $A \in L^{\mu}_{c}(\mathbb{R}^n)$ *satisfying* (2.1) *and* $u \in \mathcal{D}'(\mathbb{R}^n)$ *zero in* Ω. *Then* $(Au)_{|\Omega}$ *has sectional traces of any order on* $\partial\Omega$.

PROOF: By partition of unity and local charts we can reduce to the case $\Omega = \mathbb{R}^n_+$. We can assume u to have compact support; we then know that there exist c and ℓ such that $|\hat{u}(\xi)| \leq c(1 + |\xi|)^{\ell}$ for $\xi \in \mathbb{R}^n$, and it can easily be shown that $A'u \in C^k(\mathbb{R}^n)$ when $k < -\mu' - n - \ell$ and $A' \in L^{\mu'}(\mathbb{R}^n)$. Hence A'u can be taken to be as regular as one wishes in $\mathbb{R}^n$ when μ' is sufficiently small, and it is sufficient to establish the theorem in the case $A = \tilde{a}(x, D)$, $\tilde{a}(x, \xi) = \chi(\xi)\, a(x, \xi)$, with a homogeneous of degree μ with respect to ξ, where $\chi(\xi) = 0$ for ξ near 0 and $\chi(\xi) = 1$ for $|\xi| \geq 1$. For $\Phi \in C_0^\infty(\mathbb{R}^n)$, we have (cf. (1.3.3), Chapter IV)

$$< Au, \Phi > = \int_{|\xi|\geq 1} \hat{u}(\xi)\, đ\xi \int e^{ix\xi} a(x, \xi) \Phi(x) dx + < Ru, \Phi >$$

where R is the regularising operator with kernel

$$(x, y) \longmapsto \int_{|\xi|\leq 1} e^{i(x-y)\xi} \chi(\xi) a(x, \xi)\, đ\xi \quad .$$

Putting B = A - R, we thus have

$$(2.2.1.) \qquad < Bu, \Phi > = \int_{|\xi|\geq 1} \hat{u}(\xi) F(\xi)\, đ\xi \quad ,$$

with

$$(2.2.2.) \qquad F(\xi) = \int e^{ix\xi} a(x, \xi) \Phi(x) dx \quad .$$

By hypothesis, $a(x, \xi)$ is a rational fraction in ξ, so that it follows that the poles ξ_n remain within a compact subset of $\mathbb{C}$ when x describes a compact subset of $\mathbb{R}^n$ and ξ' describes a compact subset of $\mathbb{R}^{n-1}$.

For ξ' fixed, $F(\xi', \xi_n)$ is thus holomorphic in ξ_n for $|\xi_n|$ sufficiently large. Since

$$\xi^\alpha F(\xi) = \int e^{ix\xi} (- D_x)^\alpha (a(x, \xi) \Phi(x)) dx \quad ,$$

we see that $F(\xi)$ is of rapid decrease with respect to ξ for

$\xi' \in \mathbb{R}^{n-1}$ and $\operatorname{Im} \xi_n \geq 0$ when $\operatorname{supp} \Phi \subset \{ x_n > 0 \}$. Since $\hat{u}(\xi)$ is slowly increasing with respect to ξ for $\xi' \in \mathbb{R}^{n-1}$ and $\operatorname{Im} \xi_n \geq 0$, we therefore obtain by contour deformation at ξ_n in (2.2.1):

$$(2.2.3.) \quad \langle Bu, \Phi \rangle = \int_{\mathbb{R}^{n-1}} đ\xi' \int_{\Gamma_{\xi'}} \hat{u}(\xi', \xi_n)\, F(\xi', \xi_n)\, đ\xi_n$$

where $\Gamma_{\xi'}$ completed by the segment $[-(1-|\xi'|^2)^{1/2}, (1-|\xi'|^2)^{1/2}]$ in the case $|\xi'| < 1$, is a positive contour of the half-plane $\{ \operatorname{Im} \xi_n \geq 0 \}$ enclosing the poles ξ_n of $a(x ; \xi', \xi_n)$ such that $\operatorname{Im} \xi_n > 0$ (for x describing $\operatorname{supp} \Phi$)

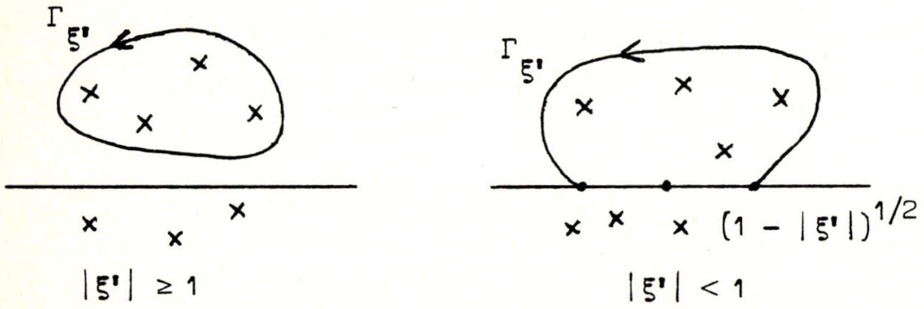

It may be noted, in view of the homogeneity of a, that we can choose $\Gamma_{\xi'} = |\xi'|\Gamma$ for $|\xi'| \geq 1$, where Γ is a fixed contour of $\{ \operatorname{Im} \xi_n > 0 \}$. We then have $|\xi_n| \leq c(1 + |\xi'|)$ for $\xi_n \in \Gamma_{\xi'}$, $\xi' \in \mathbb{R}^{n-1}$.

We take in particular $\Phi(x) = \varphi(x')\,\psi(x_n)$ with $\operatorname{supp}\psi \subset \{x_n > 0\}$. From (2.2.3), (2.2.2), we obtain

$$< Bu, \varphi \otimes \psi > = \int_{R^{n-1}} đ\xi' \int_{\Gamma_{\xi'}} \hat{u}(\xi', \xi_n)\, đ\xi_n \int e^{ix.\xi}\, a(x, \xi)\, \varphi(x)\, \psi(x_n)\, dx'\, dx_n .$$

It can be shown that the order of integration can be changed, so that we can write:

$$(2.2.4.) \quad < Bu, \varphi \otimes \psi > = \int \psi(x_n)\, dx_n \int_{R^{n-1}} G(x_n, \xi')\, đ\xi' ,$$

where

$$(2.2.5.) \quad G(x_n, \xi') = \int_{\Gamma_{\xi'}} \hat{u}(\xi', \xi_n)\, đ\xi_n \int e^{ix.\xi}\, a(x, \xi)\, \varphi(x')\, dx' .$$

(By integrating by parts with respect to x', we see that $G(x_n, \xi')$ is of rapid decrease with respect to ξ', and hence that the right-hand side of (2.2.4) is meaningful). The equality (2.2.4) is clearly true for $u \in C_0^\infty(R^n)$, and is deduced for $u \in \mathcal{E}'(R^n)$ by proceeding to the limit, approximating u by its regularised forms $\rho_k * u$, with ρ having support in $\{x_n \leq 0\}$ (see (1.1.1), Chapter I).

Finally, (2.2.4) and (2.2.5) show that, in $\{x_n > 0\}$,

Bu coincides with the function from $C^\infty(\bar{\mathbb{R}}^+, \mathcal{D}'(\mathbb{R}^{n-1}))$ defined by

$$(2.2.6.) \qquad x_n \longmapsto \left(\varphi \longmapsto \int_{\mathbb{R}^{n-1}} đ\xi' \iint_{\Gamma_{\xi'} \times \mathbb{R}^{n-1}} e^{ix.\xi}\, a(x, \xi)\, \varphi(x')\, \hat{u}(\xi', \xi_n)\, đ\xi_n\, dx'.\right.$$

REMARK 2.3 : The above proof shows that the mapping $u \longrightarrow \gamma_0(Au)_{|\Omega}$ is continuous from $\{u \in \mathcal{D}'(\mathbb{R}^n) \mid u = 0 \text{ in } \Omega\}$ into $\mathcal{D}'(\partial\Omega)$.

THEOREM 2.4 :(Regularity of the surface potentials at the boundary)

Suppose we have $A \in L^\mu_c(\mathbb{R}^n)$ *satisfying* (2.1). *Then*

i) *The operator* $K : v \longmapsto [\,A(v \otimes \delta)\,]_{|\Omega}$ *is continuous from* $C^\infty(\partial\Omega)$ *into* $C^\infty(\bar{\Omega})$.

If $v \in \mathcal{D}'(\partial\Omega)$, Kv *has sectional traces of any order on* $\partial\Omega$

ii) *The operator* $v \longmapsto \gamma_0(Kv)$ *is in* $L^{\mu+1}_c(\partial\Omega)$.

After transport by a local chart of $(\bar{\Omega}, \gamma)$, *its principal symbol is* $(x', \xi') \longrightarrow \int_\Gamma a_0(x', 0 ; \xi', \xi_n)\, đ\xi_n$, *where* a_0 *is the principal symbol of* A, *and* Γ *is a contour enclosing the poles* ξ_n *of* $a_0(x', 0 ; \xi', \xi_n)$ *such that* $\operatorname{Im} \xi_n > 0$.

iii) *For all* $s \in \mathbb{R}$, K *is continuous from* $H^s_{loc}(\partial\Omega)$ *into* $H^{s-\mu-1/2}_{loc}(\bar{\Omega})$.

PROOF: For (i), we can replace A by the operator $B = A - R$ used in the previous proof; taking (2.2.4), (2.2.5) in the particular case $u = v(x') \otimes \delta(x_n)$, $v \in C_0^\infty(\mathbb{R}^{n-1})$, we see that for $x_n > 0$:

$$(2.4.1.) \qquad (Bu)(x) = \int e^{ix'\xi'} k(x, \xi') \hat{v}(\xi') đ\xi'$$

where

$$(2.4.2.) \qquad k(x, \xi') = \int_{\Gamma_{\xi'}} e^{ix_n\xi_n} a(x ; \xi', \xi_n) d\xi_n$$

so that (i) then follows as the right-hand side of (2.4.1) is C^∞ with respect to $x \in \mathbb{R}^n$.

For (ii), we reduce to $\Omega = \mathbb{R}^n_+$ and we note first that $A'(v \otimes \delta) \in C^0(\mathbb{R}^n)$ for $v \in C_0^\infty(\mathbb{R}^{n-1})$, $A' \in L^{\mu'}(\mathbb{R}^n)$, $\mu' < -1$, and then that $A'(v \otimes \delta)(x', 0) = A''(x', D_{x'})v$ where $A'' \in L^{\mu'+1}(\mathbb{R}^n)$ has as complete symbol $a''(x', \xi') = \int_{-\infty}^{+\infty} a'(x', 0 ; \xi', \xi_n) d\xi_n$. We can therefore again replace A by B as above. Using (2.4.1), (2.4.2), we obtain

$$(Bv)(x', 0) = \int e^{ix'\xi'} k(x', 0 ; \xi') \hat{v}(\xi') đ\xi' ,$$

with

$$k(x', 0 ; \xi') = \int_{\Gamma_{\xi'}} a(x', 0 ; \xi', \xi_n) đ\xi_n ,$$

which gives (ii) since $k(x', 0 ; \xi')$ is homogeneous in ξ' of degree $\mu + 1$ for $|\xi'| \geq 1$.

For (iii), we reduce to the operator B by noting that for $A' \in L^{\mu'}(R^n)$, $v \in C_0^\infty(R^{n-1})$ and $\varphi \in C_0^\infty(R^n)$, we have:

$$\|\varphi A'(v \otimes \delta)\|_{s-\mu-1/2} \leq C \|v \otimes \delta\|_{s-\mu-1/2+\mu'} \leq C' \|v\|_{s-\mu+\mu'}$$

$$\leq C' \|v\|_s \text{ if } s-\mu+\mu' < -1/2 \text{ and } \mu' \leq \mu$$ ((4.6.1), Chap. 2).

We shall thus show that there exists a constant C such that $\|(Bu)_{|\Omega}\|_{s-\mu-1/2} \leq C \|v\|_s$ for all $v \in C_0^\infty(R^{n-1})$, where $(Bu)_{|\Omega}$ is given by (2.4.1), (2.4.2), assuming that $a(x,\xi)$ has compact support in x .

LEMMA 2.4.3: *For all multi-indices* α, β, *there exists a constant* C *such that*

$$|x^\beta D_x^\alpha k(x, \xi')| \leq C (1 + |\xi'|)^{\mu+1+\alpha_n-\beta_n} \quad \text{for } x_n \geq 0 .$$

In fact, from (2.4.2) and from Leibniz's formula, $D_x^\alpha k$ is a finite linear combination of terms of the form

$$\int_{\Gamma_{\xi'}} e^{ix_n\xi_n} \xi_n^p D_{x'}^{\alpha'} a(x ; \xi', \xi_n) d\xi_n \quad \text{with } 0 \leq p \leq \alpha_n .$$

By integrating by parts with respect to ξ_n, we then deduce that $x^\beta D_x^\alpha k$ is a finite linear combination of terms of the form

$$x'^{\beta'} \int e^{ix_n\xi_n} b(x, \xi', \xi_n) d\xi_n ,$$

with $b = D_{\xi_n}^{\beta_n}(\xi_n^p D_{x'}^{\alpha'} a)$. We then have (2.4.3) since b is of degree $\mu + p - \beta_n \leq \mu + \alpha_n - \beta_n$.

We now extend $k(x, \xi')$ for $x_n \leq 0$ such that the extension, again denoted by k, still has compact support in x and satisfies the upper bounds in Lemma (2.4.3) for all $x \in \mathbb{R}^n$. We can, for example, use a *SEELEY* extension by putting for $x_n \leq 0$:

$$k(x, \xi') = \sum_{p=1}^{\infty} \lambda_p \, k(x', -2^p x_n, \xi')$$

where (λ_p) is a sequence as in (9.2.1), Chapter I.

We put

$$\hat{k}(\zeta, \xi') = \int_{\mathbb{R}^n} e^{-ix\zeta} k(x, \xi')\, dx \quad .$$

LEMMA 2.4.4 : *For all* $q, r \in \mathbb{N}$, *there exists a constant* C *such that*

$$|\hat{k}(\zeta, \xi')| \leq C(1 + |\xi'|)^{\mu} (1 + |\zeta'|)^{-r} \left(1 + \frac{|\zeta_n|}{1 + |\xi'|}\right)^{-q} \quad .$$

From 2.4.3 we have, for given $p \in \mathbb{N}$ and for any multi-index α

$$|D_x^{\alpha} k(x, \xi')| \leq$$
$$C_{\alpha}(1 + |\xi'|)^{\mu+1-\alpha_n} (1 + |x'|)^{-q} (1 + |x_n| (1 + |\xi'|))^{-p}$$

so that, choosing $p > 1$ and supposing $q > n-1$:

$$|\zeta^{\alpha} \hat{k}(\zeta, \xi')| = \left| \int e^{ix\cdot\zeta} D_x^{\alpha} k(x, \xi')\, dx \right|$$

$$\leq C (1 + |\xi'|)^{\mu+1-\alpha_n} \int_{-\infty}^{+\infty} (1 + |x_n| (1 + |\xi'|))^{-p} dx_n$$

$$= C'(1 + |\xi'|)^{\mu+\alpha_n}$$

which readily gives Lemma 2.4.4.

We put $U(x) = \int e^{ix'\xi'} k(x, \xi')\, \hat{v}(\xi')\, đ\xi'$ for $x \in \mathbb{R}^n$.

Since U extends $(Bu)_{|\Omega}$, it will suffice to show that

(2.4.5.) $\|U\|_{s-\mu-1/2} \leq C \|v\|_s$,

i.e. that

(2.4.6.) $|< U, \varphi >| \leq C \|v\|_s \|\varphi\|_{-s+\mu+1/2}$ for all $\varphi \in C_0^\infty(R^n)$.

Now $\hat{U}(\eta) = \int \hat{k}(\eta - \xi', \xi') \hat{v}(\xi') đ\xi'$, so that

$$< U, \varphi > = \int \hat{k}(\eta - \xi', \xi') \hat{v}(\xi') \hat{\varphi}(-\eta) đ\eta\, đ\xi' ,$$

and hence, from Schwarz's inequality

$$| < U, \varphi > | \leq C \|\varphi\|_{-s+\mu+1/2} \|V\|_0 , \text{ where}$$

$$V(\eta) = \int |\hat{k}(\eta - \xi', \xi')| (1 + |\eta|)^{s-\mu-1/2} |\hat{v}(\xi')| d\xi' .$$

Lemma 2.4.4 shows that $\|V\|_0 \leq \|W\|_0$, where

$$W(\eta) = \int X(\eta, \xi')(1 + |\xi'|)^s |\hat{v}(\xi')| d\xi' , \text{ with}$$

$$X(\eta, \xi') = (1 + |\xi'|)^{\mu-s} (1 + |\eta|)^{s-\mu-1/2} (1 + |\eta' - \xi'|)^{-r} \times \left(1 + \frac{|\eta_n|}{1 + |\xi'|}\right)^{-q}$$

(r and q will be chosen later).

To conclude, it will be sufficient, therefore, to show that

(2.4.8.) $\|W\|_0 \leq C \|v\|_s$.

The change of variables $\eta' = \tilde{\eta}'$, $\eta_n = (1 + |\tilde{\eta}'|)\tilde{\eta}_n$ shows that

(2.4.9.) $\|W\|_0^2 = \int \left[(1 + |\eta'|)^{1/2} W(\eta', (1 + |\eta'|)\eta_n)\right]^2 d\eta'\, d\eta_n$.

Now Peetre's inequality (see Chapter II) shows that

$$(1 + |\eta'|)^{1/2}\, X(\eta', (1 + |\eta'|)\eta_n ; \xi') \le C\, (1 + |\eta_n|)^{s-\mu+1/2}$$

$$\times \quad (1 + |\eta' - \xi'|)^{-r+|\mu-s|}\, \left(1 + \frac{|\eta_n|}{1 + |\eta' - \xi'|}\right)^{-q}$$

so that, from (2.4.7), (2.4.9) we have:

$$\|w\|_0^2 \le C \int (1 + |\eta_n|^2)^{s-\mu-1/2}\, \|Y(\cdot, \eta_n)\|_0^2\, d\eta_n \tag{2.4.10.}$$

where

$$Y(\eta', \eta_n) = \int Z(\eta' - \xi', \eta_n)(1 + |\xi'|)^s\, |\hat{v}(\xi')|\, d\xi' \tag{2.4.11.}$$

and

$$Z(\eta', \eta_n) = (1 + |\eta'|)^{-r+|\mu-s|}\, \left(1 + \frac{|\eta_n|}{1 + |\eta'|}\right)^{-q} \tag{2.4.12.}$$

$$= (1 + |\eta'|)^{-r+|\mu-s|+q}\, (1 + |\eta|)^{-q} .$$

From (2.4.11), we have (cf. (2.1.2), Chapter I)

$$\|Y(\cdot, \eta_n)\|_0^2 \le C\, \|v\|_s^2\, \Sigma^2(\eta_n) \quad , \tag{2.4.13.}$$

where

$$\Sigma(\eta_n) = \int Z(\eta', \eta_n)\, d\eta' \quad .$$

Finally, we choose $q > n - 1$, $q > s - \mu + n - 1/2$, $r \ge |\mu - s| + q$.

Then, from (2.4.12) we have:

$$\Sigma(\eta_n) \leq \int (1 + |\eta|)^{-q} \, d\eta' = (1 + |\eta_n|)^{-q+n-1} \int (1 + |\eta'|)^{-q} \, d\eta'$$
$$= C\,(1 + |\eta_n|)^{-q+n-1}$$

and hence, from (2.4.10), (2.4.13)

$$\|w\|_0^2 \leq C\,\|v\|_s^2 \int (1 + |\eta_n|^2)^{s-\mu-1/2-q+n-1} \, d\eta_n \quad ,$$

which establishes (2.4.8) and completes the proof of Theorem 2.4.

THEOREM 2.5 (Regularity of the volume potential at the boundary):

Suppose we have $A \in L_C^{\mu}(\mathbb{R}^n)$ *satisfying* (2.1). *Then the operator* $f \to [\,A(f^{o})]_{|\Omega}$ *is continuous from* $C^{\infty}(\overline{\Omega})$ *into* $C^{\infty}(\overline{\Omega})$.

PROOF: Suppose $f \to \tilde{f}$ is an extension operator which is continuous from $C^{\infty}(\overline{\Omega})$ into $C^{\infty}(\mathbb{R}^n)$; we have $f^{o} = \tilde{f} - u$, where u is the extension by 0 into Ω of $g = \tilde{f}_{|\Omega'}$, $\Omega' = \mathbb{R}^n \setminus \overline{\Omega}$. We shall therefore show that $g \to (Au)_{|\Omega}$ is continuous from $C^{\infty}(\overline{\Omega}')$ into $C^{\infty}(\overline{\Omega})$; for this, we reduce to the operator B used in the proof of Theorem 2.2, and we use (2.2.3). By integrating by parts with respect to x', we see that for any ℓ there exists a constant C such that $|\hat{u}(\xi)| \leq C\,(1 + |\xi'|)^{-\ell}$ for $\xi' \in \mathbb{R}^{n-1}$ and $\operatorname{Im} \xi_n \geq 0$; (2.2.3) and (2.2.2) thus show that we have, for $x_n > 0$:

$$(2.5.1.) \qquad (Bu)(x) = \int_{R^{n-1}} e^{ix'\xi'} \, đ\xi' \int_{\Gamma_{\xi'}} e^{ix_n\xi_n} \, a(x\,;\,\xi',\,\xi_n)\,\hat{u}(\xi',\,\xi_n)\,đ\xi_n \,,$$

so that

$$Bu_{|\Omega} \in C^\infty(\bar{\Omega}) \,.$$

REMARK 2.6: Under the conditions of the preceding theorem, we can show that $f \longmapsto [\,A(f^\circ)\,]_{|\Omega}$ extends continuously from $H^s_{loc}(\bar{\Omega})$ to $H^{s-\mu}_{loc}(\bar{\Omega})$ when $s > -1/2$ (see BOUTET DE MONVEL [1]). By analogy with the operators which occur in the solution of the Dirichlet problem, the operators K in Theorem 2.4 are called Poisson operators; by analogy with the case in which A is differential, the operators $f \longmapsto \gamma_0\,[\,A(f^\circ)_{|\Omega}\,]$ are called trace operators. All the previous results and the definitions of the Poisson operators and the trace operators are in fact valid under a weaker assumption than (2.1): the transmission assumption for A (see BOUTET DE MONVEL [2]).

PROPOSITION 2.7: *Suppose we have* $A \in L^\mu_c(R^n)$ *satisfying* (2.1). *Then the transpose of the Poisson operator*

$$K \;:\; \mathcal{E}'(\partial\Omega) \longmapsto \mathcal{E}'(\bar{\Omega})$$
$$v \longmapsto ([\,A(v \otimes \delta)]_{|\Omega})^\circ$$

is the operator

$$T' \;:\; C^\infty(\bar{\Omega}) \longmapsto C^\infty(\partial\Omega)$$
$$f \longmapsto \gamma_0([\,{}^tA(f^\circ)]_{|\Omega'}) \,,$$

where $\Omega' = \mathbb{R}^n \setminus \bar{\Omega}$.

PROOF: We need to prove that $< f, Kv > = < T'f, v >$ for $f \in C_0^\infty(\bar{\Omega})$, $v \in C_0^\infty(\partial\Omega)$. We reduce to the case $\Omega = \mathbb{R}^n_+$; the equality is immediate when $f \in C_0^\infty(\Omega)$. When $f \in C_0^\infty(\bar{\Omega})$, we consider a regularising sequence $\rho_\epsilon(x_n) = \frac{1}{\epsilon}\rho(\frac{x_n}{\epsilon})$ with $\mathrm{Supp}\,\rho \subset \{ x_n > 0 \}$. We have

$$\begin{aligned} < f, Kv > &= < f^o, [A(v \otimes \delta)]_{|\bar{\Omega}} > \\ &= \lim_{\epsilon \to 0} < \rho_\epsilon * f^o, [A(v \otimes \delta)_{|\bar{\Omega}}] > \\ &= \lim_{\epsilon \to 0} < \rho_\epsilon * f^o, A(v \otimes \delta) > \\ &= \lim_{\epsilon \to 0} < \gamma_0 \, {}^tA(\rho_\epsilon * f^o), v > . \end{aligned}$$

To conclude, we show that $\lim_{\epsilon \to 0} \gamma_0 \, {}^tA(\rho_\epsilon * f^o) = \gamma_0([{}^tA(f^o)]_{|\Omega'})$ in $C^\infty(\Omega)$. This is obvious if A is a pseudo-differential operator of degree less than -n, so that the required equality follows by a straightforward passage to the limit in (2.5.1).

COROLLARY 2.8: *Under the conditions of Proposition* 2.7, *the operator* T' *extends continuously from* $\mathcal{D}'(\bar{\Omega})$ to $\mathcal{D}'(\partial\Omega)$. *The same is true for the trace operator* $T : f \longrightarrow \gamma_0([A(f^o)]_{|\Omega})$ *when the degree* μ *of* A *is less than* - 1/2.

PROOF: Since $T' = {}^tK$ and since K is continuous from $C_0^\infty(\partial\Omega)$ into $C_0^\infty(\bar{\Omega})$, T' extends continuously from $\mathcal{D}'(\bar{\Omega})$ to $\mathcal{D}'(\partial\Omega)$.

Since $f^o \in L^2_{loc}(\mathbb{R}^n)$, $A(f^o)$ is in $H^{-\mu}_{loc}(\mathbb{R}^n)$, and thus $T'f = Tf$ if $\mu < -1/2$.

We conclude this section with a property relating to the existence of sectional traces.

THEOREM 2.9: *Suppose* $P(x, D_x)$ *is an elliptic differential operator of degree* m *in* $\mathbb{R}^n$, *and suppose that* $u \in \mathcal{D}'(\Omega)$ *is extendable to* $\mathbb{R}^n$.

(i) *If* $Pu \in C^\infty(\overline{\Omega})$, u *has sectional traces of any order on* $\partial\Omega$.

(ii) *If* $Pu \in H^s_{loc}(\overline{\Omega})$, u *has sectional traces up to order* k *on* $\partial\Omega$ *if* $k < s + m - 1/2$.

PROOF: For (i), let $\tilde{u} \in \mathcal{D}'(\mathbb{R}^n)$ be an extension of u; we put $Pu = f$ and $P\tilde{u} - f^o = g$; let Q be a proper parametrix of P; we have $u = (Q(f^o) + Qg + R\tilde{u})_{|\Omega}$ with $R \in L^{-\infty}(\mathbb{R}^n)$; we then have (i) from Theorems 2.2 and 2.5, since Q satisfies (2.1) and since g vanishes in Ω.

For (ii), we put $Pu = f$ and we consider $\tilde{f} \in H^s_{loc}(\mathbb{R}^n)$ extending f ; let $v = u - (Q\tilde{f})_{|\Omega}$; we have $Pv \in C^\infty(\overline{\Omega})$, and hence (ii) since $Q\tilde{f} \in H^{s+m}_{loc}(\mathbb{R}^n)$.

REMARK 2.10: Suppose $M \subset \mathcal{D}'(\Omega)$ is a subspace of extendable distributions, endowed with a topology such that there exists a a continuous extension operator: $M \to \mathcal{D}'(\mathbb{R}^n)$. Remark (2.3) and the preceding proof show that, in Theorem 2.9, the traces of u depend continuously (as distributions on $\partial\Omega$) on

$(u, Pu) \in \mathbf{M} \times H^s_{loc}(\overline{\Omega})$.

REMARK 2.11 : By using the result of Exercise 6.3, given later in this chapter, Theorem 2.9 and its proof may readily be generalised to the case in which the assumption of ellipticity of P is replaced by the assumption that $\partial\Omega$ is noncharacteristic for P.

3. THE CALDERON PROJECTOR

Let $Q \in L_c^{-m}(\mathbb{R}^n)$ be a proper parametrix of the elliptic differential operator P of degree m. We have $PQ = I + R$, $QP = I + R'$ with $R, R' \in L^{-\infty}$. We recall that Q satisfies (2.1). Suppose $u \in C^\infty(\overline{\Omega})$. Applying Q on the left to both sides of (1.3), we obtain, putting $Pu = f$:

$$u = (Q\tilde{P}\gamma u)_{|\Omega} + (Q(f)^o)_{|\Omega} - [R'(u^o)]_{|\Omega}$$

so that, taking the traces:

$$\gamma u = \gamma (Q\tilde{P}\gamma u)_{|\Omega} + \gamma(Q(f^o))_{|\Omega} - \gamma R'(u^o) .$$

DEFINITION 3.1 : *The operator* C :

$C^\infty(\partial\Omega, \mathbb{C}^m) \longrightarrow C^\infty(\partial\Omega, \mathbb{C}^m)$ *defined by* $Cv = \gamma\left[(Q\tilde{P}v)_{|\Omega}\right]$

(see (1.3), (1.4), (1.5) *and Theorem* 2.4) *is termed the Calderón projector associated with* P, Ω, γ. *By construction, we have* $\gamma u = C(\gamma u) + \gamma(Q(f^{o}))_{|\Omega} - \gamma R'(u^{o})$ *for* $u \in C^{\infty}(\bar{\Omega})$, $f = Pu$.

PROPOSITION 3.2 :

i) *The operator* $v \longmapsto (Q\tilde{P}v)_{|\Omega}$ *is, for all* $\sigma \in \mathbb{R}$, *continuous from* $\prod_{\ell=0}^{m-1} H_{loc}^{\sigma-\ell-1/2}(\partial\Omega)$ *into* $H_{loc}^{\sigma}(\Omega)$.

ii) *We have* $C = (C_{j,\ell})_{j,\ell=0,\dots,m-1}$, *with* $C_{j,\ell} \in L_c^{j-\ell}(\partial\Omega)$, *proper.*

iii) C *is a projector in the sense that* $C^2 - C \in L^{-\infty}(\partial\Omega)$.

PROOF: By transport using a local chart of ($\bar{\Omega}$, ν), (1.4) and (1.5) show that

$$(3.2.1.)\quad (Q\tilde{P}v)_{|\Omega} = \frac{1}{i}\sum_{\ell+k+1\leq m}\left[QP_{\ell+k+1}(x, D_{x'})D_{x_n}^{k}(v_\ell \otimes \delta)\right]_{|\Omega},$$

$$\gamma_j\left[(Q\tilde{P}v)_{|\Omega}\right] = \frac{1}{i}\sum_{\ell+k+1\leq m}\gamma_0\left(\left[D_{x_n}^{j}\,QP_{\ell+k+1}(x, D_{x'})D_{x_n}^{k}(v_\ell \otimes \delta)\right]_{|\Omega}\right).$$

Since the operators $QP_{\ell+k+1}D_{x_n}^{k}$, $D_{x_n}^{j}\,QP_{\ell+k+1}D_{x_n}^{k}$ satisfy (2.1) and are of degree $-\ell-1$, $j-\ell-1$ respectively, (i) and (ii) follow from Theorem 2.4; this also shows that $C_{j,\ell}$ has as principal symbol (of degree $j-\ell$):

$$(3.2.2.)\qquad c_{j,\ell}(x', \xi') = \frac{1}{2i\pi}\int_\Gamma p^{-1}(x', 0; \xi', \xi_n) \sum_{\ell+k+1\leq m} p_{\ell+k+1}(x', 0; \xi')\, \xi_n^{j+k}\, d\xi_n ,$$

where Γ is a contour of $\{ \operatorname{Im} \xi_n > 0 \}$ enclosing the zeros ξ_n of $p(x', 0; \xi', \xi_n)$ such that $\operatorname{Im} \xi_n > 0$.

To establish (iii), we consider $v \in C^\infty(\partial\Omega, \mathbb{C}^m)$, and put $u = (Q \tilde{P} v)_{|\Omega}$; by construction of C, we have $\gamma u = Cv$; since $Pu = (R \tilde{P} v)_{|\Omega}$, we also have

$$\gamma u = C(\gamma u) + \gamma Q \left[(R \tilde{P} v)_{|\Omega} \right]^o - \gamma R'(u^o) ,$$

so that $C - C^2 = r$, with

$$rv = \gamma Q \left[(R \tilde{P} v)_{|\Omega} \right]^o - \gamma R' \left[(Q \tilde{P} v)_{|\Omega} \right]^o .$$

Thus $C^2 - C \in L^{-\infty}(\partial\Omega)$ since r extends continuously from $\mathcal{D}'(\partial\Omega)$ to $C^\infty(\partial\Omega)$.

Let c be the principal symbol of C, i.e. the matrix $(c_{j,\ell})_{j,\ell=0,\dots,m-1}$, where $c_{j,\ell}$ is the principal symbol (of degree $j-\ell$) of $C_{j,\ell}$. For $(x', \xi') \in T^*(\partial\Omega) \setminus 0$, $c(x', \xi')$ is a projector from $\mathbb{C}^m$, the image of which we shall make explicit; following transport by a local chart of $(\overline{\Omega}, \nu)$ we denote by $p(x', x_n; \xi', \xi_n)$ the principal symbol of P; we denote by $S^+(x', \xi')$ (resp : $S^-(x', \xi')$) the space of solutions U bounded for $x_n \geq 0$ (resp : $x_n \leq 0$) of the ordinary differential equation $p(x', 0; \xi', D_{x_n})U = 0$.

We put $\gamma U = (D^j U(0))_{j=0,\ldots,m-1}$ and

$$D^{\pm}(x', \xi') = \{ \gamma U \mid U \in S^{\pm}(x', \xi')\} .$$

From the ellipticity of P, the equation $p(x', 0; \xi', \xi_n) = 0$ has no real root ξ_n, and hence $C^m = (D^+(x', \xi') \oplus D^-(x', \xi')$ furthermore, the number of roots ξ_n, for which $\operatorname{Im} \xi_n > 0$, is locally constant and therefore independent of $(x', \xi') \in T^*(R^{n-1}) \setminus 0$ when $n \geq 3$; since this is changed into its complement in m when ξ' is replaced by $-\xi'$, we see that this number is equal to m/2 when $n \geq 3$. In the case n = 2, we hypothesise that this number is equal to m/2. We thus have $\dim D^{\pm}(x', \xi') = m/2$.

PROPOSITION 3.3 : *Using the preceding notation, we have* $\operatorname{Im} c(x', \xi') = D^+(x', \xi')$, *and thus* $\dim D^{\pm}(x', \xi') = m/2$.

PROOF: Let $U \in S^+(x', \xi')$ and U^o be its extension by 0 for $x_n < 0$; as in (1.3), (1.4), (1.5), we have the jump formula:

$$p(x', 0; \xi', D_{x_n})U^o = \frac{1}{i} \sum_{\ell+k+1 \leq m} p_{\ell+k+1}(x', 0; \xi')\, \gamma_\ell U \otimes D^k \delta$$

where p_j is the principal symbol of P_j; since $U^o \in \mathcal{S}'(R)$, we thereby deduce that

$$p(x', 0; \xi', \xi_n)\,\widehat{U^o}(\xi_n) = \frac{1}{i} \sum_{\ell+k+1\leq m} p_{\ell+k+1}(x', 0; \xi')\, \xi_n^k\, \gamma_\ell U$$

and, by a straightforward contour deformation, we have

$$(3.3.1.) \quad U(x_n) = \frac{1}{2i\pi} \int_\Gamma e^{ix_n\xi_n}\, p^{-1}(x', 0; \xi', \xi_n) \sum_{\ell+k+1\leq m} p_{\ell+k+1}(x', 0; \xi')\, \xi_n^k\, \gamma_\ell U \; d\xi_n \qquad \text{for } x_n \geq 0$$

where Γ is taken as in (3.2.2). We obtain

$$\gamma_j U = \frac{1}{2i\pi} \int_\Gamma p^{-1}(x', 0; \xi', \xi_n) \sum_{\ell+k+1\leq m} p_{\ell+k+1}(x', 0; \xi')\, \xi_n^{j+k}\, \gamma_\ell U \; d\xi_n$$

i.e. from (3.2.2), we have $\gamma U = c(x', \xi')\, \gamma U$, and therefore $\gamma U \in \operatorname{Im} c(x', \xi')$.

Conversely, let $w \in \mathbb{C}^m$ be such that $w \in \operatorname{Im} c(x', \xi')$. For $x_n \in \mathbb{R}$, we define $U(x_n)$ by the equality (3.3.1) with $\gamma_\ell U$ replaced by w_ℓ. Then $U \in S^+(x', \xi')$ and $\gamma U = c(x', \xi')w = w$, so that $w \in D^+(x', \xi')$.

REMARK 3.4 : The above proposition shows that the rank of $c(x', \xi')$ remains constant, and therefore that the subspaces $\operatorname{Im} c(x', \xi')$ are organised into a vector sub-bundle of rank $m/2$ of the trivial bundle $\mathbb{C}^m \times T^*(\partial\Omega) \setminus 0$ above

$T^*(\partial\Omega) \setminus 0$. We denote this sub-bundle by Im c.

4. APPLICATION TO ELLIPTIC BOUNDARY-VALUE PROBLEMS

Suppose we have

$$\wp : C^\infty(\bar{\Omega}) \longrightarrow C^\infty(\bar{\Omega}) \times C^\infty(\partial\Omega, \mathbb{C}^\mu)$$
$$u \longmapsto (Pu, B\gamma u) .$$

We denote by b the principal symbol of B, i.e. the matrix $(b_{j,k})$ where $b_{j,k}$ is the principal symbol (of degree d_j-k) of $B_{j,k}$. For $(x', \xi') \in T^*(\partial\Omega)\setminus 0$, $b(x', \xi')$ defines a linear mapping of $\mathbb{C}^m$ into $\mathbb{C}^\mu$ whose restriction to $\mathrm{Im}\, c(x', \xi')$ we denote by $\tilde{b}(x', \xi')$. We denote by $\tilde{b} : \mathrm{Im}\, c \longrightarrow \mathbb{C}^\mu \times T^*(\partial\Omega) \setminus 0$ the corresponding morphism of vector bundles on $T^*(\partial\Omega)\setminus 0$.

THEOREM 4.1 : *Assume that* $\tilde{b}$ *is surjective. Then* $\wp$ *admits a right parametrix* $\mathcal{Q}$. *The orthogonal of* $\mathrm{Im}\,\wp$ (*in* $\mathcal{E}'(\bar{\Omega}) \times \mathcal{E}'(\partial\Omega, \mathbb{C}^\mu)$) *is contained in* $C^\infty(\bar{\Omega}) \times C^\infty(\partial\Omega, \mathbb{C}^\mu)$. *If moreover* Ω *is bounded, then* $\mathrm{Im}\,\wp$ *is closed and of finite codimension.*

PROOF: BC is a $\mu \times m$ matrix of operators $(BC)_{j,k} \in L_c^{d_j-k}(\partial\Omega)$, and the surjectivity of its principal symbol is equivalent to that of $\tilde{b}$. BC is therefore right-elliptic in the sense of Agmon-Douglis-Nirenberg, thus there exists $A = (A_{k,j})$, with $A_{k,j} \in L_c^{k-d_j}(\partial\Omega)$ for $k = 0,\ldots, m-1$ and $j = 0,\ldots, \mu-1$, such that $BC.A = I + r$, $r \in L^{-\infty}(\partial\Omega, \mathbb{C}^\mu)$. We define $\mathcal{Q}$ by:

$$(4.1.1.) \quad \mathcal{Q}(f, g) = \{ Q(f^o) + Q\,\tilde{P}\,A(g - B\gamma[Q(f^o)]_{|\Omega}) \}_{|\Omega} .$$

$\mathcal{Q}$ is continuous from $C^\infty(\bar{\Omega}) \times C^\infty(\partial\Omega, \mathbb{C}^\mu)$ into $C^\infty(\bar{\Omega})$ in view of Theorems 2.4, 2.5, and it can easily be shown that $\rho\,\mathcal{Q} = I + \mathcal{R}$, with

$$(4.1.2.) \quad \mathcal{R}(f, g) = (\{ R\,\tilde{P}\,A(g - B\,\gamma[Q(f^o)]_{|\Omega} + R(f^o)\}_{|\Omega} \quad r(g - B\,\gamma[Q(f^o)]_{|\Omega}) \quad) .$$

Now the operator $f \longrightarrow \gamma[Q(f^o)]_{|\Omega}$ extends continuously from $\mathcal{D}'(\bar{\Omega})$ to $\mathcal{D}'(\partial\Omega, \mathbb{C}^m)$ in view of Corollary 2.8, since $\gamma^j \circ Q$ is of degree -1 at most for $0 \leq j \leq m-1$. Therefore its left composition with r B is continuous from $\mathcal{D}'(\bar{\Omega})$ into $C^\infty(\partial\Omega, \mathbb{C}^\mu)$ and its left composition with $R\,\tilde{P}\,A\,B$ is continuous from $\mathcal{D}'(\bar{\Omega})$ to $C^\infty(\mathbb{R}^n)$. The operator $f \to [R(f^o)]_{|\Omega}$ has its kernel in $C^\infty(\bar{\Omega} \times \bar{\Omega})$ and the other operators which appear in (4.1.2) are clearly regularising. $\mathcal{R}$ is therefore regularising. The orthogonal of $\mathrm{Im}\,\rho$ coincides with $\mathrm{Ker}\,{}^t\rho$; now ${}^tQ\,{}^t\rho = I + {}^t\mathcal{R}$, with ${}^t\mathcal{R}$ regularising; hence if (f, g) is in the orthogonal of $\mathrm{Im}\,\rho$, we have $(f, g) = -\,{}^t\mathcal{R}(f, g) \in C^\infty(\bar{\Omega}) \times C^\infty(\partial\Omega, \mathbb{C}^\mu)$. Finally, if Ω is bounded, we know that $\mathcal{R}$ is a compact operator from $C^\infty(\bar\Omega) \times C^\infty(\partial\Omega, \mathbb{C}^\mu)$. Thus $\mathrm{Im}(I + \mathcal{R})$ is closed and of finite codimension, and the same applies for $\mathrm{Im}\,\rho$ which contains $\mathrm{Im}\,(I + \mathcal{R})$.

REMARK 4.2 : We can show that ${}^t\rho$: $\mathcal{E}'(\bar\Omega) \times \mathcal{E}'(\partial\Omega, \mathbb{C}^\mu) \longrightarrow \mathcal{E}'(\bar\Omega)$ is given by ${}^t\rho(F, G) = {}^tPF + \sum_{j,k} {}^tB_{j,k}\, G_j \otimes D^k \delta$. Using the jump formula for ${}^tP$:

$$ {}^tP(f^\circ) = ({}^tPf)^\circ + \tilde\pi(\gamma f) $$

and we therefore deduce that $(f, g) \in C_0^\infty(\bar\Omega) \times C_0^\infty(\partial\Omega, \mathbb{C}^\mu)$ is in Ker ${}^t\rho$ if and only if

$$ (4.2.1.)\quad {}^tP\, f = 0 \ , \quad \tilde\pi(\gamma f) + \sum_{j,k} {}^tB_{j,k}\, g_j \otimes D^k \delta = 0 $$

We now proceed to the case of Sobolev spaces. We denote by m' the smallest integer $\leq m$ such that $B_{j,k} = 0$ for $k \geq m'$, $j = 0,\dots, \mu-1$. Then $B\,\gamma u$ is well defined when u has sectional traces on $\partial\Omega$ up to order $m' - 1$; from Theorem 2.9, this is the case, in particular, when u is an extendable distribution such that $Pu \in H^s_{loc}(\bar\Omega)$ with $s > m' - m - 1/2$.

Let $s, \sigma \in \mathbb{R}$ with $s > m' - m - 1/2$. We consider the operator (not necessarily bounded)

$$\mathcal{P}_{s,\sigma} \quad : \quad H^{\sigma}_{loc}(\bar{\Omega}) \longrightarrow H^{s}_{loc}(\bar{\Omega}) \times H^{\sigma-d-1/2}_{loc}(\partial\Omega)$$

$$u \longmapsto (Pu,\ B\,\gamma u)$$

in which we have put $H^{\sigma-d-1/2}_{loc}(\partial\Omega) = \prod_{j=0}^{\mu-1} H^{\sigma-d_j-1/2}_{loc}(\partial\Omega)$,

and where the domain of $\mathcal{P}_{s,\sigma}$ is

$$\mathcal{D}_{s,\sigma} = \{\ u \in H^{\sigma}_{loc}(\bar{\Omega}) \mid Pu \in H^{s}_{loc}(\bar{\Omega})\ ,\quad B\gamma u \in H^{\sigma-d-1/2}_{loc}(\partial\Omega)\ \}.$$

This domain is dense in $H^{\sigma}_{loc}(\bar{\Omega})$; using Theorem 2.9 and Remark 2.10, it can be shown that $\mathcal{P}_{s,\sigma}$ is closed.

THEOREM 4.3 : *Suppose that* $\tilde{b}$ *is surjective. Let* $s,\sigma \in \mathbb{R}$ *with* $s > m'-m-1/2$, $\sigma \leq s+m$. *Then* $\mathcal{P}_{s,\sigma}$ *admits a right quasi-inverse* $\mathcal{Q}_{s,\sigma}$ *in the sense that*

$$\mathcal{Q}_{s,\sigma} \quad : \quad H^{s}_{loc}(\bar{\Omega}) \times H^{\sigma-d-1/2}_{loc}(\partial\Omega) \longrightarrow H^{\sigma}_{loc}(\bar{\Omega})$$

is continuous, with $\mathcal{P}_{s,\sigma}\mathcal{Q}_{s,\sigma} = I + \mathcal{R}_{s,\sigma}$, $\mathcal{R}_{s,\sigma}$ *continuous from* $H^{s}_{loc}(\bar{\Omega}) \times H^{\sigma-d-1/2}_{loc}(\partial\Omega)$ *into* $C^{\infty}(\bar{\Omega}) \times C^{\infty}(\partial\Omega, \mathbb{C}^{\mu})$. *The orthogonal of* $\operatorname{Im}\mathcal{P}_{s,\sigma}$ (*in* $H^{-s}_{\bar{\Omega},comp}(\mathbb{R}^n) \times H^{-\sigma+d+1/2}_{comp}(\partial\Omega)$) *is the same as the orthogonal of* $\operatorname{Im}\mathcal{P}$. *If, furthermore,* Ω *is bounded, then* $\operatorname{Im}\mathcal{P}_{s,\sigma}$ *is closed and has finite codimension (equal to that of* $\operatorname{Im}\mathcal{P}$).

PROOF: Let $f \longmapsto \tilde{f}$ be an extension operator, continuous from $H^{s}_{loc}(\bar{\Omega})$ into $H^{s}_{loc}(\mathbb{R}^n)$; we define $\mathcal{Q}_{s,\sigma}(f, g)$

as being the right-hand side of formula (4.1.1) defining $\mathcal{Q}(f,g)$, but with f° replaced by $\tilde{f}$. The continuity of $\mathcal{Q}_{s,\sigma}$ follows from Proposition 3.2 and from the assumption $\sigma \leq s+m$. It can be shown that $P_{s,\sigma}\mathcal{Q}_{s,\sigma} = I + R_{s,\sigma}$, where $R_{s,\sigma}(f, g)$ is defined by the right-hand side of (4.1.2) with f° in this formula replaced by $\tilde{f}$; $R_{s,\sigma}$ obviously has values in $C^{\infty}(\bar{\Omega}) \times C^{\infty}(\partial\Omega, \mathbb{C}^{\mu})$. If (F,G) belongs to the domain $\mathcal{D}'_{s,\sigma}$ of ${}^tP_{s,\sigma}$, it follows immediately that ${}^tP_{s,\sigma}(F, G) = {}^tP(F, G)$. Since $\operatorname{Ker} {}^tP \subset C^{\infty}_0(\bar{\Omega}) \times C^{\infty}_0(\partial\Omega, \mathbb{C}^{\mu}) \subset \mathcal{D}'_{s,\sigma}$, we obtain $\operatorname{Ker} {}^tP_{s,\sigma} = \operatorname{Ker} {}^tP$, which gives the stated equality between the orthogonal of the images of $P_{s,\sigma}$, P. Finally, when Ω is bounded, $R_{s,\sigma}$ is a compact operator from $H^s(\bar{\Omega}) \times H^{\sigma-d-1/2}(\partial\Omega)$, and therefore $\operatorname{Im} P_{s,\sigma}$ is closed, with finite codimension.

THEOREM 4.4 (Elliptic Regularity): *Suppose that* $\tilde{b}$ *is injective. Let* $s,\sigma \in \mathbb{R}$ *with* $s > m'-m-1/2$, $\sigma \leq s+m$.

i) *Let* $u \in \mathcal{D}'(\Omega)$ *be extendable to* $\mathbb{R}^n$ *such that* $Pu \in H^s_{loc}(\bar{\Omega})$, $B\gamma u \in H^{\sigma-d-1/2}_{loc}(\partial\Omega)$. *Then* $u \in H^{\sigma}_{loc}(\bar{\Omega})$ *and* $\gamma_j u \in H^{\sigma-j-1/2}_{loc}(\partial\Omega)$ *for* $0 \leq j < s+m-1/2$.

ii) *Suppose* $\sigma' < \sigma$ *and let* K *be a compact subset of* $\bar{\Omega}$; *then there exists a constant* C *such that*

(4.4.1.) $$\|u\|_{\sigma} + \sum_{0\leq j<s+m-1/2} \|\gamma_j u\|_{\sigma-j-1/2}$$

$$\leq C\,(\|Pu\|_s + \|B\gamma u\|_{\sigma-d-1/2} + \|u\|_{\sigma'})$$

for all u *with support in* K *satisfying the assumptions of* (i).

iii) *We have* $\operatorname{Ker} \rho_{s,\sigma} = \operatorname{Ker} \rho \subset C^\infty(\overline{\Omega})$.

iv) *If, furthermore,* Ω *is bounded, then* $\operatorname{Ker} \rho_{s,\sigma} = \operatorname{Ker} \rho$ *is of finite dimension.*

PROOF: For (i), we put Pu = f and $v = u - (Q\tilde{f})_{|\Omega}$; we have Pv = φ , with $\varphi = -(R\tilde{f})_{|\Omega}$. In view of Theorem 2.9, v has sectional traces of all orders on $\partial\Omega$, and we have $P(v^0) = \varphi^0 + \tilde{P}(\gamma v)$. Applying Q on the left to this equality, we obtain

$$v = [Q(\varphi^0)]_{|\Omega} + (Q\,\tilde{P}\,\gamma v)_{|\Omega} - [R'(v^0)]_{|\Omega} \tag{4.4.2.}$$

so that

$$(I - C)\gamma v = \gamma\,[Q(\varphi^0)]_{|\Omega} - \gamma\,R'(v^0) \in C^\infty(\partial\Omega, \mathbb{C}^m) ,$$

and therefore

$$\begin{pmatrix} I - C \\ B \end{pmatrix} \gamma v \in C^\infty(\partial\Omega, \mathbb{C}^m) \oplus H^{\sigma-d-1/2}_{\mathrm{loc}}(\partial\Omega) .$$

Now the injectivity of the principal symbol of $\begin{pmatrix} I - C \\ B \end{pmatrix}$ is equivalent to that of $\tilde{b}$; therefore $\begin{pmatrix} I - C \\ B \end{pmatrix}$ is left-elliptic in the sense of Agmon-Douglis-Nirenberg, so that

$$\gamma v \in \prod_{j=0}^{m-1} H^{\sigma-j-1/2}_{\mathrm{loc}}(\partial\Omega) .$$

From (4.4.2) and from Proposition 3.2, we deduce that

$v \in H^{\sigma}_{loc}(\overline{\Omega})$, which then gives the stated regularity properties for u and $\gamma_j u$ since $u - v = (Q\tilde{f})_{|\Omega} \in H^{s+m}_{loc}(\overline{\Omega})$.

We prove (ii) in the same way by using the elliptic inequality for $\begin{pmatrix} I - C \\ B \end{pmatrix}$. (iii) follows from (i). If Ω is bounded, (4.4.1) shows that the topologies induced on Ker $\mathcal{P}$ by $H^{\sigma}(\overline{\Omega})$ and $H^{\sigma'}(\overline{\Omega})$ are coincident, which gives (iv).

4.5 LOPATINSKI'S DETERMINANT AND CONDITION: The boundary-value problem (1.1) is said to be elliptic if $\tilde{b}$ is bijective. The number μ of scalar boundary conditions must then be equal to m/2, and the preceding theorems show that $\mathcal{P}$, $\mathcal{P}_{s,\sigma}$ are indexed operators (with the same index) when Ω is bounded. We conclude this section by giving invertibility criteria for $\tilde{b}$ in the case $\mu = m/2$. Following transport by a local chart of $(\overline{\Omega}, \nu)$, we denote by $p(x, \xi)$, $b_{j,k}(x', \xi')$ the principal symbols of P, $B_{j,k}$ respectively, and we put

$$b_j(x', \xi) = \sum_{k=0}^{m-1} b_{j,k}(x', \xi')\, \xi_n^k \qquad (0 \le j \le \mu-1) .$$

We factorise $p(x', 0, \xi)$ into

$p(x', 0; \xi', \xi_n) = p^+(x', \xi', \xi_n)\, p^-(x', \xi', \xi_n)$ where $p^{\pm}(x'; \xi', \xi_n)$ are polynomials in ξ_n whose zeros ξ_n are such that $\pm \operatorname{Im} \xi_n > 0$; $p^{\pm}$ is of degree m/2 in ξ_n. Let

$$b_j(x'; \xi', \xi_n) = \sum_{k=0}^{\mu-1} b'_{j,k}(x', \xi')\, \xi_n^k$$ denote the remainder

of the Euclidian division of $b_j(x'; \xi', \xi_n)$ by $p^+(x'; \xi_n)$
From Proposition 3.3, $\tilde{b}(x', \xi')$ is invertible if and only if the only bounded solution U, for $x_n \geq 0$, of

$$\begin{cases} p(x', 0; \xi', D_{x_n})\, U(x_n) = 0 \\ (b_j(x'; \xi', D_{x_n})U)_{|x_n=0} = 0 \quad (j = 0, \ldots, \mu-1) \end{cases}$$

is U = 0.

Since the bounded solutions U, for $x_n \geq 0$ of the equation $p(x', 0; \xi', D_{x_n})\, U(x_n) = 0$ are the solutions U of the equation $p^+(x'; \xi', D_{x_n})\, U(x_n) = 0$, and since

$b_j(x'; \xi', D_{x_n})U = b'_j(x'; \xi', D_{x_n})U$ when

$p^+(x'; \xi', D_{x_n})U = 0$, we see that the invertibility of $\tilde{b}(x', \xi')$ is equivalent to that of the matrix

$(b'_{j,k}(x', \xi'))_{j,k=0,\ldots,\mu-1}$. The determinant of this matrix is called the Lopatinski determinant of (1.1) relative to the local chart considered. We say that (1.1) satisfies the Lopatinski condition in (x', ξ') if this determinant is non-zero; this condition is equivalent to the invertibility of $\tilde{b}(x', \xi')$.

5. EXAMPLES

The Dirichlet problem in Ω for the elliptic differential operator P of degree m is

$$(5.1.)\quad \begin{cases} Pu = f & \text{in } \Omega \\ \gamma_j u = g_j & \text{in } \partial\Omega \quad (j = 0,\ldots, m/2-1)\ . \end{cases}$$

We thus have in this case $\mu = m/2$, $B_{j,k} = \delta_{j,k} I$, $m' = m/2$.

THEOREM 5.2: *The Dirichlet problem* (5.1) *is elliptic. The corresponding operators* $\mathcal{P}$, $\mathcal{P}_{s,\sigma}$ *(with* $s > -\frac{m+1}{2}$, $\sigma \le s+m$) *are therefore indexed (with the same index) when* Ω *is bounded; this index is zero when* P *is formally self-adjoint or strongly elliptic.*

PROOF: Using the notation of 4.5 , we have $b_j(x', \xi) = \xi_n^j = b'_j(x', \xi)$ for $j = 0,\ldots, \mu-1$ (putting $\mu = m/2$); therefore $b'_{j,k}(x', \xi') = \delta_{j,k}$ for $j,k = 0,\ldots, \mu-1$, and this shows that (5.1) is elliptic. In view of Remark 4.2, $\operatorname{Ker} {}^t\mathcal{P}$ is the set of the $(f, g) \in C^\infty(\overline{\Omega}) \times C^\infty(\partial\Omega, \mathbb{C}^\mu)$ such that

$$(5.2.1.)\qquad {}^tP f = 0\ , \qquad \tilde{\pi}(\gamma f) + \sum_{j=0}^{\mu-1} g_j \otimes D^j \delta = 0\ .$$

Now, we have locally

$$\tilde{\pi}(\gamma f) = \frac{1}{i} \sum_{\ell+k+1\le m} \pi_{\ell+k+1}(x, D_{x'})\gamma_\ell u \otimes D^k \delta \text{ with}$$

$\pi_m = \pi_m(x) \neq 0$ since ${}^tP$ is elliptic. We therefore deduce that (5.2.1) is equivalent to

$${}^tPf = 0 \quad , \quad \gamma_j f = 0 \text{ for } 0 \le j \le \mu-1 \quad ,$$

g_j being equal to the coefficient of $D^j \delta$ in $-\tilde{\pi}(\gamma f)$ and it therefore follows that

$$\text{(5.2.2.)} \qquad \operatorname{Ker} {}^t\mathcal{P} \approx \{ f \in C^\infty(\bar{\Omega}) | {}^tPf = 0 , \gamma_j f = 0 \text{ for } 0 \le j \le \mu-1 \}$$

$$\approx \{ f \in C^\infty(\bar{\Omega}) | P^* f = 0 , \gamma_j f = 0 \text{ for } 0 \le j \le \mu-1 \}$$

In particular $\operatorname{Ker} {}^t\mathcal{P} \approx \operatorname{Ker} \mathcal{P}$ if $P = P^*$, so that $\mathcal{P}$ then has zero index in the case where Ω is bounded. We now suppose that P is strongly elliptic and that Ω is bounded. Gårding's inequality ensures the existence of $c > 0$ and $\lambda \in \mathbb{R}$ such that

$$\text{(5.2.3.)} \qquad \operatorname{Re}(Pu, u) + \lambda(u, u) \ge c \|u\|_\mu^2 \quad \text{for all} \quad u \in H_0^\mu(\Omega) \quad .$$

We therefore deduce that $\operatorname{Ker} \mathcal{P}' = \{ 0 \}$, where $\mathcal{P}'$ is associated with $P' = P + \lambda$. In view of (5.2.2), we similarly have $\operatorname{Ker} {}^t\mathcal{P}' = \{ 0 \}$, and hence the index of $\mathcal{P}'$ is equal to zero. Let $s > -\frac{m+1}{2}$ and $\sigma = s+m$; then $\mathcal{P}'_{s,\sigma} - \mathcal{P}_{s,\sigma}$ is the operator

$$H^{s+m}(\bar{\Omega}) \longrightarrow H^s(\bar{\Omega}) \times H^{\sigma-d-1/2}(\partial\Omega)$$

$$f \longmapsto (\lambda f, 0) \qquad .$$

As this operator is compact, we have:

$$\operatorname{Index} \mathcal{P} = \operatorname{Index} \mathcal{P}_{s,\sigma} = \operatorname{Index} \mathcal{P}'_{s,\sigma} = \operatorname{Index} \mathcal{P}' = 0 \quad .$$

REMARK 5.3: In the case in which $P = -\Delta$ and Ω is bounded, we have Green's formula

$$\text{(5.3.1.)} \int_\Omega -\Delta u . v \, dx = \int_\Omega \operatorname{grad} u . \operatorname{grad} v \, dx - \int_{\partial\Omega} \frac{\partial u}{\partial N} . v \, d\sigma ,$$

where N is the inward unit normal. Gårding's inequality (5.2.3) thus holds for $\lambda = 0$. Therefore

$$\rho : C^\infty(\bar\Omega) \longrightarrow C^\infty(\bar\Omega) \times C^\infty(\partial\Omega)$$

$$u \longmapsto (-\Delta u, \gamma_0 u) \quad \text{is an isomorphism,}$$

and

$$\rho_{s,\sigma} : \mathcal{D}_{s,\sigma} \subset H^\sigma(\bar\Omega) \longrightarrow H^s(\bar\Omega) \times H^{\sigma-1/2}(\partial\Omega)$$

$$u \longmapsto (-\Delta u, \gamma_0 u)$$

is bijective, with continuous inverse (for $s > -3/2$, $\sigma \leq s+2$) .

Now let P be elliptic of degree 2; we consider the problem

$$(5.4.) \quad \begin{cases} Pu = f & \text{in } \Omega \\ \gamma_1 u = g & \text{in } \partial\Omega . \end{cases}$$

In this case $\mu = m/2 = 1$, $m' = m$ and, using the notation of 4.5 , $b(x', \xi) = \xi_n$, $b'(x', \xi') = \xi_n^+(x', \xi')$, where $\xi_n^+(x', \xi')$ denotes the root ξ_n of $p(x', 0; \xi', \xi_n) = 0$ with positive imaginary part. Hence (5.4) is elliptic; in the particular case of the Neumann problem, i.e. of problem (5.4) with $P = -\Delta$, $\nu = N$, Green's formula (5.3.1) shows that $\mathrm{Ker}\,\rho$ is constituted by locally constant functions in Ω, and also that $\mathrm{Ker}\,{}^t\rho$ is constituted by the $(f, g) \in C_0^\infty(\bar\Omega) \times C_0^\infty(\partial\Omega)$ such that $\Delta f = 0$, $\gamma_1 f = 0$, $g = -\gamma_0 f$, i.e. such that f is locally constant in Ω , $g = -f_{|\partial\Omega}$.

We now return to the Dirichlet problem for an elliptic operator

P of degree m; suppose we have

$$\mathbb{P} : D \subset L^2(\Omega) \longrightarrow L^2(\Omega)$$
$$u \longmapsto Pu \quad ,$$

where the domain D of $\mathbb{P}$ is $D = H^m(\Omega) \cap H^{\mu}_0(\Omega)$, with $\mu = m/2$. For this boundary-value problem, we have the analogue of Theorem 8.14, Chapter IV:

THEOREM 5.5 : *Suppose that Ω is bounded.*

(i) *$\mathbb{P}$ is a closed operator with dense domain. Its adjoint $\mathbb{P}^*$ has the same domain D as $\mathbb{P}$ and is defined by $\mathbb{P}^* u = P^* u$ for $u \in D$. $\mathbb{P}$ is self-adjoint if and only if P is formally self-adjoint.*

(ii) *$\mathbb{P}$ is an indexed operator.*

PROOF: From (4.4.1), there exists C such that

$$\|u\|_m \leq C\left(\|Pu\|_0 + \sum_{j=0}^{\mu-1} \|\gamma_j u\|_{m-j-1/2} + \|u\|_0\right) \text{ for } u \in H^m(\Omega).$$

We therefore deduce, using Remark 2.10, that $\mathbb{P}$ is closed. Theorems 4.3, 4.4 show that Ker $\mathbb{P}$ is of finite dimension and contained in $C^\infty(\bar{\Omega})$, and that Im $\mathbb{P}$ is closed. We thus have $\text{Im}\,\mathbb{P} = (\text{Ker}\,\mathbb{P}^*)^\perp$. We now investigate $\mathbb{P}^*$; its domain D^* is the set of the $v \in L^2(\Omega)$ such that (Pu, v) depends continuously on $u \in D$ for the topology induced by $L^2(\Omega)$ and, for $v \in D^*$, $\mathbb{P}^* v$ is characterised by: $(Pu, v) = (u, \mathbb{P}^* v)$ for $u \in D$. Taking, in particular, $u \in C_0^\infty(\Omega)$, it follows that, for $v \in D^*$, we then have $P^* v = \mathbf{P}^* v \in L^2(\Omega)$, and therefore, from Theorem 2.9, that v has sectional traces on $\partial\Omega$ up to order m-1; it can easily be shown that we still have the jump formula $P^*(v^\circ) = (P^* v)^\circ + \tilde{P}(\gamma v)$ for the operator P^*.

Furthermore, for $u \in H^m(\Omega)$ and $v \in L^2(\Omega)$, we have $(Pu, v) = (u, P^*(v^\circ))$, where the bracket on the right-hand side is that of the duality between $H^m(\Omega)$ and $H^{-m}_{\bar{\Omega}}(\mathbb{R}^n)$. Thus, for $u \in H^m(\Omega)$ and $v \in D^*$, we have:

$$(Pu, v) = (u, P^*v) + (u, \tilde{\pi}(\gamma v)) ,$$

which shows that $v \in L^2(\Omega)$, $(u, \tilde{\pi}(\gamma v)) = 0$ for $u \in D$. It can be shown that this last condition is equivalent to $\gamma_j v = 0$ for $0 \leq j \leq \mu-1$, and hence that $D^* = \{v \in L^2(\Omega) \mid P^*v \in L^2(\Omega) , \gamma_j v = 0$ for $0 < j < \mu - 1\} = D$ in view of Theorem 4.4.

6. ADDITIONAL NOTES AND EXERCISES

6.1 EXAMPLE OF A PSEUDO-DIFFERENTIAL OPERATOR WHICH DOES NOT SATISFY THE REGULARITY PROPERTY OF THEOREM 2.5

Let $\alpha > -1$; let x_+^α denote the function on $\mathbb{R}$ which is equal to x^α for $x > 0$ and equal to 0 for $x \leq 0$. Show that the operator A of convolution with x_+^α is in $L^\mu(\mathbb{R})$ with $\mu = -1-\alpha$ (see Exercise 11.4, Chapter IV).

Let $f \in C_0^\infty(\overline{\mathbb{R}}^+)$ such that $f(x) = 1$ for $0 \leq x \leq 1$. Calculate the restriction of $A(f^0)$ to $]0,1[$, and deduce from this that $A(f^0) \notin C^\infty(\overline{\mathbb{R}}^+)$ when $\alpha \notin \mathbb{N}$.

6.2 CAUCHY'S FORMULA AND THE HILBERT OPERATOR

a) Let Ω be the open unit disc in $\mathbb{R}^2$, and let $P = \frac{1}{2}\left(\frac{\partial}{\partial x} + i\frac{\partial}{\partial y}\right)$ be the Cauchy-Riemann operator in $\mathbb{R}^2$. We recall (see 10.4, Chapter I) that $E = \frac{1}{\pi z}$ (with $z = x + iy$) is a fundamental solution of P. We adopt the notation of Section 1.

Let $v \in C^\infty(\partial\Omega)$. Show that the function $V = (E * Pv)_{|\Omega}$ is given by

$$V(z) = \frac{1}{2i\pi}\int_{\partial\Omega} \frac{v(t)}{t-z}\,dt \qquad (z \in \Omega).$$

Let $z_0 \in \partial\Omega$. Show that

$$V(z) = v(z_0) + \frac{1}{2i\pi}\int_{\partial\Omega} \frac{v(t)-v(z_0)}{t-z}\,dt$$

$$= v(z_0) + \frac{1}{2i\pi}\lim_{\epsilon\to 0}\int_{C_\epsilon} \frac{v(t)-v(z_0)}{t-z}\,dt,$$

where C_ε is obtained by excluding from $\partial\Omega$ the arc with centre z_0 and with length 2ε. Deduce that the Calderón projector

$$C : C^\infty(\partial\Omega) \longrightarrow C^\infty(\partial\Omega)$$

$$v \longmapsto \gamma[(E * \widetilde{P}v)_{|\Omega}]$$

is given by $C = \frac{1}{2}(I + H)$, with

$$(H\,v)(z) = \frac{1}{i\pi}\ \text{p.v.} \int_{\partial\Omega} \frac{v(t)}{t - z}\,dt \qquad (z \in \partial\Omega) ,$$

where we have put

$$\text{p.v.} \int_{\partial\Omega} \frac{v(t)}{t - z_0}\,dt = \lim_{\epsilon \to 0_+} \int_{C_\epsilon} \frac{v(t)}{t - z_0}\,dt .$$

Deduce that the boundary value problem

$$\begin{cases} u \in C^\infty(\overline{\Omega}) \\ u \quad \text{holomorphic in} \quad \Omega \\ u|_{\partial\Omega} = v \end{cases}$$

has a solution u if and only if the data v satisfies the equation $v = Hv$, and that u is then unique and given by the Cauchy formula

$$u(z) = \frac{1}{2i\pi} \int_{\partial\Omega} \frac{v(t)}{t - z}\,dt \quad (z \in \Omega) .$$

The operator $H : C^\infty(\partial\Omega) \to C^\infty(\partial\Omega)$ is termed the Hilbert operator (for the unit circle). Show that $H^2 = I$. Show that when we set $z = e^{i\theta}(z \in \partial\Omega\ , \theta \in \mathbb{R})$, H becomes identical to the operator of convolution (in the sense of periodic distributions with period 2π) with the distribution $\mathcal{H} = \frac{1}{2\pi} - \frac{1}{2i\pi}\ \text{p.v.}\ \cot\frac{\theta}{2}$. Calculate the Fourier coefficients of $\mathcal{H}$ and show that H extends to a unitary operator in $L^2(\partial\Omega)$.

6.3 WEAKENING OF ASSUMPTION (2.1)

By appropriate choice of the contour $\Gamma_{\xi'}$ in the equality (2.2.3), show that all the results of Section 2 still hold if assumption (2.1) is replaced by: A is proper, and each term of the complete symbol $\sum_{j \geq 0} a_j(x, \xi)$ of A is a rational function with respect to ξ in a symmetric conic neighbourhood (i.e. invariant under homothetic transformations with non-zero real ratio) of the fibred space conormal to $\partial\Omega$.

6.4 LOCAL REGULARITY IN THE NEIGHBOURHOOD OF A POINT OF $\partial\Omega$.

We adopt the notation of Section 4. Let $x_0 \in \partial\Omega$. Suppose that $\tilde{b}(x, \xi')$ is injective for all $\xi' \in T^*_{x_0}(\partial\Omega) \setminus \{0\}$, and that $u \in \mathcal{D}'(\Omega)$, extendable, such that $Pu \in H^s_{loc}(\bar{\Omega})$ with $s > m' - m - \frac{1}{2}$. Assume that in some neighbourhood of x_0 (respectively in $\bar{\Omega}$ and in $\partial\Omega$) we have $Pu \in H^{s'}$ and $B\gamma u \in H^{\sigma'-d-\frac{1}{2}}$ with $\sigma' \leq s' + m$. By repeating the proof of part (i) of Theorem 4.4, prove that $u \in H^{\sigma'}$ and that $\gamma_j u \in H^{\sigma'-j-\frac{1}{2}}$ in some neighbourhood of x_0 for $0 \leq j < s+m-\frac{1}{2}$.

6.5 GLOBAL REGULARITY

We adopt the notation of Section 4, assuming Ω to be unbounded, and $\partial\Omega$ bounded. Suppose that the differential operator P is elliptic in $BL^m(\mathbb{R}^n)$. Show (cf Section 11.1, Chapter IV) that this is equivalent to the following conditions: the coefficients of P, together with all their derivatives, are bounded in $\mathbb{R}^n$; moreover, there exists $c > 0$ such that $|p_m(x,\xi)| \geq C$ for $x \in \mathbb{R}^n$, $\xi \in \mathbb{R}^n$, $|\xi| = 1$, where p_m is the homogeneous principal symbol of degree m of P. Suppose that $\tilde{b}$ is injective, and that we have $u \in H^{-\infty}(\bar{\Omega})$ such that $Pu \in H^s(\bar{\Omega})$, $B\gamma u \in H^{\sigma-d-\frac{1}{2}}(\partial\Omega)$, with $s > m' - m - \frac{1}{2}$, $\sigma \leq s + m$. Using Theorem 4.4 and a parametrix of P in $BL^{-m}(\mathbb{R}^n)$, show that $u \in H^\sigma(\bar{\Omega})$ and that an a priori estimate of the type (4.4.1) is valid for all u satisfying the above assumptions. Deduce that part (i) of Theorem 5.5 remains true for P and Ω.

6.6 CONVERSE OF THE REGULARITY THEOREM

Assume that the boundary-value problem (1.1) satisfies part (ii) of the statement of Theorem 4.4 (a priori estimate). By applying the inequality (4.4.1) for $K \subset \Omega$ and by using Proposition 8.13, Chapter IV, show that P is elliptic in $\bar{\Omega}$

By applying the inequality (4.4.1) for $u = (QPv)_{|\Omega}$, where $v \in \mathcal{D}_{K'}(\partial\Omega, \underset{m}{\oplus} E)$, show that:

$$\sum_{k=0}^{m-1} |v_k| \leq C\left(|BCv| + \sum_{k=0}^{m-1} |(I - C)v| + \sum_{k=0}^{m-1} |v_k|\right)$$

for all $v \in \mathcal{D}_{K'}(\partial\Omega, \underset{m}{\oplus} E)$. Then deduce that the operator

$$\begin{pmatrix} I - C \\ B \cdot C \end{pmatrix}$$ is left-elliptic, and that $\tilde{b}$ is therefore

injective.

6.7 CASE OF SYSTEMS WHICH ARE ELLIPTIC IN THE SENSE OF DOUGLIS-NIRENBERG

Suppose $\bar{\Omega}$ is a manifold with boundary, imbedded in a manifold M (see Remark 9.6, Chapter I). Let

$E = \bigoplus_{j=1}^{J} E_j$, $F = \bigoplus_{i=1}^{I} F_i$, where the E_j, F_i are vectorial bundles above M.

Suppose we have an elliptic differential operator P: $C^\infty(M, E) \to C^\infty(M, F)$ of type $(s, t) \in Z^I \times Z^J$, i.e. (see Section 7.6, Chapter IV) an operator P of the form

$(P_{ij})_{\substack{i=1,\dots,I \\ j=1,\dots,J}}$, where $P_{i,j} : C^\infty(M, E_j) \to C^\infty(M, F_i)$ is a differential operator of degree $t_j - s_i$, with $\sigma_{s,t}(P) = (\sigma_{t_j - s_i}(P_{i,j}))$ bijective.

Let $G = \bigoplus_{j=1}^{d} G_j$, where G_j are vectorial bundles above $\partial\Omega$. Let $\mathcal{B} : C^\infty(M, E) \to C^\infty(M, G)$ be a differential operator of type $(r, t) \in \mathbb{R}^d \times \mathbb{R}^J$.

Consider the boundary-value problem

$$\begin{cases} Pu = f & \text{in } \Omega \\ (\mathcal{B}u)_{|\partial\Omega} = g & \text{in } \partial\Omega \end{cases} .$$

In order to allow this problem to be studied by means of a formalism analogous to that in Chapter V, it is convenient to introduce a real field ν transversal to $\partial\Omega$, and to put

$$\gamma_j u = \left(\left(\frac{1}{i}\frac{\partial}{\partial\nu}\right)^j u\right)_{|\partial\Omega} \quad \text{for} \quad u \in C^\infty(\overline{\Omega}, E) ,$$

and $\gamma u = (\gamma_0 u, \ldots, \gamma_{m-1}u) \in C^\infty(\partial\Omega, \underset{m}{\oplus} E)$, where m is an integer large enough to ensure that the following two conditions are satisfied:

(6.7.1.) the boundary condition $(\mathcal{B}u)_{|\partial\Omega} = g$ is expressed in the form $B(\gamma u) = g$, with $B = (B_0, B_1 \ldots B_{m-1})$, where $B_k : C^\infty(\partial\Omega, E) \to C^\infty(\partial\Omega, G)$ is a differential operator of type $(r , t-k)$ for $k = 0, \ldots, m-1$ (we put $(t-k)_j = t_j-k$ for $j = 1,\ldots,J$).

(6.7.2.) The jump formula for P does not involve any terms $\gamma_j u_k$ such that $j \geq m$. We can therefore write this formula in the form

$$P(u^0) = (Pu)^0 + \tilde{P}(\gamma u) \quad \text{for} \quad u \in C^\infty(\overline{\Omega}, E) ,$$

where $\tilde{P}(\gamma u)$ is a distribution carried by $\partial\Omega$ which depends only on P and γu.

It is then a simple matter to adapt Sections 3 and 4, taking account of the following remarks:

a) We take a parametrix Q of P such that $Q = (Q_{j,k})$ is of type (t,s) and each operator constituting $Q_{j,k}$ satisfies (2.1). The existence of such a parametrix is immediate: we take Q_o of type (t,s) having as principal symbol the inverse matrix of $\sigma_{s,t}(P)$; putting, for example, $PQ_o = I - R$, we have R of degree -1, and we take $Q \sim \sum_{k \geq 0} Q_o R^k$.

b) We define the Calderón projector $C = \gamma Q \widetilde{P}$. We have $C : C^{\infty}(\partial\Omega , \underset{m}{\oplus} E) \to C^{\infty}(\partial\Omega , \underset{m}{\oplus} E)$, $C = (C_{k,\ell})$, $C_{k,\ell}$ a pseudo-differential operator of type $(t - k, t - \ell)$.

Let $\sigma_o(C) = (\sigma_{t-k,t-\ell} \quad (C_{k,\ell}))_{k,\ell=0,\ldots,m-1}$ be the principal symbol of C . It can be shown, following transport by a local chart to the boundary of $(\bar{\Omega}, \nu)$ and trivialisations of E, F , that $\sigma_o(x' , \xi')$ is a projector, and that the mapping $U \to \gamma U$ is an isomorphism of $S^+(x', \xi')$ onto the image of $\sigma_o(C)(x', \xi')$, where $S^+(x', \xi')$ is the space of solutions U, bounded for $x_n \geq 0$, of the differential system $\sigma_{s,t}(P)(x' , 0 ; \xi' , D_{x_n}) U(x_n) = 0.$ We then deduce that the image of $\sigma_o(C)(x' , \xi')$ does not depend on the choice (to within an isomorphism) of the integer m, and also (see for example INCE [1]) that the rank of $\sigma_o(C)(x', \xi')$ is equal to one half of the degree in ξ_n of $\det \sigma_{s,t}(P)(x', 0 ; \xi' , \xi_n)$, i.e. to $\frac{1}{2} (\sum_{j=1}^{J} t_j \dim E_j - s_j \dim F_j)$, with, in the case n=2, a restriction analogous to that indicated prior to the statement of Proposition 3.3.

c) Consider the principal symbol b of B:

$b = (\sigma_{r,t-k}(B_k))_{k=0,\ldots,m-1}$, and the restriction $\tilde{b}$ of b to Image $\sigma_o(C)$. Thus $\tilde{b}$ is a morphism of Image $\sigma_o(C)$ into G. It can be shown that the property of $\tilde{b}$ being injective or surjective depends neither on the integer m nor on the field ν chosen.

d) For $j = 1,\ldots,J$ let m_j' denote the smallest integer for which $(\mathcal{B}u)_{|\partial\Omega}$ can be expressed by means of the traces $\gamma_k u_j$ such that $j = 1,\ldots,J$ and $0 \le k \le m_j' - 1$. Under the assumption $m_j' - \frac{1}{2} < t_j - s_i$ for $i = 1,\ldots,I$ and $j = 1,\ldots,J$, generalise Theorem 4.1 to the case of the operator

$$\mathcal{P} : C^\infty(\overline{\Omega}, E) \longrightarrow C^\infty(\overline{\Omega}, F) \times C^\infty(\partial\Omega, G)$$

$$u \longmapsto (Pu, (\mathcal{B}u)_{|\partial\Omega}) .$$

e) Put for example $H^s_{loc}(\overline{\Omega}, F) = \prod_{i=1}^{I} H^{s_i}_{loc}(\overline{\Omega}, F)$.
Suppose that $m_j' - \frac{1}{2} < t_j$ for $j = 1,\ldots,J$. If $u \in \mathcal{D}'(\Omega, E)$ is extendable and satisfies $P u \in H^s_{loc}(\overline{\Omega}, F)$, show that $(\beta u)_{|\partial\Omega}$ is meaningful. For $\rho \le 0$, consider the operator

$$\mathcal{P}_{s,t,\rho} : H^{t+\rho}_{loc}(\overline{\Omega}, E) \longrightarrow H^s_{loc}(\overline{\Omega}, F) \times H^{r+\rho-\frac{1}{2}}_{loc}(\partial\Omega, G)$$

$$u \longmapsto (Pu, (\mathcal{B}u)_{|\partial\Omega})$$

with domain

$$\mathcal{D}_{s,t,\rho} = \{u \in H^{t+\rho}(\bar{\Omega}, E) \mid Pu \in H^s_{loc}(\bar{\Omega}, F),$$
$$(\mathcal{B}u)_{|\partial\Omega} \in H^{r+\rho-\frac{1}{2}}_{loc}(\partial\Omega, G)\}.$$

Generalise Theorem 4.3 to the case of the operator $\mathcal{P}_{s,t,\rho}$. Generalise Theorem 4.4 similarly (in particular, show that if $u \in \mathcal{D}'(\Omega, E)$ is extendable and satisfies $Pu \in H^s_{loc}(\bar{\Omega}, F)$, $(\mathcal{B}u)_{|\partial\Omega} \in H^{r+\rho-\frac{1}{2}}_{loc}(\partial\Omega, G)$, then $u \in H^{t+\rho}_{loc}(\bar{\Omega}, F)$ and $\gamma_k u_j \in H^{t_j+\rho-k-\frac{1}{2}}(\partial\Omega, G)$ for $0 \le k < t_j - \frac{1}{2}$, $j = 1, \ldots, J$).

f) Give properties analogous to those established in Exercises 6.4, 6.5, 6.6.

6.8 THE OBLIQUE DERIVATIVE PROBLEM

We consider a regular open domain Ω of $\mathbb{R}^n$, and we denote by N the inward unit normal to $\partial\Omega$. We prescribe on $\partial\Omega$ a real vector field $x' \longmapsto \nu(x') \in \mathbb{R}^n$, of class C^∞, such that $\nu(x') \neq 0$ for all $x' \in \partial\Omega$. We put $\nu(x') = \tau(x') + \lambda(x')N$, where $\tau(x') \in T_{x'}(\partial\Omega)$ is the tangential component of $\nu(x')$ and $\lambda(x') \in \mathbb{R}$ is the normal component of $\nu(x')$. We study the oblique derivative problem

$$\text{(6.8.1)} \qquad \begin{cases} \Delta u = f & \text{in } \Omega \\ \dfrac{1}{i}\dfrac{\partial u}{\partial \nu} = g & \text{in } \partial\Omega \end{cases}$$

a) Let $x'_0 \in \partial\Omega$. Suppose we have a local chart $x' \to y' = \chi(x')$ from $\partial\Omega$ to the neighbourhood of x'_0, with $\chi(x'_0) = 0$. Show that the mapping $y = (y', y_n) \mapsto x = x' + y_n N(x')$, where $x' = \chi^{-1}(y')$, is a diffeomorphism of a neighbourhood of 0 into $\mathbb{R}^n$, and that its inverse is a local chart $\tilde{\chi}$ from $\bar{\Omega}$ into x'_0 transporting N into $\frac{\partial}{\partial y_n}$. Show that the principal symbols of the transportations under $\tilde{\chi}$ of the operators Δ and $\frac{1}{i}\frac{\partial}{\partial\nu}$ are respectively $|{}^t\chi'(x')\eta'|^2 + \eta_n^2$ and $< \chi'(x')\,\tau(x')\,,\ \eta' > + \lambda(x')\,\eta_n$. Show that the Lopatinski determinant of (6.8.1) in the chart $\tilde{\chi}$ is equal to

$$< \chi'(x')\tau(x')\,,\ \eta' > + i\lambda(x')\,|{}^t\chi'(x')\eta'|$$

b) Deduce from this that the boundary-value problem (6.8.1) is always elliptic when n = 2, and that for $n \geq 3$ it is elliptic if and only if the vector $\nu(x')$ is nowhere tangential to $\partial\Omega$.

c) Suppose that Ω is bounded; let $u = Kh$ denote the unique solution of the Dirichlet problem

$$\begin{cases} \Delta u = 0 & \text{in} \quad \Omega \\ u_{|\partial\Omega} = h & \text{in} \quad \partial\Omega . \end{cases}$$

Show that u is a solution of the oblique derivative

problem

$$\begin{cases} \Delta u = 0 & \text{in} \quad \Omega \\ \frac{1}{i}\frac{\partial u}{\partial \nu} = g & \text{in} \quad \partial\Omega \end{cases}$$

if and only if

$$\begin{cases} u = K\,h \\ Th = g & \text{in} \quad \partial\Omega \end{cases}$$

where $T = \frac{1}{i}\frac{\partial}{\partial \nu} \circ K$ is a (scalar) p.d.o. of degree 1 in $\partial\Omega$, of which the principal symbol, following transport by a local chart $x' \overset{\chi}{\longmapsto} y'$ of $\partial\Omega$, is

$$< \chi'(x')\ \nu(x')\ ,\ \eta' > + \, i\lambda(x')\ |{}^t\chi'(x')\eta'|\ .$$

d) Investigate the problem (6.8.1) in the case $n = 2$, $\Omega = \{z = x + iy \in \mathbb{C}\ |z| < 1\}$, $\nu(z) = z^p$ for $z \in \partial\Omega$, where $p \in \mathbb{Z}$ (identify $\mathbb{R}^2$ and $\mathbb{C}$). More precisely, consider the operator

$$\mathcal{P} : C^\infty(\overline{\Omega}) \longrightarrow C^\infty(\overline{\Omega}) \oplus C^\infty(\partial\Omega)$$

defined by

$$\mathcal{P} u = (\Delta u\ ,\ (\frac{\partial u}{\partial \nu})_{|\partial\Omega})\ .$$

Suppose we have $u \in C^\infty(\overline{\Omega})$ with real values. Put $U = \frac{\partial u}{\partial z} = \frac{1}{2}(\frac{\partial u}{\partial x} - i\frac{\partial u}{\partial y})$. Show that u is in the kernel of $\mathcal{P}$ if and only if we have

$$\begin{cases} U \in C^\infty(\bar{\Omega}) \\ U \quad \text{holomorphic in} \quad \Omega \\ \operatorname{Re}\,(z^p U(z)) = 0 \quad \text{for} \quad |z| = 1 . \end{cases}$$

Deduce from this that $\dim \operatorname{Ker} \mathcal{P} = 1$ when $p > 0$, $\dim \operatorname{Ker} \mathcal{P} = 2$ when $p = 0$.

In the case $p < 0$, let $U(z) = \sum_{j=0}^{+\infty} a_j z^j$ be the entire-series expansion of U. Show that

$$z^p U(z) = a_{-p} + 2i \ \operatorname{Im} \sum_{j=0}^{-p-1} a_j z^{j+p} + h(z)$$

for $0 < |z| < 1$, with h holomorphic for $|z| < 1$, $h(0) = 0$. Deduce from this that u is in the kernel of $\mathcal{P}$ if and only if $h = 0$, and a_{-p} is purely imaginary. Conclude that $\dim \operatorname{Ker} \mathcal{P} = 2 - 2p$ when $p < 0$. Now let $g \in C^\infty(\partial\Omega)$ with real values. We know that there exists a unique function $G \in C^\infty(\bar{\Omega})$ holomorphic in Ω such that $\operatorname{Re} G = g$ in $\partial\Omega$, with $G(0) \in \mathbb{R}$ (in fact, if $g(e^{i\theta}) = \sum_{n=-\infty}^{+\infty} a_n e^{in\theta}$ is the Fourier series expansion of g, we have

$$G(z) = a_0 + 2 \sum_{n=1}^{+\infty} a_n z^n \quad \text{for} \quad |z| \le 1) .$$

Show that the conditions $\Delta u = 0$ in Ω, $\frac{\partial u}{\partial \nu} = g$ in $\partial\Omega$ are equivalent to

$$(6.8.2) \quad \begin{cases} U \in C^\infty(\overline{\Omega}) \\ U \text{ holomorphic in } \Omega \\ \operatorname{Re}\,(z^p U(z) - 2G(z)) = 0 \quad \text{for} \quad |z| = 1 \,. \end{cases}$$

Deduce from this that $\mathcal{P}$ is surjective when $p \leq 0$. When $p > 0$, show that (6.8.2) is equivalent to: U holomorphic in Ω, $z^p U(z) = 2G(z)$ for $|z| < 1$. Deduce from this that the image of $\mathcal{P}$ is of codimension $2p - 1$. Conclude that the index of the operator $\mathcal{P}$ is in all cases equal to $2(1 - p)$.

6.9 ELLIPTIC BOUNDARY-VALUE PROBLEMS WITH INTERFACES: Suppose Ω_1, Ω_2 are two regular open domains in $\mathbb{R}^n$ such that $\partial\Omega_1 = \Gamma$, $\partial\Omega_2 = \Gamma \cup \Gamma'$, with Γ, Γ' disjoint (and Γ' possibly empty).

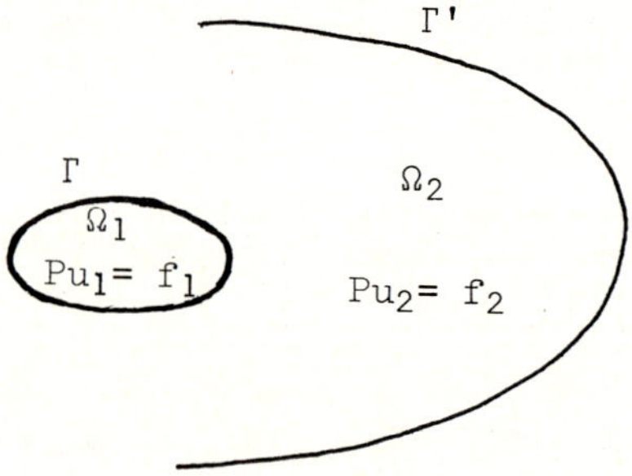

Let P_1, P_2 be two elliptic differential operators in $\mathbb{R}^n$, of degrees m_1, m_2 respectively. Suppose that in the neighbourhood of $\partial\Omega_2$ there is a field of vectors ν transversal to $\partial\Omega_2$. For $u_1 \in C^\infty(\overline{\Omega}_1)$, we put

$\gamma^{(1)} u_1 = (\gamma_0 u_1, \dots, \gamma_{m_1-1} u_1) \in C^\infty(\Gamma, \mathbb{C}^{m_1})$, with $\gamma_j u_1 = ((\frac{1}{i}\,\nu)^j u_1)_{|\Gamma}$. For $u_2 \in C^\infty(\overline{\Omega}_2)$, we similarly put

$\gamma^{(2)}u_2 = (\gamma_0 u_2, \ldots, \gamma_{m_2-1}u_2) \in C^\infty(\Gamma, \mathbb{C}^{m_2})$, and we put

$\gamma' u_2 = (\gamma'_0 u_2, \ldots, \gamma'_{m_2-1} u_2) \in C^\infty(\Gamma', \mathbb{C}^{m_2})$, with

$\gamma'_j u_2 = ((\frac{1}{i} \nu)^j u_2)_{|\Gamma'}$.

Suppose we have p.d.o. matrices $B^{(1)} = \left(B^{(1)}_{j,k}\right)_{\substack{j=0,\ldots,\mu-1\\ k=0,\ldots,m_1-1}}$,

$$B^{(2)} = \left(B^{(2)}_{j,k}\right)_{\substack{j=0,\ldots,\mu-1\\ k=0,\ldots,m_2-1}} , \quad B' = \left(B'_{j,k}\right)_{\substack{j=0,\ldots,\mu'-1\\ k=0,\ldots,m_2-1}},$$

with

$$B^{(1)}_{j,k} \in L^{d_j^{(1)}-k}(\Gamma) , \quad B^{(2)}_{j,k} \in L^{d_j^{(2)}-k}(\Gamma) , \quad B'_{j,k} \in L^{d'_j-k}(\Gamma') .$$

Consider the boundary-value problem

$$(6.9.1) \qquad \begin{cases} P_1 u_1 = f_1 & \text{in } \Omega_1 \\ P_2 u_2 = f_2 & \text{in } \Omega_2 \\ B^{(1)}\gamma^{(1)}u_1 + B^{(2)}\gamma^{(2)}u_2 = g & \text{in } \Gamma \\ B'\gamma' u_2 = g' & \text{in } \Gamma' . \end{cases}$$

We adopt the following assumption with regard to the degrees of the operators:

$$m_1 - d_j^{(1)} = m_2 - d_j^{(2)} \qquad \text{for} \quad j = 0,\ldots,\mu-1 .$$

a) Suppose we have, for $\ell = 1,2$, a parametrix Q_ℓ of P_ℓ. We consider the three Calderón projectors $C^{(1)}$, $C^{(2)}$, C' defined by

$$C^{(1)}v_1 = \gamma^{(1)}[(Q_1\tilde{P}_1 v_1)_{|\Omega_1}] \; , \quad C^{(2)}v_2 = \gamma^{(2)}[(Q_2\tilde{P}_2 v_2)_{|\Omega_2}] \; ,$$

$$C'v_2 = \gamma'[(Q_2\tilde{P}_2 v_2)_{|\Omega_2}] \; .$$

Let c_1, c_2, c' denote their principal symbols, and let b_1, b_2, b' denote the principal symbols of $B^{(1)}$, $B^{(2)}$, B'. By again using the method of reduction to the boundary, we are led to considering the morphism

$$\tilde{b} : (\operatorname{Im} c_1 \oplus \operatorname{Im} c_2) \oplus \operatorname{Im} c' \longrightarrow \mathbb{C}^{\mu} \oplus \mathbb{C}^{\mu'}$$

$$((w_1 \; , \; w_2) \; , \; w') \longmapsto (b_1w_1 + b_2w_2 \; , \; b'w') \; .$$

b) Show that Theorem 4.1 remains valid when $C^{\infty}(\bar{\Omega})$ is replaced by $C^{\infty}(\bar{\Omega}_1) \oplus C^{\infty}(\bar{\Omega}_2)$, $C^{\infty}(\partial\Omega \, , \mathbb{C}^{\mu})$ by $C^{\infty}(\Gamma, \mathbb{C}^{\mu}) \oplus C^{\infty}(\Gamma, \mathbb{C}^{\mu'})$, and the assumption that Ω is bounded is replaced by: Ω_1, Ω_2 bounded. [In particular, it may be noted that the matrix of operators $(B^{(1)}C^{(1)}, B^{(2)}C^{(2)})$ is right-elliptic in the sense of Douglis-Nirenberg when its rows are given the indices $m_1 - d_j^{(1)} = m_2 - d_j^{(2)}$ $(j = 0,\ldots,\mu-1)$, the columns of $B^{(1)} C^{(1)}$ are given the indices $m_1 - k$ $(k = 0,\ldots, m_1 - 1)$ and the columns of $B^{(2)} C^{(2)}$ the indices $m_2 - k$ $(k = 0,\ldots, m_2 - 1)$].

c) Let m_1' denote the smallest integer $\leq m$ such that $B_{j,k}^{(1)} = 0$ for $k \geq m_1'$, $j = 0,\ldots,\mu-1$. We define m_2' and m' similarly.

Suppose $s_1, \sigma_1, s_2, \sigma_2 \in \mathbb{R}$ are such that $s_1 > m_1' - m_1 - \frac{1}{2}$,

$$s_2 > m_2' - m_2 - \tfrac{1}{2}, \qquad s_2 > m' - m_2 - \tfrac{1}{2},$$

$$\sigma_1 - m_1 \le s_1, \ \sigma_2 - m_2 \le s_2, \ \sigma_1 - m_1 = \sigma_2 - m_2 .$$

Adapt Theorem 4.3 as before. Replace $H^{s}_{loc}(\bar{\Omega})$ by $H^{s_1}_{loc}(\bar{\Omega}_1) \times H^{s_2}_{loc}(\bar{\Omega}_2)$, $H^{\sigma}_{loc}(\bar{\Omega})$ by $H^{\sigma_1}_{loc}(\bar{\Omega}_1) \times H^{\sigma_2}_{loc}(\bar{\Omega}_2)$ and $H^{\sigma-d-\frac{1}{2}}_{loc}(\partial\Omega)$ by $H^{\theta}_{loc}(\Gamma) \times H^{\sigma_2-d'-\frac{1}{2}}_{loc}(\Gamma')$ where $\theta = \sigma_1 - d^{(1)} - \frac{1}{2} = \sigma_2 - d^{(2)} - \frac{1}{2}$.

Adapt Theorem 4.4 in similar fashion. It may be noted that the injectivity of $\tilde{b}$ implies the left-ellipticity, in the sense of Douglis-Nirenberg, of the matrix of operators

$$\left(\begin{array}{c|c} I - C^{(1)} & 0 \\ \hline 0 & I - C^{(2)} \\ \hline B^{(1)} & B^{(2)} \end{array}\right)$$

when the first block of columns is given the indices $m_1 - k\,(k=0,\dots,m_1-1)$, the second block of columns $m_2 - k\,(k=0,\dots,m_2-1)$, the first block of rows the indices $m_1 - j\ (j=0,\dots,m_1-1)$, the second the indices $m_2 - j\ (j=0,\dots,m_2-1)$, and the third the indices

$$m_1 - d_j^{(1)} = m_2 - d_j^{(2)} \quad (j=0,\dots,\mu-1) .$$

d) Give properties analogous to those established in Exercises

6.4, 6.5, 6.6. Generalise everything done earlier in the context of manifolds with boundary, bundles, and systems which are elliptic in the sense of Douglis-Nirenberg (see Exercise 6.7)

e) Consider problem (6.9.1) in the case

$\Omega_1 = \{ x \in R^n \mid x_n > 0 \}$, $\Omega_2 = \{ x \in R^n \mid x_n < 0 \}$, $\Gamma = R^{n-1}$;

$\Gamma' = \emptyset$. Put $u = (u_1, \tilde{u}_2)$, with

$\tilde{u}_2(x', x_n) = u_2(x', -x_n)$. By this means reduce the problem to one which is elliptic in the sense of Douglis-Nirenberg (see Exercise 6.7) for u in the open domain Ω_1.

CHAPTER 6

EVOLUTION EQUATIONS

1. THE CAUCHY-KOVALESKI AND HOLMGREN THEOREMS

1.1 INTRODUCTION: Suppose $P(x, D_x)$ is a differential operator of degree m with C^∞ coefficients in $\mathbb{R}^n$.

One natural idea for attempting to find a solution u in the neighbourhood of 0 of the Cauchy problem

$$(1.1.1.) \qquad P(x, D_x)u = f(x)$$

$$(1.1.2.) \qquad D_n^j u\Big|_{x_n=0} = g_j(x') \quad j = 0,\dots,m-1 \quad x = (x', x_n)$$

consists of calculating the derivatives $u^{(\alpha)}(0)$ from the equations.

If $\alpha_n < m$, we immediately obtain the $u^\alpha(0)$ by using (1.1.2). Then, to calculate $D_n^m u(0)$ from (1.1.1), we assume that this equation can be written in the form

$$(1.1.3.) \qquad a_m(x)\, D_n^m u = \sum_{\substack{|\alpha| \leq m \\ \alpha_n < m}} a_\alpha(x)\, D^\alpha u + f(x)$$

with $a_m(0) \neq 0$. In this case, it can be shown that there exists a unique formal solution of (1.1.1) and (1.1.2); furthermore, when the data a_α, f, g_j are analytic in a neighbourhood of 0, the solution u is itself analytic.

We now give a somewhat more general formulation of this result.

THEOREM 1.2: *Suppose* $\ell(\alpha) = \sum_{j=1}^{n} c_j \alpha_j$ *is a linear form on* $\mathbb{R}^n$ *with coefficients* $c_j \geq 0$. *Consider an equation of the form*

$$D^\gamma u = \sum_\alpha a_\alpha(x) D^\alpha u + f(x) \tag{1.2.1.}$$

with $a_\alpha = 0$ *if* $|\alpha| > |\gamma|$ *or* $\ell(\alpha) \geq \ell(\gamma)$.

We impose the following supplementary conditions:

$$D_j^k(u - \varphi) = 0 \quad \text{on} \quad x_j = 0 \quad \text{for} \quad j = 1,\dots,n \quad \text{and} \quad k = 0,\dots,\gamma_j - 1 . \tag{1.2.2.}$$

Then, given f, a_α, φ *analytic in the neighbourhood of* 0, *there exists a unique function* u *which is analytic in the neighbourhood of* 0 *and which satisfies* (1.2.1) *and* (1.2.2).

For example, if we take $\ell(\alpha) = \alpha_n$ and $\gamma = (0,\dots, 0, m)$, we obtain the Cauchy problem (1.1.3), (1.1.2).

PROOF OF UNIQUENESS: Suppose that φ is equal to zero; then condition (1.2.2) implies:

(1.2.3.) $D^{\mu} u(0) = 0$ for any multi-index μ such that $\mu_j < \gamma_j$ for at least one index j.

For indices of the form $\gamma + \delta$ with $\delta \geq 0$, equation (1.2.1) and Leibniz's formula give

$$(1.2.4.)\quad D^{\gamma+\delta} u(0) = \sum_{\alpha} \sum_{\beta \leq \delta} \binom{\delta}{\beta} D^{\beta} a_{\alpha}(0)\, D^{\alpha+\delta-\beta} u(0) + D^{\delta} f(0).$$

If f is zero, it can be shown by recurrence over the set of numbers $\ell(\gamma+\delta)$ that $D^{\gamma+\delta} u(0) = 0$ for all $\delta \geq 0$. In fact, the multi-indices which appear in the right-hand side of (1.2.4) satisfy the recurrence assumption since

$$(1.2.5.)\quad \ell(\alpha+\delta-\beta) = \ell(\alpha) + \ell(\delta-\beta) < \ell(\gamma) + \ell(\delta) = \ell(\gamma+\delta).$$

In order to prove the existence, we define u by its Taylor series expansion about the origin. Replacing u by $u - \varphi$, we obtain the case $\varphi = 0$, so that we force the derivatives of u to satisfy (1.2.3).

Next, we successively determine the derivatives of order greater than γ by imposing the equality (1.2.4). In order to prove the analyticity of u in the neighbourhood of 0, we shall show that there exist constants $M_0 \in \mathbb{R}^+$ and $M = (M_1, \ldots, M_n) \in \mathbb{R}^n_+$ such that

(1.2.6.) $|D^{\alpha} u(0)| \leq M_0 M^{\alpha} |\alpha|!$ for any multi-index $\alpha \geq 0$,

where $M^{\alpha} = M_1^{\alpha_1} \times \ldots \times M_n^{\alpha_n}$.

We know, by hypothesis, that there exist constants L and L_0 such that

$$|D^{\beta} a_{\alpha}(0)| \leq L^{|\beta|+1} |\beta|!$$

(1.2.7.) for all α, β

$$|D^{\beta} f(0)| \leq L_0 L^{|\beta|} |\beta|! \quad .$$

By means of a suitable choice of the M_j, we prove (1.2.6) by recurrence over the numbers $\ell(\alpha)$. We assume that it is true for all the $\alpha = \gamma+\mu$ such that $\ell(\gamma+\mu) < \ell(\gamma+\delta)$ and we show that it is satisfied for $\alpha = \gamma+\delta$. Substituting the upper bounds (1.2.7) into (1.2.4) we obtain, taking account of (1.2.5) :

$$(1.2.8.)\ |D^{\gamma+\delta} u(0)| \leq \Big(\sum_{\beta_\alpha \leq \delta} \binom{\delta}{\beta} L\, L^{|\beta|} |\beta|!\, M_0 M^{\alpha+\delta-\beta} |\alpha+\delta-\beta|! + L_0 L^{|\delta|} |\delta|! \Big) \quad .$$

The combinatorial interpretation of the binomial coefficients shows that $\binom{\delta}{\beta} \leq \binom{|\delta|}{|\beta|}$ and consequently $\binom{\delta}{\beta} \cdot \frac{|\alpha+\delta-\beta|!\ |\beta|!}{|\gamma+\delta|!} \leq 1$. Using this inequality, (1.2.8) implies that

$$(1.2.9.)\quad |D^{\gamma+\delta}\, u(0)| \le M_0\, M^{\gamma+\delta} |\gamma+\delta|! \left[L\Big(\sum_{\beta \ge 0} \frac{L^{|\beta|}}{M^{\beta}}\Big)\Big(\sum_{\ell(\alpha) < \ell(\gamma)} \frac{1}{M^{\gamma-\alpha}}\Big) + \frac{1}{M^{\gamma}} \cdot \frac{L^{|\delta|}}{M^{\delta}} \cdot \frac{L_0}{M_0} \right] .$$

Without reducing generality, we can assume that $L \ge 1$; we put $M_j = 2L\, A^{c_j}$ $j = 1, \ldots, n$ with $A \ge 1$. Then the bracket in the right-hand side of (1.2.9) is bounded above by the quantity

$$(1.2.10.)\qquad L.2^n \sum_{\substack{\alpha \\ \ell(\alpha) < \ell(\gamma)}} \frac{1}{A^{\ell(\gamma)-\ell(\alpha)}} + \frac{L_0}{M_0} .$$

It is sufficient to take A and M_0 sufficiently large that the quantity (1.2.10) is smaller than unity; this then allows us to argue by recurrence and cóncludes the proof of the thoerem.

REMARK 1.3: The statement and the proof of Theorem 1.2 remain valid word for word when the coefficients $a_\alpha(x)$ are N x N matrices and the functions φ , f, u have values in $\mathbb{C}^N$.

REMARK 1.4: Expression (1.2.10) shows that the radius of convergence of the Taylor series expansion for u is bounded below by a number > 0 which depends only on the constant L appearing in the upper bounds (1.2.7), i.e. on the smallest radius of convergence ρ of the Taylor series expansions about 0 of the functions a_α, f, φ and on the $\sup_{\substack{z \in \mathbb{C}^n \\ |z| \le \rho/2}} |a_\alpha(z)|$.

As a special case of this theorem we deduce:

THEOREM 1.5 (Cauchy-Kovaleski): *Suppose* $P(x, D_x)$ *is an operator with coefficients analytic in some neighbourhood of a point* $x^0 \in \mathbb{R}^n$. *We take functions* f, φ, ψ *which are analytic in the neighbourhood of* x_0. *Suppose that* ψ *is real-valued and satisfies*

$$(1.5.1.) \qquad P_m(x^0, d\psi(x^0)) \neq 0 \quad .$$

We denote by S *the hypersurface defined in the neighbourhood of* x^0 *by the equation* $\psi(x) = \psi(x^0)$. *Then in some neighbourhood of* x^0 *there exists a unique analytic solution of the Cauchy problem*

$$(1.5.2.) \qquad \begin{cases} P(x, D_x)u = f & \text{in the neighbourhood of } x^0 \\ D^\alpha(u-\varphi)|_S = 0 & \text{if } |\alpha| \leq m-1 \end{cases} \quad .$$

PROOF: Suppose for example that $\frac{\partial\psi}{\partial x_n}(x^0) \neq 0$; then the change of variables

$$(1.5.3.) \qquad \begin{aligned} y_n &= \psi(x) - \psi(x^0) \\ y_j &= x_j - x_j^0 \qquad j = 1, \ldots, n-1 \end{aligned}$$

is an analytic diffeomorphism of a neighbourhood of x^0 onto a neighbourhood of $y = 0$; it transforms problem (1.5.2) into

$$\text{(1.5.4.)} \quad \begin{aligned} \tilde{P}(y, D_y)\tilde{u} &= \tilde{f} \\ D^\alpha(\tilde{u}-\tilde{\varphi})\big|_{y_n=0} &= 0 \qquad |\alpha| \le m-1 \ . \end{aligned}$$

Formula (7.3.5) in Chapter I for the transformation of the principal parts shows that $\tilde{P}_m(0, \ldots, 0, 1) \neq 0$. Consequently we can isolate the term in D_n^m in (1.5.4), and we therefore come down to the particular case of the example given after the statement of Theorem 1.2.

REMARK 1.6: In the case where P is an N x N system, assumption (1.5.1) has to be replaced by the condition

$$\det P_m(x^0, d\psi(x^0)) \neq 0 \ .$$

This assumption leads us to introduce:

DEFINITION 1.7: *A hypersurface* S *in* $\mathbb{R}^n$, *of class* C^1, *is said to be characteristic at* x^0 *for the operator* $P(x, D_x)$ *of degree* m *if* $P_m(x^0, d\psi(x^0)) = 0$, *where* $\psi(x) = 0$ *is an equation of* S *in the neighbourhood of* x^0. *(P is an* N x N *system, this condition must obviously be replaced by*

$$\det P_m(x^0, d\psi(x^0)) = 0 \ ) \ .$$

For example, the hyperplane $x_n = 0$ is characteristic at 0 if and only if the coefficient of D_n^m in $P(x, D_x)$ vanishes at 0.

REMARK 1.8: In the case of operators P(D) with constant coefficients, the uniqueness of the solution of the Cauchy problem

$$(1.1.7.)\qquad \begin{aligned} P(D)u &= 0 \\ D_n^j u_{|x_n=0} &= 0 \qquad j = 0,\ldots, m-1 \end{aligned}$$

with m = degree of P(D)

is directly related to the fact that the hyperplane $x_n = 0$ is non-characteristic.

In fact, suppose that $P_m(0, \ldots, 0, 1) = 0$. Then, if P is homogeneous of degree m, the function $u(x) = x_n^{m+1}$ is a non-zero analytic solution of problem (1.1.7). More generally, if we drop the assumption of homogeneity, we can find in HORMANDER [1], Chapter V, the construction of a non-zero C^∞ solution of (1.1.7).

Nonetheless, in certain cases there is a substitute theorem for local existence and uniqueness with respect to a characteristic surface; this is as follows:

THEOREM 1.9 (Goursat-Beudon): *Consider the equation*

$$(1.9.1.)\qquad D_1 D_n^{m-1} u = \sum_{\substack{|\alpha| \leq m \\ \alpha_n < m \\ \alpha \neq (1,0,0,m-1)}} a_\alpha(x)\, D^\alpha u(x) + f(x)$$

with the conditions

$$(1.9.2.)\qquad \begin{aligned} D_n^k(u-\varphi)_{|x_n=0} &= 0 \qquad k = 0, \ldots, m-2 \\ (u-\varphi)_{|x_1=0} &= 0 \end{aligned}$$

.

Suppose that the functions a_α, f, φ *are analytic in the neighbourhood of* 0; *then there exists a unique solution* u *of* (1.9.1), (1.9.2), *which is analytic in the neighbourhood of* 0.

To prove this, it is sufficient to apply Theorem 1.2 with
$\gamma = (1, 0, \ldots, 0, m-1)$ and $\ell(\alpha) = \alpha_n + m\,\alpha_1$.

There is, in fact, no need to assume the analyticity of the solution in order to have local uniqueness in the Cauchy problem (1.5.2). We have:

THEOREM 1.10 (Holmgren): *Suppose* $P(x, D_x)$ *is a differential operator of degree* m *with coefficients analytic in a neighbourhood* V *of* x^0. *Suppose that* S *is a hypersurface containing* x_0 *and that* S *is defined by an analytic equation* $\psi(x) = 0$. *We put* $V^+ = \{x \in V \mid \psi(x) \geq 0\}$. *Suppose that* S *is not characteristic for* P *at* x^0. *Then there exists a neighbourhood* W *of* x^0 *such that any function* $u \in C^m(V^+)$ *which satisfies*

$$(1.10.1.)\qquad \begin{array}{llll} P\,u = 0 & \text{in} & V^+ & \\ D^\alpha u_{|S} = 0 & \text{in} & V \cap S & |\alpha| \leq m-1 \end{array}$$

is necessarily equal to zero in $W^+ = \{x \in W \mid \psi(x) \geq 0\}$.

PROOF: To prove this uniqueness result we proceed by duality, using the existence of a solution of a transposed problem. After carrying out the analytic change of variables (1.5.3), we can assume that S is defined by the equation $y_n = 0$. Next, we apply the analytic change of variables

$$(1.10.2.)\quad \begin{cases} x_n = y_n + (y_1^2 + \dots + y_{n-1}^2) \\ x_j = y_j \qquad\qquad j = 1, \dots, n-1 \end{cases}$$

which transforms the hyperplane $y_n = 0$ into the paraboloid $\mathcal{P}$ with equation $x_n = x_1^2 + \dots + x_{n-1}^2$. We put

$$\mathcal{P}^+ = \left\{ x \mid x_n \geq x_1^2 + \dots + x_{n-1}^2 \right\}, \quad \tilde{V}^+ = \mathcal{P}^+ \cap \tilde{V},$$

where $\tilde{V}$ corresponds to the open set V after the change of variables, and we denote by $\tilde{P}$ the operator obtained by transport of P. We now have to show that there exists a neighbourhood $\tilde{W}$ of 0 such that if $\tilde{u} \in C^m(\tilde{V}^+)$ satisfies

$$(1.10.3.)\quad \begin{cases} \tilde{P}\tilde{u} = 0 \quad \text{in } \tilde{V}^+ \\ D^\alpha \tilde{u}|_{\mathcal{P}} = 0 \quad \text{in } \tilde{V} \cap \mathcal{P}, \quad |\alpha| \leq m-1 \end{cases}$$

then it is zero in $\tilde{W}^+ = \tilde{W} \cap \mathcal{P}^+$.

Suppose $r > 0$; we define the sets

$$B_r = \left\{ x \in R^n \mid \max_j |x_j| \leq \sqrt{r} \right\}; \quad \mathcal{P}_r^+ = \left\{ x \in \mathcal{P}^+ \mid x_n \leq r \right\};$$

for r sufficiently small, we have the inclusions

$$\mathcal{P}_r^+ \subset B_r \subset \tilde{V}.$$

The Cauchy-Kovaleski Theorem 1.5 and Remark 1.4 show that it is possible to find an r_o sufficiently small that for any $h \in]0, r_o[$ and for any polynomial $q(x_1, \dots, x_{n-1})$ the Cauchy problem

$$(1.10.4.)\quad {}^t\tilde{P}(x, D_x)v = 0 \quad \text{in} \quad B_r$$
$$D^j_{x_n} v|_{x_n=0} = 0 \quad j = 0,\ldots, m-2 \quad , \quad D_n^{m-1} v|_{x_n=0} = q$$

admits an analytic solution v in B_r.

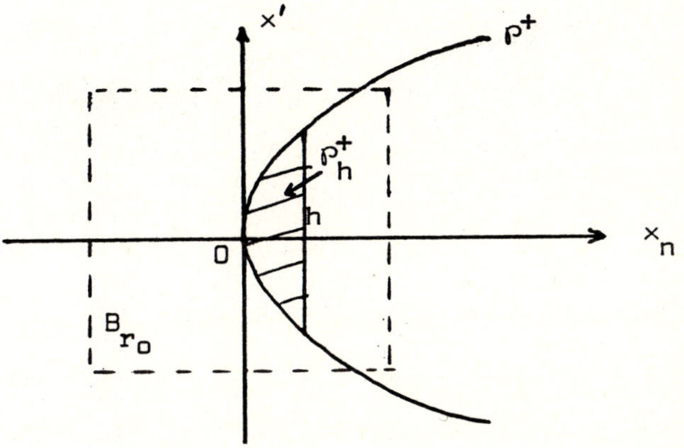

If we apply Green's formula in the compact domain P^+_h , then taking account of (1.10.3) and (1.10.4), we obtain

$$(1.10.5.)\int_{P^+_h}(v.\tilde{P}\,\tilde{u} - \tilde{u}.{}^t\tilde{P}\,v)dx = \int_{|x'|\leq\sqrt{h}} C(x')\,\tilde{u}(x', h)\,q(x')\,dx'$$

where $C(x')$ is a function which depends only on the coefficients of $\tilde{P}$ and which does not vanish.

Equations (1.10.3) and (1.10.4) imply that the left-hand side of (1.10.5) is equal to zero; consequently $u(x', h) = 0$ for $|x'| \leq \sqrt{h}$ since q is an arbitrary polynomial; this shows that $u = 0$ in $P^+_{r_o}$. It is therefore sufficient to take $\tilde{W} = B_{r_o}$.

REMARK 1.11: The determination of r_o shows that the open set W depends only on the open set V, on the radius of convergence of the Taylor series expansions at x° of the coefficients of P and of ψ and on the modulus of the coefficients as in Remark 1.4.

REMARK 1.12: In all the cases for which there is an existence theorem for problem (1.10.4), the above proof shows that there is also a uniqueness theorem (1.10.1).

In the case where P has constant coefficients and S is a hyperplane, Holmgren's Theorem 1.10 and Remark 1.8 show that in order to have local uniqueness, it is necessary and sufficient that S be non-characteristic. We then also have global uniqueness:

PROPOSITION 1.13: *Suppose* P(D) *is a differential operator with constant coefficients; suppose that the hyperplane* $x_n = 0$ *is non-characteristic for* P. *Let* $u \in \mathcal{D}'(\mathbb{R}^n)$ *be such that* $Pu = 0$ *in* $\mathbb{R}^n_+$ *and* $\operatorname{supp} u \subset \overline{\mathbb{R}}^n_+$. *Then* $u = 0$.

PROOF: Let ρ_k be a regularising sequence (see (1.1.1), Chapter I) with $\rho \in C_0^{\infty}(\mathbb{R}^n_+)$. The function $u_k = u * \rho_k$ is C^{∞}, with support in $\overline{\mathbb{R}}^n_+$, and satisfies $P u_k = 0$ in $\overline{\mathbb{R}}^n_+$. As P has constant coefficients, Holmgren's Theorem 1.10 and Remark 1.11 imply that $u_k = 0$ and consequently that $u = \lim\limits_{k\,\infty} u_k = 0$.

2. NECESSARY CONDITION FOR THE CAUCHY PROBLEM TO BE WELL POSED

The Cauchy-Kovaleski and Holmgren theorems are essentially local results; in this section we give a necessary condition for the Cauchy problem to be well posed globally.

Let $x = (x_0, x_1, \ldots, x_n) \in \mathbb{R}^{n+1}$; we put $x = (x_0, x')$. We denote by $P(x, D_x)$ a differential operator of degree m with coefficients in $C^\infty(\mathbb{R}^{n+1})$.

DEFINITION 2.1 : *Let* $T \in \mathbb{R}$*; we say that the Cauchy problem is well posed for* P *in the half-space* $x_0 \geq T$, *if for any* $f \in C^\infty(\mathbb{R}^{n+1})$, $g_j \in C^\infty(\mathbb{R}^n)$ $j = 0, .., m-1$, *there exists a unique* $u \in C^\infty(\mathbb{R}^{n+1})$ *such that*

$$(2.1.1.) \qquad Pu = f \quad \text{for} \quad x_0 > T \;, \quad D_0^j u = g_j$$

$$\text{for} \quad x_0 = T \quad \text{and} \quad j = 0, \ldots, m-1 \;.$$

We say that the Cauchy problem is well posed in the direction $N = (1,0 \ldots, 0)$ *if it is well posed in each half space* $x_0 \geq T$.

It is possible to obtain another formulation by replacing the initial conditions by a condition on the support. To simplify the notation we take $T = 0$.

PROPOSITION 2.2 : *Suppose that the hyperplane* $x_o = 0$ *is noncharacteristic for* P. *Then the Cauchy problem is well posed for* P *in the half space* $x_o \geq 0$ *if and only if for any* $f_o \in C^\infty(R^{n+1})$ *with* $\text{supp } f_o \subset \bar{R}_+^{n+1} = \{x \mid x_o \geq 0\}$ *there exists a unique* $u_o \in C^\infty(R^{n+1})$ *such that*

$$P u_o = f_o \quad \textit{and} \quad \text{supp } u_o \subset \bar{R}_+^{n+1} . \tag{2.2.2.}$$

PROOF: The proof of necessity is immediate, since all the derivatives of the solution u of (2.1.1) with $g_j = 0$, $j = 0,\ldots, m-1$ are equal to zero on the noncharacteristic hyperplane $x_o = 0$.

Conversely, Borel's Theorem allows us to construct a function $v \in C^\infty(R^{n+1})$ of which the traces on $x_o = 0$ satisfy $D_o^j v = g_j$ for $j = 0,\ldots, m-1$ and $f_o = Pv - f$ flat on $x_o = 0$.

Then if u_o is the solution of (2.2.2), the function $u_o + v$ is a solution of (2.1.1).

The problem of giving a general characterisation of operators for which the Cauchy problem is well posed remains open; nonetheless, a large number of partial results exist. We give below a necessary condition on the principal part of P.

THEOREM 2.3 : (Lax-Mizohata) *Suppose that for any* $T \in \mathbb{R}$, *the hyperplane* $x_0 = T$ *is noncharacteristic for* P, *and that the Cauchy problem is well posed in the direction* $N = (1, 0, \ldots, 0)$. We then have the implication

$$\left.\begin{array}{c} P_m(x, \zeta_0, \xi') = 0 \\ \\ x \in \mathbb{R}^{n+1}, \quad \zeta_0 \in \mathbb{C}, \quad \xi' \in \mathbb{R}^n \end{array}\right\} \Longrightarrow \operatorname{Im} \zeta_0 = 0. \tag{2.3.1.}$$

We start by giving the following lemma:

LEMMA 2.4 : *Suppose* K *is a compact subset of* $\mathbb{R}^{n+1}$; *then there exists a number* C *and an integer* $k \geq 0$ *such that*

$$|u|_0^- \leq C\, |Pu|_k^-, \quad \textit{for all} \quad u \in C_K^\infty \tag{2.4.1.}$$

$$\textit{with} \quad |v|_j^- = \sup_{\substack{|\alpha| \leq j \\ x_0 \leq 0}} |D^\alpha v(x)| .$$

PROOF: Let $f \in C_K^\infty$; the Seeley extension (Proposition 9.2, Chapter I) of $f_- = f|_{\mathbb{R}_-^{n+1}}$ to $\mathbb{R}^{n+1}$ is a function in $C_0^\infty(\mathbb{R}^{n+1})$ which we denote by Sf_-. Suppose V is a compact neighbourhood of K; we can force Sf_- to have support in V for any f. The continuity of the Seeley extension shows that for any integer $k \geq 0$ there exist C and q such that

$$|Sf_-|_k \leq C\, |f_-|_q^-, \quad \text{for any } f \in C_K^\infty . \tag{2.4.2.}$$

For $u \in C_K^\infty$, we put $f = Pu$, $g = Sf_-$ and we define, with

$\varepsilon > 0$, the function

$$f_\varepsilon = \theta_\varepsilon f + (1 - \theta_\varepsilon)g \in C_V^\infty ,$$

where $\theta_\varepsilon(x) = \theta(\frac{x_0}{\varepsilon})$ and $\theta \in C^\infty(R)$ is equal to zero for $x_0 \geq 1$ and equal to unity for $x_0 \leq 0$.

Let $u_\varepsilon \in C^\infty(\mathbb{R}^{n+1})$ be the unique solution of

$$P u_\varepsilon = f_\varepsilon , \quad \text{supp } u_\varepsilon \subset \left\{ x \mid x_0 \geq a \right\} = H_a$$

(where $a = \inf_{x \in V} x_0$) .

Since $f_\varepsilon = f$ on $\mathbb{R}_-^{n+1}$, we have $u_\varepsilon = u$ on $\mathbb{R}_+^{n+1}$ and

$$|u|_0^- = |u_\varepsilon|_0^- \leq |u_\varepsilon|_0 . \tag{2.4.3.}$$

By hypothesis, P is a continuous bijection of the Fréchet space $E = \left\{ u \mid u \in C^\infty(R^{n+1}), \text{ supp } u \subset H_a \right\}$; consequently there exist C, an integer k and a compact set K' such that

$$\sup_{x \in K} |u(x)| \leq \underset{\substack{|\alpha| \leq k \\ x \in K'}}{C} |D^\alpha Pu(x)| , \quad \text{for any } u \in E. \tag{2.4.4.}$$

Applying (2.4.4) to u_ε and taking account of (2.4.3), we obtain

$$|u|_0^- \leq C |f_\varepsilon|_k , \quad \text{for any } u \in C_K^\infty , \ \varepsilon > 0. \tag{2.4.5.}$$

Furthermore, we have $|f_\varepsilon|_k \xrightarrow[\varepsilon \to 0]{} |g|_k$ since the function $(f_\varepsilon - g) = \theta_\varepsilon(f - g)$ has support in the strip

$0 \le x_o \le \varepsilon$ and is flat for $x_o = 0$. We therefore deduce from (2.4.5) that

$$|u|_o^- \le C\, |g|_K ,$$

and taking account of (2.4.2) we finally obtain

$$|u|_o^- \le C\, |f_-|_q^- = C\, |Pu|_q^- , \text{ for any } u \in C_K^\infty .$$

In order to prove the theorem, we consider the point $x = 0$ and we use a method adapted from IVRII - PETKOV [1] and HORMANDER [7].

Let $\zeta_o \in C$ be such that

$$P_m(0\ ;\ \zeta_o, \xi') = 0 \quad \text{with} \quad \xi' \in R^n .$$

We denote its multiplicity by r. If Im $\zeta_o \ne 0$, we can assume for example that Im $\zeta_o < 0$, so that we necessarily have $\xi' \ne 0$. The vector $(\mathrm{Re}\ \zeta_o, \xi')$ is transversal to the hyperplane $x_o = 0$; hence we can apply a linear change of variables which preserves the variable x_o and which transforms $(\mathrm{Re}\ \zeta_o, \xi')$ into $(0,\dots, 0, 1)$. We thereby reduce to $\zeta_o = - i$, $\xi' = e_n = (0, \dots, 0, 1)$ and we have

(2.4.6.) $$P_m(0; D_o, 0,\dots, D_n) = (D_o + i\, D_n)^r\, Q(D_o, D_n)$$

where Q is homogeneous of degree $m - r$ and $Q(- i, 1) \ne 0$.

In order to localise the operator at $x = 0$ and to get rid of terms of lower order, *Ivrii* and *Petkov* introduce the change of variables

$$y_j = x_j \rho^{s_j} \qquad j = 0, \ldots, n \ , \text{ with } \rho > 0 \ , \ s_j > 0 \ .$$

The operator $P(x, D_x)$ is transformed into

$$P_\rho(y, D_y) = P(\ldots, y_j \rho^{-s_j}, \ldots, \rho^{s_j} D_{y_j}, \ldots) \quad .$$

The inequality (2.4.1) implies that for any compact set K_1 there exists ρ_o such that

$$(2.4.7.) \quad |v|_o^- \leq C \rho^{kS} |P_\rho(y, D_y)v|_k^- \ , \quad v \in C_{K_1}^\infty \ , \quad \rho \geq \rho_o$$

where $S = \max_j s_j$.

The idea behind the proof consists of constructing an asymptotic solution u_ρ of $P_\rho u_\rho = O(\rho^{-\infty})$ in a neighbourhood of 0 so as to contradict the inequality (2.4.7). We put

$$s_j = 2r \quad \text{if} \quad j = 1, \ldots, n-1 \quad \text{and} \quad s_o = s_n = 4r \quad ;$$

then

$$\rho^{-4mr} P_\rho(y, D_y) = P_m(0; D_o, \ldots, D_n) + \begin{pmatrix} \text{differential operator of} \\ \text{degree } m \text{ with coefficients} \\ O(\rho^{-2r}) \end{pmatrix}$$

We seek an asymptotic solution in the form

$$u_\rho(y) = \sum_{j \geq o} e^{i\rho\, \varphi(y)} \frac{v_j(y)}{\rho^j}$$

with $v_o(0) = 1$.

Expression (2.4.6) leads us to take for the phase φ the function $\varphi(y) = -iy_0 + y_n$, since this satisfies

$(D_0 + iD_n)\varphi = 0$, $Q(D_0\varphi, D_n\varphi) = q \neq 0$; moreover

$$\operatorname{Im}\varphi(y) \geq 0 \quad \text{for} \quad y_0 \leq 0 .$$

With this choice of φ the asymptotic expansion of $P_\rho u_\rho$ starts with a term of degree m - r, and setting this to zero leads to the equation $q(D_0 + c D_n)^r v_0(y) = 0$. It is sufficient to take $v_0 \in C^\infty_{K_1}$ equal to unity in the neighbourhood of 0. The equations which follow are of the form

$$q(D_0 + i D_n)^r v_j(y) = F_j(y) \tag{2.4.8.}$$

where $F_j \in C^\infty_{K_1}$ is determined from the $v_0, .., v_{j-1}$.

We solve (2.4.8) in the neighbourhood of 0 by convolution with the fundamental solution (6.1.3) of Chapter I of the Cauchy-Riemann operator $\frac{i}{2}(D_0 + i D_n)$. We can thus find an integer J such that the partial sum

$$u^J_\rho = \sum_{j=0}^{J} \rho^{i\rho\varphi} v_j \rho^{-j}$$

satisfies

$$|P_\rho u^J_\rho|^-_k = O\left(\frac{1}{\rho^{4kr+1}}\right) \quad \text{and} \quad |u^J_\rho|^- \geq 1$$

for ρ sufficiently large; this contradicts (2.4.7) and concludes the proof of the theorem.

REMARK 2.5 : A completely analogous theorem exists for

differential systems, it being sufficient to replace P_m by the determinant of the principal part in condition (2.3.1) (see IVRII-PETKOV [1]).

REMARK 2.6 : Apart from the case of operators with constant coefficients, a characterisation of operators for which the Cauchy problem is well posed is known only in a number of special cases. For example, if we assume that the roots at ζ_o of (2.3.1) are of constant multiplicity, we have a condition which is necessary (FLASCHKA-STRANG [1]) and sufficient (CHAZARAIN [1]) and which involves lower-order terms of P, i.e. the *Levi* condition. If we suppose that the roots at ζ_o are of multiplicity less than or equal to 2, again we have a number of very complete results (see IVRII-PETKOV [1], HORMANDER [7]). When the coefficients are analytic, BONY and SCHAPIRA [1] have shown that condition (2.3.1) is sufficient for problem (2.1.1) to admit a unique analytic solution when the data are analytic.

In Section 4, we shall show in particular that if the roots at ζ_o are simple, then condition (2.3.1) is necessary and sufficient.

3. HYPERBOLIC OPERATORS WITH CONSTANT COEFFICIENTS

In the case where P has constant coefficients, Gårding

has given an algebraic characterisation of the operators P(D) for which the Cauchy problem is well posed. The subject is discussed in depth in HORMANDER [1]; however, for completeness, we shall reproduce a part of this discussion here, since certain results will be needed in Section 4 in connection with questions relating to speed of propagation.

Throughout this section, we shall be in $\mathbb{R}^n$.

PROPOSITION 3.1 : *Assume that the Cauchy problem is well posed for the operator* P(D) *in the direction* $N \in \mathbb{R}^n \setminus 0$. *Then there exists a real number* γ_o *such that*

$$(3.1.1.)\quad P(\xi + \tau N) \neq 0 \textit{ for } \xi \in \mathbb{R}^n,\ \tau \in C \textit{ and } \operatorname{Im} \tau < \gamma_o .$$

PROOF: The generality will not be restricted if we take $N = (0, .., 0, 1)$. The argument used in proving (2.4.4) applies particularly in this case; consequently, for any compact set K there exist C, an integer k and a compact set K' such that

$$(3.1.2.)\quad \sup_{x \in K} |u(x)| \leq C \sup_{\substack{|\alpha| \leq k \\ x \in K'}} |D^\alpha(P(D)\, u(x))| , \quad u \in C^\infty(\mathbb{R}^n)$$
$$\text{and } \operatorname{supp} u \subset \overline{\mathbb{R}}^n_+ .$$

Let $\tau \in C$ be such that $P(\xi', \tau) = 0$ with $\xi' \in \mathbb{R}^{n-1}$, and put $\zeta = (\xi', \tau)$; we shall show that there exists γ_o

independent of ξ' such that $\operatorname{Im}\tau \geq \gamma_0$. We can therefore suppose that $\operatorname{Im}\tau \leq 0$. To show this, we apply the inequality (3.1.2) (with K reduced to a point $a \in \mathbb{R}^n_+$,) to the function

$$u(x) = e^{i(x-a).\zeta}\,\theta(x_n)$$

where $\theta \in C^\infty(\mathbb{R})$ is equal to unity for $x_n \geq \frac{a_n}{2}$ and zero for $x_n \leq 0$. Since $P(D)\,u(X) = 0$ for $x_n \geq \frac{a_n}{2}$ and $u(a) = 1$, we obtain

$$1 \leq C\,(1 + |\zeta|)^{k+m}\, e^{\frac{1}{2}a_n.\operatorname{Im}\tau} \quad ,$$

or, with real constants b and c,

$$(3.1.3.) \quad b\,\operatorname{Log}(1 + |\zeta|) + C \leq \operatorname{Im}\tau \ , \text{ when } P(\xi', \tau) = 0 ,$$

$$\operatorname{Im}\tau \leq 0 \ , \ \xi' \in \mathbb{R}^{n-1} \ .$$

However, a result due to Seidenberg-Tarski (see HORMANDER [1], Appendix) shows that the logarithm in (3.1.3) may be replaced by a constant, and this proves the proposition.

This proposition leads to

DEFINITION 3.2 : *An operator with constant coefficients is said to be correct in the sense of Petrovski in the direction* N = (0, ..., 1) *if*

i/ *the coefficient of the term of highest degree in* ξ_n *in* $P(\xi)$ *is independent of* ξ'

ii/ *condition* (3.1.1) *is satisfied for a certain* γ_0.

EXAMPLES : The heat operator $\partial_{x_n} - \Delta_{x'}$ is correct in the direction N. The Schrödinger operator $i\partial_{x_n} - \Delta_{x'}$ is correct in the directions N and -N.

Remark 1.8 shows that in general we cannot expect to have uniqueness of the Cauchy problem for a correct operator; nonetheless we have

PROPOSITION 3.3 : *Suppose* P *is a correct operator in the direction* N; *then it admits a fundamental solution* E *with support in* $\overline{\mathbf{R}^n_+}$.

PROOF: By hypothesis, the polynomial $P(\xi', \tau)$ factorises into $\text{Const.} \prod_{j=1}^{p} (\tau - \lambda_j(\xi'))$, consequently there exists $c > 0$ such that

$$(3.3.1.) \quad |P(\tau, \xi')| \geq c\, |\operatorname{Im} \tau - \gamma_0|^p , \quad \text{for } \operatorname{Im} \tau < \gamma_0 \text{ and } \xi' \in \mathbf{R}^{n-1} .$$

This inequality allows us to define a distribution $F \in \mathcal{D}'$ by

$$(3.3.2.) \quad < F, \varphi > = \int \frac{\hat{\varphi}(\xi', \tau)}{P(\xi', \tau)} \, d\!\!\!/\xi , \qquad \varphi \in C_0^\infty$$

with $\tau = \xi_n + i\gamma$ and $\gamma < \gamma_0$.

We have

$$< P(-D)F, \varphi > = \int \hat{\varphi}(\xi', \tau) \, d\!\!\!/\xi = \varphi(0)$$

and therefore $P(D)\, E = \delta_o$ if $E = \check{F}$.

In order to show that $\operatorname{supp} F \subset \overline{\mathbb{R}^n_-}$, we take $\varphi \in C_0^\infty(\mathbb{R}^n_+)$ in (3.3.2). The Paley-Wiener theorem applied to φ in conjunction with the Cauchy theorem, shows that expression (3.3.2) is independent of γ for $\gamma < \gamma_o$ and that there exists a constant C such that

$$(3.3.3.) \quad |< F, \varphi >| \leq C \frac{\exp \gamma a}{|\gamma - \gamma_o|^p} \cdot \int \frac{d\xi}{(1 + |\xi|)^{n+2}} \quad , \quad \gamma < \gamma_o$$

where $a = \inf_{x \in \operatorname{supp} \varphi} x_n$. Making $\gamma \longrightarrow -\infty$, the upper bound (3.3.3) shows that $< F, \varphi > = 0$ and consequently that $\operatorname{supp} E \subset \overline{\mathbb{R}^n_+}$.

REMARK 3.4 : We say that P(D) is an evolution operator in the direction N if it admits a fundamental solution with support in $\overline{\mathbb{R}^n_+}$. An operator which is correct in the sense of Petrovski is an evolution operator; the converse does not hold, however, and we refer to HORMANDER [4] for a characterisation of evolution operators.

Remark 1.8 and Proposition 3.1 show that a necessary condition for the Cauchy problem to be well posed for P(D) in the direction N is that P be hyperbolic in the sense of the following definition:

DEFINITION 3.5 : *We say that* P(D) *is hyperbolic in the direction* N *if*

(i) $P_m(N) \neq 0$

(ii) *The condition* (3.1.1) *is satisfied for some* γ_o .

The last part of this section will be devoted to proving that this condition is sufficient; to this end, we require certain algebraic results relating to hyperbolic operators.

First, we have:

PROPOSITION 3.6 : *Suppose* P(D) *is a hyperbolic operator in the direction* N; *then it is hyperbolic in the direction* - N *and its principal part* $P_m(D)$ *is hyperbolic in the direction* N.

PROOF: Since $P_m(N) \neq 0$, the sum of the roots in τ of $P(\xi + \tau N)$ is an affine function of ξ of which the imaginary part is bounded above by $m\,\gamma_o$; it is therefore constant. Consequently, there exists some γ such that

$$P(\xi + \tau N) \neq 0 \quad \text{for} \quad (\xi, \tau) \in R^n \times C \text{ and } |\text{Im } \tau| \leq \gamma$$

and this proves the first assertion.

We have $\lambda^{-m} P(\lambda\xi + \lambda\,\tau N) \xrightarrow[\lambda \to +\infty]{} P_m(\xi + \tau N)$; now the roots in τ of the left-hand side are located within a strip $|\text{Im } \tau| \leq \frac{\gamma}{\lambda}$; therefore the roots in τ of the limit are real.

If the principal part P_m of P is hyperbolic, this does not in general imply that P is itself hyperbolic (see SVENSSON [1]); however, we have

PROPOSITION 3.7 : *Suppose that all the roots of the equation in* τ :

$$P_m(\xi + \tau N) = 0 \tag{3.7.1.}$$

are real and distinct for ξ *not proportional to* N. *Then* P(D) *is hyperbolic in the direction* N. *We say that it is strictly hyperbolic.*

For example, the wave operator $\partial^2_{x_n} - \Delta_{x'}$ is strictly hyperbolic in the direction (0, , ..., 0, 1). To prove the proposition, we take N = (0, ...,0, 1); we start by first proving

LEMMA 3.8 : *If* P *is strictly hyperbolic in the direction* N, *then there exists* c > 0 *such that*

$$|P(\xi', \tau)| \geq c\,|\mathrm{Im}\,\tau|\,(\,|\xi'| + |\tau|)^{m-1} \tag{3.8.1.}$$

$$\text{for} \quad \xi' \in \mathbb{R}^{n-1},\ \tau \in \mathbb{C} \quad \text{and} \quad |\mathrm{Im}\,\tau| > \frac{1}{c}\,.$$

PROOF: Let $\lambda_h(\xi')$ be the roots in τ of $P_m(\xi', \tau) = 0$; factorising P_m, we obtain

$$P_m(\xi', \tau) \cdot \overline{\frac{\partial P_m}{\partial \tau}(\xi', \tau)} = a \sum_{h=1}^{m} (\tau - \lambda_h(\xi')) \, |G_h(\xi', \tau)|^2$$

where $G_h(\xi', \tau) = \prod_{j \neq h} (\tau - \lambda_j(\xi'))$ and a is a non-zero constant.

The function $\sum_{h=1}^{m} |G_h(\xi', \tau)|^2$ does not vanish for $(\xi', \tau) \in (R^{n-1} \times C) \setminus 0$ and is homogeneous of degree $2(m - 1)$; there therefore exists $c > 0$ such that

$$(3.8.2.) \quad \left| P_m(\xi', \tau) \overline{\frac{\partial P_m}{\partial \tau} (\xi', \tau)} \right| \geq c \, |\text{Im } \tau| \, (|\tau| + |\xi'|)^{2(m-1)} .$$

Furthermore $\left|\frac{\partial P_m}{\partial \tau} (\xi', \tau)\right| \leq C \, (|\tau| + |\xi'|)^{m-1}$, hence (3.8.2) implies

$$|P_m(\xi', \tau)| \geq C' \, |\text{Im } \tau| \, (|\tau| + |\xi'|)^{m-1}$$

from which we immediately deduce (3.8.1).

We now return to the proof of the proposition. We put $P(\xi', \tau) = P_m(\xi', \tau) + Q(\xi', \tau)$. Since we have an upper bound $|Q(\xi', \tau)| \leq C \, (|\xi'| + |\tau|)^{m-1}$, the inequality (3.8.1) shows that $P(\xi', \tau) \neq 0$ when $|\text{Im } \tau| \geq \gamma$ for γ sufficiently large.

Naturally, Proposition 3.3 applies for hyperbolic operators; we shall show further that E has support in a cone of $R^n_+ \cup \{ 0 \}$.

DEFINITION 3.9: *Suppose* P *is a hyperbolic operator in the domain* N; *we denote by* $\Gamma(P, N)$ *the connected component of* N *in the open set* $\{ \xi \mid \xi \in \mathbb{R}^n$ *and* $P_m(\xi) \neq 0 \}$.

We shall now show that P is also hyperbolic in each direction of $\Gamma(P, N)$. We start with the following lemma:

LEMMA 3.10: $\Gamma(P, N)$ *is equal to the set of the* $\xi \in \mathbb{R}^n$ *such that the polynomial* $P_m(\xi + \tau N)$ *has only strictly negative roots in* τ.

PROOF: Let Γ' be this latter set; this set is clearly open and by definition $P_m(\xi) \neq 0$ for $\xi \in \Gamma'$ and further, $N \in \Gamma'$. In order to show that $\Gamma' \subset P$, it remains for us to prove that Γ' is connected. More precisely, we shall show that if $\xi \in \Gamma'$, then $\eta = \lambda\xi + \mu N \in \Gamma'$ where λ and μ are positive real numbers which sum to unity. By hypothesis , we have the factorisation

$$(3.10.1.)\ P_m(\xi + \tau N) = P_m(N) \prod_{j=1}^{m} (\tau - \tau_j(\xi)) \text{ with } \tau_j < 0$$

and consequently the polynomial

$$P_m(\eta + \tau N) = P_m(\lambda\xi + \mu N + \tau N) = P_m(N) \prod_1^r (\mu + \tau - \lambda \tau_j)$$

admits the roots $\lambda \tau_j - \mu < 0$.

Finally, if ξ belongs to the boundary of Γ', the roots of $P_m(\xi + \tau N)$ are negative or zero and there exists at least one zero root since $\xi \notin \Gamma'$. Hence $P_m(\xi) = 0$, which proves that ξ also belongs to the boundary of $\Gamma(P, N)$ and consequently $\Gamma' = \Gamma(P, N)$.

The cone $\Gamma(P, N)$ depends only on P_m; however we have

THEOREM 3.11: *Suppose* P *is hyperbolic in the direction* N. *Let* $\eta \in \Gamma(P, N)$; *then*

$$(3.11.1.) \quad P(\xi + \tau N + \mu\eta) \neq 0 \quad \textit{for} \quad \operatorname{Im} \tau < \gamma_0$$
$$\textit{and } \operatorname{Im} \mu \leq 0 \ , \ \xi \in R^n \ .$$

PROOF: For $\operatorname{Im} \mu = 0$, this is condition (3.1.1). The coefficient of μ^m is $P_m(\eta) \neq 0$; consequently the number of zeros μ such that $\operatorname{Im} \mu < 0$ is constant when the parameter τ remains in the half plane $\operatorname{Im} \tau < \gamma_0$. However, for τ large and $\mu = \lambda\tau$, the equation $P(\xi + \tau N + \mu\eta) = 0$ is equivalent to

$$(3.11.2.) \qquad \tau^{-m} P(\xi + \tau(N + \lambda\eta)) = 0$$

which converges to $P_m(N + \lambda\eta) = 0$ when $\operatorname{Im} \tau \to -\infty$.

Lemma 3.10 shows that $P_m(N + \lambda\eta) \neq 0$ for $\operatorname{Re} \lambda > 0$, consequently, the same applies for (3.11.2) when $\operatorname{Im} \tau < \gamma_0$, and this proves the theorem.

As an application, we deduce

THEOREM 3.12. *If* P *is hyperbolic in the direction* N, *it is hyperbolic in any direction of* $\Gamma(P, N)$ *and, moreover, this cone is convex.*

PROOF: Let $\eta \in \Gamma(P, N)$; we start by showing that P is hyperbolic in the direction $\eta + \varepsilon N$ with $\varepsilon > 0$. In fact, Theorem 3.11 implies

$$P(\xi + \tau(\eta + \varepsilon N)) \neq 0 \quad \text{for} \quad \mathrm{Im}\, \tau < \inf(0, \frac{\gamma_0}{\varepsilon}) \ , \ \xi \in \mathbb{R}^n$$

and Lemma 3.10 shows that $P_m(\eta + \varepsilon N) \neq 0$ for $\varepsilon > 0$. Writing $\eta = (\eta - \varepsilon N) + \varepsilon N$, we then deduce that $\eta \in \Gamma(P, N)$ by taking $\varepsilon > 0$ small enough to ensure that $\eta - \varepsilon N$ remains in the open set $\Gamma(P, N)$.

To prove that the cone $\Gamma(P, N)$ is convex we show that the segment joining two of its points η_1 and η_2 remains in its interior. In fact we showed during the proof of Lemma 3.10 that this segment is inside the cone $\Gamma(P, \eta_1) = \Gamma(P, N)$.

An analogous result holds for strictly hyperbolic operators; this is:

THEOREM 3.13: *If* P *is strictly hyperbolic in the direction* N, *then it is strictly hyperbolic in any direction of the cone* $\Gamma(P, N)$.

PROOF: For $\xi \in \mathbb{R}^n$ and not proportional to N, we have the factorisation

$$\text{(3.13.1.)} \qquad P_m(\xi + \tau N) = P_m(N) \prod_1^m (\tau - \tau_j(\xi))$$

where the $\tau_j(\cdot)$ are C^∞ functions which are mutually distinct for $\xi \in \mathbb{R}^n \setminus (\mathbb{R}N)$. Since $P_m(N) \neq 0$, the roots $\tau_j(\xi)$ tend to the single root $\tau = -\lambda$ when ξ tends towards λN; consequently, the functions $\tau_j(\xi)$ are continuous on $\mathbb{R}^n$.

Suppose we have $\eta \in \Gamma(P, N)$, not proportional to N, and $\mu \in \mathbb{R}$; we deduce from (3.13.1):

$$(3.13.2.) \qquad P_m(\xi + \mu\eta) = P_m(-N) \prod_1^m \tau_j(\xi + \mu\eta) .$$

We now show that

$$(3.13.3.) \qquad \tau_j(\xi + \mu\eta) \longrightarrow \mp \infty \quad \text{when} \quad \mu \longrightarrow \pm \infty .$$

The homogeneity and the continuity of the functions τ_j show that

$$\tau_j(\xi + \mu\,\eta) = \mu\,\tau_j\left(\frac{\xi}{\mu} + \eta\right) \underset{|\mu|\,\infty}{\sim} \mu.\,\tau_j(\eta)$$

which proves (3.13.3) since Lemma 3.10 tells us that the $\tau_j(\eta)$ are strictly negative.

The assertion (3.13.3) and the continuity of the τ_j imply that for any j there exists μ_j such that $\tau_j(\xi + \mu_j\,\eta) = 0$, and (3.13.2) shows that the μ_j are roots of $P_m(\xi + \mu\eta) = 0$. It now only remains to show that the μ_j are distinct in pairs when ξ is not proportional to η. In fact, if there existed $\mu_j = \mu_k = \mu$ with j and k distinct, we would have $\tau_j(\xi + \mu\eta) = \tau_k(\xi + \mu\eta) = 0$ and consequently $\xi + \mu\eta = \lambda\,N$ with $\lambda \in \mathbb{R}$. Then the equality $P_m(\xi + \mu\eta) = 0 = \lambda^m P_m(N)$ shows that $\lambda = 0$, i.e. that $\xi + \mu\eta = 0$, which is a contradiction.

We are now in a position to prove the principal result of this section; this is

THEOREM 3.14: *Suppose* P *is a hyperbolic operator in the direction* N; *then* P *admits a unique fundamental solution with support in the dual cone*

$$\Gamma^*(P, N) = \{ x \in \mathbb{R}^n \mid x.\xi \geq 0 \text{ for all } \xi \in \Gamma(P, N) \}$$

PROOF: Since P is, in particular, correct in the sense of Petrovski in the direction N, Proposition 3.3 shows that it admits a fundamental solution E with support in $\overline{\mathbb{R}^n_+}$ and defined by

$$(3.14.1.) \quad < \check{E}, \varphi > = \int_{R^n} \frac{\hat{\varphi}(\xi + i\gamma N)}{P(\xi + i\gamma N)} \, d\!\!/\xi \quad , \text{ for all } \quad \varphi \in C_0^\infty$$

with $\mathrm{Re}\,\gamma < \gamma_0$.

Theorem 3.11 implies that the function

$$(\xi + i\eta) \longrightarrow \frac{\hat{\varphi}(\xi + i\gamma N + i\eta)}{P(\xi + i\gamma N + i\eta}$$

is holomorphic in the open set $R^n - i\,\Gamma(P, N)$. Furthermore, the factorisation

$$P(\xi + i\eta + i\gamma N) = P_m(iN) \prod_1^m (\gamma - \tau_k(\xi + i\eta))$$

shows that

$$|P(\xi + i\eta + i\gamma N)| \geq |P_m(N)|\,|\gamma - \gamma_0|^m \tag{3.14.2.}$$

when $\operatorname{Re}\gamma < \gamma_0$. The Paley-Wiener theorem applied to φ implies the existence of C such that

$$|\hat{\varphi}(\xi + i\gamma N + i\lambda\eta)| \leq C \frac{\exp H_k(\gamma N + \lambda\eta)}{(|\xi| + |\gamma| + |\lambda|)^{n+1}} \tag{3.14.3.}$$

for $\xi \in \mathbb{R}^n$, λ and γ real. We deduce from (3.14.2), (3.14.3) that we can apply a translation of the contour of integration in (3.14.1) and obtain

$$< \check{E}, \varphi > = \int_{\mathbb{R}^n} \frac{\hat{\varphi}(\xi + i\gamma N + i\lambda\eta)}{P(\xi + i\gamma N + i\lambda\eta)} \, đ\xi$$

for all $\gamma < \gamma_0$, $\lambda < 0$ and $\eta \in \Gamma(P, N)$.

To show that E has support in $\Gamma^*(P, N)$, we take $\varphi \in C_0^\infty$ such that

$$\operatorname{supp}\varphi \subset \{ x \in \mathbb{R}^n \mid x.\eta \geq 0 \}$$

where $\eta \in \Gamma(P, N)$. When $\lambda \to -\infty$, the inequalities (3.14.2) (3.14.3) show that $< \check{E}, \varphi > = 0$. Consequently, $\operatorname{supp} E \subset \{ x \mid x.\eta \geq 0 \}$, and this shows that $\operatorname{supp} E \subset \Gamma^*(P, N)$.

Since $\Gamma(P, N)$ contains a neighbourhood of N, the dual cone $\Gamma^*(P, N)$ in included in $\mathbb{R}_n^+ \cup \{ 0 \}$. From Proposition 2.2 of Chapter I, we can therefore perform a convolution of a distribution with support in $\mathbb{R}_+^n$ and a distribution with support in $\Gamma^*(P, N)$; this immediately implies the uniqueness of a fundamental solution with support in $\mathbb{R}_+^n$ (this also follows from Proposition 1.13). Finally, Proposition 3.1, Remark 1.8 and Theorem 3.14 show that we have

THEOREM 3.15: *The following assertions are equivalent:*

i/ *the Cauchy problem is well posed for* P(D) *in the direction* N

ii/ P *is hyperbolic in the direction* N

iii/ P *admits a fundamental solution with support in a cone* C *such that* $C \cap \{ x \mid x.N \leq 0 \} = \{ 0 \}$.

In the case where P(D) is an r x r system, we have

PROPOSITION 3.16: *The following assertions are equivalent:*

i/ *The system* $P = (P_{ij}(D))_{ij=1\ldots r}$ *admits a fundamental solution with support in a cone* C *such that*

$$C \cap \{ x \mid x.N \geq 0 \} = \{ 0 \}.$$

ii/ *The scalar operator* $p = \det(P_{ij}(D))$ *is hyperbolic in the direction* N .

When this is satisfied, we say that P *is hyperbolic in the direction* N.

PROOF: i/ $\Longrightarrow$ ii/ . Let $E = (E_{hk})$ be the fundamental solution of P with support in C; we have

$(P_{ij}(D)\delta_0) * (E_{hk}) = \delta_0.I$. Taking the determinant in the sense of the convolution this equality shows that

$$p\delta_0 * \det(E_{hk}) = \delta_0 .$$

Since we have $\operatorname{supp}(\det(E_{hk})) \subset \operatorname{supp} E \subset C$, Theorem 3.15 implies that the operator p is hyperbolic in the direction N. In order to prove the implication ii/$\Longrightarrow$ i/, we introduce the matrix $Q = (Q_{s,t}(D))$ of the cofactors of $P_{ij}(D))$. Let E_0 be the fundamental solution of p; the equality

$$P(D) \quad Q(D) = I \cdot p(D)$$

shows that P admits $E = Q(D) \quad (E_0 \cdot I)$ as a fundamental solution with support in a cone of $R^n_+ \cup \{0\}$.

4. HYPERBOLIC CAUCHY PROBLEMS WITH VARIABLE COEFFICIENTS

4.1 INTRODUCTION: We consider the Cauchy problem for first-order systems of the form $P = D_t - A(t, x, D_x)$ where A is an N x N matrix of pseudo-differential operators of degree one in $x \in X$ and with C^∞ dependence on the "time" parameter t .

We now propose to prove, under certain assumptions, the existence and uniqueness of u such that

$$\text{(4.1.1.)} \quad \begin{cases} D_t u - A(t, x, D_x)u = f & \text{in } \mathbb{R} \times X \\ u|_{t=0} = g \end{cases}$$

The investigation may be carried out in analogous fashion when X is a compact manifold without boundary or when $X = \mathbb{R}^n$. We shall restrict our attention to the latter case since it requires certain additional assumptions on the behaviour of the symbol of A for large $|x|$. More precisely, we shall assume henceforth that the complete symbol $a(t, x, \xi)$ of A is in the space $BS^1(\mathbb{R}^{n+1} \times \mathbb{R}^n ; \mathcal{L}(\mathbb{C}^N, \mathbb{C}^N))$ (see 10.1, Chapter III) and that it admits a principal part $a_1(t, x, \xi)$ which is homogeneous of degree one in ξ for $|\xi| \geq 1$. Then, for $u \in C^\infty([0\ T] ; \mathcal{S})$, we have:

$$(4.1.2.)\quad (Au)(t, x) = \int e^{ix\xi}\, a(t, x, \xi)\, u(t, \hat{\xi})\, d\xi \in C^\infty([0\ T] ; \mathcal{S}).$$

In the special case where A is a differential operator with constant coefficients of the form $\sum_1^n A_j D_j$, Proposition 3.16 shows that the Cauchy problem (4.1.1) is well posed if and only if the equation

$$\det\left(\tau - \sum_1^n A_j \xi_j\right) = 0$$

has only real roots in τ for $\xi \in \mathbb{R}^n$. For example, numerous systems of physical origin lead to the case where the matrices A_j are hermitian.

In the general case of operators of the form (4.1.2), we shall prove in this section that the Cauchy problem is well posed for operators which are symmetrisable in the sense of the following definition:

DEFINITION 4.2: *The operator* $D_t - A$ *is said to be symmetrisable if there exists a* C^∞ *mapping:*

$$(t, x, \xi) \in \mathbb{R}^{n+1} \times (\mathbb{R}^n \setminus 0) \longrightarrow r_0(t, x, \xi)$$

with values in N × N *hermitian matrices, which is homogeneous of degree* 0 *in* $\xi \in \mathbb{R}^n \setminus 0$, *and which is bounded together with all its derivatives for* $|\xi| = 1$ *and such that*

(4.2.1.) $r_0(t, x, \xi) \cdot a_1(t, x, \xi)$ *is hermitian*

(4.2.2.) $r_0(t, x, \xi) \geq c I$

with $c > 0$ *independent of* $(x, \xi) \in \mathbb{R}^n \times \mathbb{R}^n \setminus 0$ *and of* t *in a compact subset of* $\mathbb{R}$.

We start by establishing an energy inequality.

THEOREM 4.3 : *Let* $P = D_t - A$ *be a symmetrisable operator. Then, for all* $s \in \mathbb{R}$ *and* $T > 0$, *there exists a* C *such that*

$$(4.3.1.)_s \quad \|u(t, .)\|_s^2 \leq C \left[\|u(0, .)\|_s^2 + \int_0^T \|Pu(t', .)\|_s^2 \, dt' \right]$$

for $t \in [0\ T]$ *and* $u \in C^0([\,0\ T\,]; H^{s+1}) \cap C^1([\,0\ T\,]; H^s)$. *Furthermore, we can replace* u(0, .) *by* u(T, .) *in the right-hand side of* $(4.3.1)_s$.

PROOF: We shall show that, under the conditions of the theorem, there exists a C' such that for $T' \in [\,0, T\,]$ we have

$$(4.3.2.)_s \quad \|u(T', .)\|_s^2 \leq C' \left[\|u(0, .)\|_s^2 + \int_0^{T'} (\|u(t', .)\|_s^2 + \|Pu(t', .)\|_s^2) \, dt' \right].$$

We shall see that $(4.3.2)_s$ implies $(4.3.1)_s$; $(4.3.2)_s$ has the advantage of allowing errors of order $O(\|u(t', .)\|_s)$ in $\|Pu(t', .)\|_s$. We first note that it is sufficient to establish $(4.3.2)_o$ since by applying $(4.3.2)_o$ to $v = \Lambda_s u$ (where Λ_s is the p.d.o. with symbol $(1 + |\xi|^2)^{s/2}$) we obtain $(4.3.2)_s$ by observing that from Section 11.1 of Chapter IV we have

$$\|Pv(t', .)\|_o = \|Pu(t', .)\|_s + O(\|u(t', .)\|_s) .$$

In order to prove $(4.3.2)_s$, we introduce a self-adjoint p.d.o. $R(t, x, D_x) \in BL^o$ with principal symbol $r_o(t, x, \xi)$ with C^∞ dependence on t. The idea is to calculate $F(t) = \frac{d}{dt}(Ru(t, .), u(t, .))$ (we denote by (,) the inner product in $L^2(\mathbb{R}^n)$) and by $\| \ \|$ the associated norm).

We obtain

$$F(t) = (R\,\partial_t u, u) + (Ru, \partial_t u) + (R_t' u, u)$$

where the symbol of the operator R_t' is obtained by differentiating the symbol of R with respect to t. Or, putting Pu = f.

$$(4.3.3.) \qquad F(t) = i(RA\,u, u) - i(Ru, Au) + i(Rf, u) + (Ru, if) + (R_t' u, u)$$

$$= i((RA - A^*R)u, u) + O(\|u(t, .)\|^2 + \|f(t, .)\|^2) .$$

The assumption (4.2.1) implies that the operator $RA - A^*R$ is of degree 0, and consequently Exercise 11.1 in Chapter IV shows that

$$((RA - A^*R)\, u(t, .), u(t, .)) = O(\|u(t, .)\|^2) \quad .$$

Substituting in (4.3.3) and integrating between 0 and T' we obtain, taking account of the L^2 continuity of the operator R,

$$(4.3.4.) \quad (Ru(T', .), u(T', .)) \leq C \left[\|u(0, .)\|^2 + \int_0^{T'} u(\|(t', .)\|^2 + \|f(t', .)\|^2)\, dt' \right]$$

with $T' \in [0\ T]$.

Furthermore, assumption (4.2.2) allows *Gårding's* inequality to be used (see (11.1.12), Chapter IV): there exist $c > 0$ and C such that

$$(4.3.5.)(Ru(T', .), u(T')) \quad \geq \quad c\, \|u(T', .)\|^2 - C\, \|u(T', .)\|_{-1}^2 .$$

Replacing R by $R + C\, \Lambda_{-2}$, we can assume that (4.3.5) holds with $C = 0$. This gives $(4.3.2)_s$ by substituting into (4.3.4).

In order to show that $(4.3.2)_s$ implies $(4.3.1)_s$, it is sufficient to apply the following lemma to $y(t) = \|u(t, .)\|_s^2$ and $\|Pu(t, .)\|_s^2 = \varphi(t)$.

LEMMA 4.4 : *Suppose* y *and* φ *are positive continuous functions in* $[0\ T]$ *which, with constants* α, $C \geq 0$, *satisfy the inequality*

$$y(t) \leq C\left[\alpha + \int_0^t (y(t') + \varphi(t'))\,dt'\right] \quad .$$

Then

$$(4.4.1.) \qquad y(t) \leq C\left[\alpha\, e^{Ct} + \int_0^t e^{C(t-t')}\,\varphi(t')\,dt'\right] .$$

PROOF: The right-hand side of (4.4.1) is the solution z of $z' = Cz + C\varphi$, $z(0) = C\alpha$. We have

$$z(t) = C\left[\alpha + \int_0^t (z(t') + \varphi(t'))\,dt'\right]$$

and consequently the function $\rho = y - z$ satisfies

$$\rho(t) \leq C \int_0^t \rho(t')\,dt' \quad .$$

Putting $\psi(t) = \int_0^t \rho(t')\,dt'$, we obtain:

$$\psi'(t) \leq C\,\psi(t) \quad , \quad \text{i.e.} \quad {}^{d}/_{dt}\,(e^{-Ct}\,\psi(t)) \leq 0 \quad .$$

We thus have $\psi(t) \leq 0$ since $\psi(0) = 0$. As $\rho(t) \leq C\,\psi(t)$, we also have $\rho(t) \leq 0$, that is $y(t) \leq z(t)$.

Finally, replacing t by -t, we see that we can replace $u(0, .)$ by $u(T, .)$ in $(4.3.1)_s$.

By using a functional analysis argument, we shall now deduce from the inequalities $(4.3.1)_s$ a theorem concerning existence, uniqueness and regularity.

THEOREM 4.5 : *Suppose* $P = D_t - A$ *is symmetrisable. Let* $s \in \mathbb{R}$, $f \in L^2([\,0\ T\,];\, H^s)$ *and* $g \in H^s$. *Then there exists a unique* $u \in C^0([\,0\ T\,];\, H^s)$ *such that*

$$(4.5.1.)\quad \begin{cases} Pu = f & (\text{ in the sense of } \mathcal{D}'(]\,0\ T\,[\,\times \mathbb{R}^n) \\ u(0,\,.) = g & . \end{cases}$$

Moreover, the energy inequality $(4.3.1)_s$ *remains valid for* u. *If* $f \in C^\infty([\,0\ T\,];\, H^{+\infty})$ *and* $g \in H^{+\infty}$ *then* $u \in C^\infty([\,0\ T\,];\, H^{+\infty})$.

PROOF: The proof of uniqueness is immediate; in fact if $u \in C^0([\,0\ T\,];\, H^s)$ satisfies $D_t\, u = Au$, we deduce that $u \in C^1([\,0\ T\,];\, H^{s-1})$ and the inequality $(4.3.1)_{s-1}$ then shows that $u = 0$.

In order to prove the existence, we introduce the space $E = \{\varphi \in C^\infty([\,0\ T\,];\, H^{-\infty}) \mid \varphi(T) = 0\}$. We shall now show that it is possible to define a linear form ℓ on P^*E by

$$P^*\varphi \longrightarrow \ell(P^*\varphi) = \int_0^T (f(t,\,.),\varphi(t,\,.))\,dt + \frac{1}{i}(g,\,\varphi(0,\,.)) \quad .$$

For this we note that the energy inequality $(4.3.1)_{-s}$ holds for the Cauchy problem relating to the operator P^* with $\|u(T,\,.)\|_s$ on the right-hand side of $(4.3.1)_{-s}$ (note that P^* is symmetrisable by r_0^{-1}). We thus have, for $\varphi \in E$:

$$\|\varphi(t, .)\|_{-s}^2 \leq C \int_0^T \|P^* \varphi(t', .)\|_{-s}^2 \, dt' \quad , \quad t \in [\, 0 \; T\,] \quad ,$$

so that

$$|\ell(P^* \varphi)|^2 \leq C' \int_0^T \|P^* \varphi(t', .)\|_{-s}^2 \, dt' \qquad .$$

This shows that ℓ is well defined and is, moreover, continuous for the topology induced on P^*E by $L^2([\, 0 \; T\,]; H^{-s})$. Consequently the Hahn-Banach theorem implies the existence of an element $u \in (L^2([\, 0 \; T\,]; H^{-s}))' = L^2([\, 0 \; T\,]; H^{s})$ such that

$$(4.5.2.) \quad (u, P^* \varphi) = \int_0^T (f(t, .), \varphi(t, .))\, dt + \frac{1}{i} (g, \varphi(0, .))$$

for all $\varphi \in E$. In particular, if $\varphi \in C_0^\infty(]\, 0 \; T\,[\times R^n)$, (4.5.2) shows that $Pu = f$ in $\mathcal{D}'([\, 0 \; T\,] \times R^n)$. Thus $D_t u = Au + f \in L^2([\, 0 \; T\,]; H^{s-1})$, so that $u \in C^0([\, 0 \; T\,]; H^{s-1})$. After integrating by parts with respect to t, (4.5.2) implies $(u(0, .), \varphi(0, .)) = (g, \varphi(0, .))$ for all $\varphi \in E$ and consequently $u(0, .) = g$. If $f \in C^\infty([\, 0 \; T\,]; H^{+\infty})$ and $g \in H^{+\infty}$, the above argument shows that $u \in C^0([\, 0 \; T\,]; H^{+\infty})$ and since $D_t u = Au + f$ we then deduce step by step that $u \in C^k([\, 0 \; T\,]; H^{+\infty})$ for all $k \geq 0$.

It now remains to show that $u \in C^0([\, 0 \; T\,]; H^s)$ and that it satisfies the energy inequality $(4.3.1)_s$.

Suppose we have sequences (f_j), (g_j) such that

$f_j \in C_0^\infty([\,0\;T\,] \times \mathbb{R}^n)$ and $f_j \longrightarrow f$ in $L^2([\,0\;T\,]; H^s)$,

and that $g_j \in C_0^\infty(\mathbb{R}^n)$ and $g_j \to g$ in H^s. Let

$u_j \in C^\infty([\,0\;T\,]; H^{+\infty})$ be the solution of $Pu_j = f_j$, $u_j(0, .) = g_j$.

The inequality $(4.3.1)_s$ applied to $u_j - u_k$ shows that u_j is a Cauchy sequence in $C^0([\,0\;T\,]; H^s)$ so that $u_j \longrightarrow \tilde{u}$ in $C^0([\,0\;T\,]; H^s)$. In the limit, we have $P\tilde{u} = f$, $\tilde{u}(0, .) = g$; consequently, the uniqueness shows that $\tilde{u} = u$. Furthermore, the inequality $(4.3.1)_s$ for u is obtained by proceeding to the limit in this inequality applied to the u_j.

We now give some examples of symmetrisable operators.

EXAMPLE 4.6 : When the principal symbol a_1 of A is hermitian, the operator is obviously symmetrisable. For example, the Maxwell equations in a vacuum are written classically as

$$\partial_t E = \text{curl}\, H \qquad\qquad \partial_t H = -\,\text{curl}\, E$$

where E and H are functions in $\mathbb{C}^3$, defined over $\mathbb{R}^4$.

Putting $u = \begin{pmatrix} E \\ H \end{pmatrix}$, this is equivalent to the symmetric system

$$\partial_t u = \begin{bmatrix} 0 & \begin{matrix} 0 & 0 & 0 \\ 0 & 0 & 1 \\ 0 & -1 & 0 \end{matrix} \\ \begin{matrix} 0 & 0 & 0 \\ 0 & 0 & -1 \\ 0 & 1 & 0 \end{matrix} & 0 \end{bmatrix} \frac{\partial u}{\partial x_1} + \begin{bmatrix} 0 & \begin{matrix} 0 & 0 & -1 \\ 0 & 0 & 0 \\ 1 & 0 & 0 \end{matrix} \\ \begin{matrix} 0 & 0 & 1 \\ 0 & 0 & 0 \\ -1 & 0 & 0 \end{matrix} & 0 \end{bmatrix} \frac{\partial u}{\partial x_2} + \begin{bmatrix} 0 & \begin{matrix} 0 & 1 & 0 \\ -1 & 0 & 0 \\ 0 & 0 & 0 \end{matrix} \\ \begin{matrix} 0 & -1 & 0 \\ 1 & 0 & 0 \\ 0 & 0 & 0 \end{matrix} & 0 \end{bmatrix} \frac{\partial u}{\partial x_3}$$

EXAMPLE 4.7 : The operator $P = D_t - A$ is said to be strictly hyperbolic if the eigenvalues of $a_1(t, x, \xi)$ are real and distinct when $(t, x, \xi) \in \mathbb{R}^{n+1} \times (\mathbb{R}^n \setminus 0)$. Suppose further that a_1 is independent of (t, x) for $|x|$ sufficiently large. The projector onto the eigen-subspace of $a_1(t, x, \xi)$ associated with the eigenvalue $\lambda_j(t, x, \xi)$ is given by

$$(4.7.1.)\quad p_j(t, x, \xi) = \frac{1}{2i\pi} \oint_{|\lambda - \lambda(t, x, \xi)| = \rho} (a_1(t, x, \xi) - \lambda)^{-1} \, d\lambda$$

with
$$\rho = \frac{1}{2} \inf_{\substack{|\xi| = 1 \\ (t, x) \in \mathbb{R}^{n+1} \\ k \neq h}} |\lambda_k(t, x, \xi) - \lambda_h(t, x, \xi)| \quad .$$

Expression (4.7.1) shows that p_j is independent of (t, x) for $|x|$ sufficiently large, and is homogeneous of degree 0 in $\xi \in \mathbb{R}^n \setminus 0$. The same therefore applies for the function

$$r_o(t, x, \xi) = \sum_{j=1}^{N} p_j^*(t, x, \xi) \, p_j(t, x, \xi) \quad .$$

For $v \in \mathbb{C}^N$, we have

$$(r_o a_1 v, v) = \sum_1^N (p_j a_1 v, p_j v) = \sum_1^N (a_1 p_j v, p_j v)$$

$$= \sum \lambda_j \|p_j v\|^2 \quad .$$

Thus, $r_o a_1$ is hermitian. Finally, $r_o(t, x, \xi)$ is positive-definite since $(r_o(t, x, \xi)v, v) = \sum_j \|p_j v\|^2 > 0$ for $v \in C^N \setminus 0$.

EXAMPLE 4.8 : Strictly hyperbolic scalar operator of order m.

Suppose we have

$$P(t, x, D_t, D_x) = D_t^m - \sum_{j=0}^{m-1} A_{m-j} D_t^j \tag{4.8.1.}$$

where $A_{m-j} = a_{m-j}(t, x, D_x)$ with $a_{m-j} \in BS^{m-j}(R^{n+1} \times R^n)$; suppose that a_{m-j} admits a principal part $a^o_{m-j}(t, x, \xi)$ which is homogeneous of degree m-j for $\xi \in R^n \setminus 0$ and which is independent of (t, x) for |x| sufficiently large.

The Cauchy problem for P is written formally as follows:

$$Pu = f \quad , \quad D_t^k u_{|t=0} = g_k \qquad k = 0,\dots, m-1 \quad . \tag{4.8.2.}$$

We shall now reduce this to a first-order system; to do this, we put $u_j = D_t^{j-1} \Lambda_{m-j} u$ and $U = (u_1,\dots, u_m)$. Then, (4.8.2) is equivalent to

$$D_t U - AU = F \quad , \quad U_{|t=0} = G \tag{4.8.3.}$$

where

$$A = \begin{bmatrix} 0 & \Lambda & 0 & \cdots & 0 \\ \vdots & \ddots & \ddots & \ddots & \vdots \\ \vdots & & \ddots & \ddots & 0 \\ 0 & \cdots & \cdots & 0 & \Lambda \\ B_1 & \cdots & \cdots & \cdots & B_m \end{bmatrix} \qquad \text{with } B_j = A_{m-j+1}\,\Lambda_{-m+j}$$

$$F = (0, \ldots, 0, f)$$

$$G = (\Lambda_{m-j}\, g_j) \quad .$$

Consequently B_j is a p.d.o. of degree 1 with symbol $a_{m-j+1}(t, x, \xi)(1 + |\xi|^2)^{-\frac{1}{2}(m-j)}$ and A has as **homogeneous** principal symbol of degree 1,

$$A_1(t, x, \xi) = \begin{bmatrix} 0 & |\xi| & 0 & \cdots & 0 \\ \vdots & \ddots & \ddots & \ddots & \vdots \\ \vdots & & \ddots & \ddots & 0 \\ 0 & \cdots & \cdots & 0 & |\xi| \\ \dfrac{a_m}{|\xi|^{m-1}}, & \cdots & \cdots & \cdots, & a_1 \end{bmatrix} .$$

It can easily be shown that $\det(\tau I - A_1(t, x, \xi)) = p_m(t, x, \tau, \xi)$ where p_m is the principal symbol of P.

The above allows us to state the following:

THEOREM AND DEFINITION 4.9 : *The operator* P *defined in* (4.8.1) *is said to be strictly hyperbolic if for all* $(t, x, \xi) \in R^{n+1} \times (R^n \setminus 0)$ *the zeros in* τ *of its principal symbol* $p_m(t, x, \tau, \xi) = \tau^m - \sum_{j=0}^{m-1} a^0_{m-j}(t, x, \xi)\,\tau^j$ *are real and distinct. We assume in addition that* p_m *is independent of* (t, x) *for* $|x|$ *sufficiently large. Let* $s \in \mathbb{R}$, $f \in L^2([0\,T]; H^s)$, $g_j \in H^{s+m-j}$ $j = 1, \ldots, m$. *Then there exists a unique*

$$u \in \bigcap_{k=0}^{m-1} C^k([0\,T]; H^{s+m-1-k})$$

such that $Pu = f$ *in* $\mathcal{D}'(]0\,T[\times \mathbb{R}^n)$, $D_t^{j-1} u(0, .) = g_j$.
Furthermore, there exists C *such that*

$$(4.9.1.)\quad \sum_{j=1}^{m} \|D_t^{j-1} u(t, .)\|^2_{s+m-j} \le C\left[\sum \|g_j\|^2_{j+m-j} + \int_0^T \|f(t, .)\|^2_s\, dt\right]$$

for all f, g_j *satisfying the above assumptions and* $t \in [0\,T]$.

Finally, if $f \in C^\infty([0\,T]; H^{+\infty})$ *and* $g_j \in H^{+\infty}$ *we have* $u \in C^\infty([0\,T]; H^{+\infty})$.

PROOF: It is sufficient to apply Theorem 4.5 to the system (4.8.3).

The standard example of a strictly hyperbolic operator is the wave operator $P = D_t^2 - c^2(x)\,\Delta_x$ where c is a strictly positive C^∞ function (which in this case is constant for $|x|$ large). Physically, $c(x)$ is interpreted as being the wave propogation speed. We shall show more generally that we have the notion of a propagation speed when P is a strictly hyperbolic differential operator.

Consider a differential operator of degree m, $P = P(t, x, D_t, D_x)$, with C^∞ coefficients in an open subset Ω of $\mathbb{R}^{n+1}$. Let p_m be its principal part; saying that P is strictly hyperbolic with respect to t is equivalent to saying that for all $(t, x) \in \Omega$ the polynomial $p_{t,x}(\tau, \xi) = p_m(t, x, \tau, \xi)$ is strictly hyperbolic with respect to $N = (1; 0)$. We denote by $\Gamma(t, x)$ the cone $\Gamma(p_{t,x}, N)$ (see Definition 3.9). We start by giving a definition of the hypersurfaces which will play a role analogous to that of the hyperplane $t = 0$.

DEFINITION 4.10 : *A hypersurface* S *of class* C^1 *in* $\mathbb{R}^{n+1}$ *is said to be spacelike for* P *at a point* $(t, x) \in S$ *if the covectors normal to* S *at* (t, x) *are in* $\Gamma(t, x) \cup (-\Gamma(t, x))$. *It is said to be spacelike for* P *if this is true at all points.*

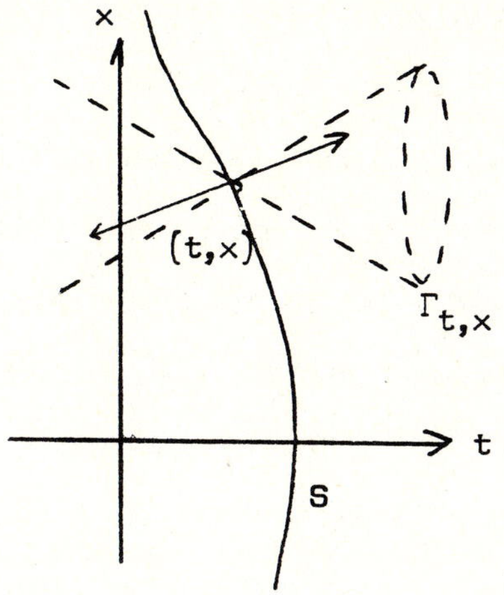

We note that with this definition, Theorem 3.13 shows that $p_{t,x}$ is strictly hyperbolic with respect to any non-zero conormal of S at (t, x). This will form the basis of the following local uniqueness result.

DEFINITION 4.11 : *Suppose* P *is a differential operator of degree* m *with* C^∞ *coefficients which is strictly hyperbolic in a neighbourhood* V *of a point* (t_0, x^0). *Let* S *be a* C^∞ *hypersurface of* V. *Suppose that* S *is spacelike for* P *at* (t_0, x^0). *Then there exists a neighbourhood* W *of* (t_0, x^0) *such that if* $u \in C^m(V)$ *satisfies* $Pu = 0$ *in* V , $D^\alpha u_{|S} = 0 \quad |\alpha| \leq m-1$, *we have* $u = 0$ *in* W.

PROOF: Let χ be a diffeomorphism defined in the neighbourhood of (t_0, x^0) which transports S into

$\tilde{S} = \{ (\tilde{t}, \tilde{x}) \mid \tilde{t} = 0 \}$ and (t_o, x^o) into 0. Let $\tilde{P}$ be the operator obtained by transporting P. Now (7.3.5), Chap. I, shows that $\tilde{p}_{0,0}(\tilde{\tau}, \tilde{\xi}) = p_{t_0,x_0}(\tau, \xi)$; consequently $\tilde{p}_{0,0}$ is strictly hyperbolic with respect to $\tilde{N} = (1; 0)$. A transversality argument shows immediately that $\tilde{p}_{\tilde{t},\tilde{x}}$ is still strictly hyperbolic with respect to $\tilde{N}'$ when $(\tilde{t}, \tilde{x})$ and $\tilde{N}'$ are respectively sufficiently near 0 and $\tilde{N}$. The above allows us to prove the existence of a neighbourhood W in which uniqueness applies, by proceeding as in the proof of Theorem 1.10 (see Remark 1.12). We are in fact then led to using an existence theorem for a Cauchy problem of the type (1.10.4). For this, it is sufficient to apply Theorem 4.9 to ${}^t\tilde{P}$ after extending it to the whole of $\mathbb{R}^{n+1}$. This is possible by virtue of:

LEMMA 4.12 : *Let* $Q = Q(t, x, D_t, D_x)$ *be a differential operator (or pseudo-differential operator with respect to* x) *of degree* m, *with* C^∞ *coefficients, which is strictly hyperbolic in a ball* B_r *of radius* r *at* 0. *Then there exists a strictly hyperbolic operator* $\tilde{Q}$ *of degree* m *which coincides with* Q *in the ball* $B_{r/2}$ *and which has constant coefficients outside the ball* B_r.

PROOF: Let $\theta(s) \in C^\infty(\mathbb{R})$ be equal to unity for $s \le \frac{r}{2}$, zero for $s \ge r$, and lying between 0 and 1. The mapping

$$(t, x) \in \mathbb{R}^{n+1} \longrightarrow \rho(t, x) = \theta(|t| + |x|) \begin{pmatrix} t \\ x \end{pmatrix} \in \mathbb{R}^{n+1}$$

coincides with the identity map in the ball $B_{r/2}$ and is zero outside the ball B_r.

Then the operator $\tilde{Q}$ with complete symbol $Q(\rho(t, x), \tau, \xi)$ meets the requirements.

This local uniqueness result can be globalised. For example, the following theorem establishes the existence of a finite wave propagation speed for strictly hyperbolic differential operators:

THEOREM 4.13 : *Suppose* P *is a differential operator of degree* m, *strictly hyperbolic in* $[\,0, T\,] \times R^n$ *and with constant coefficients for* $|x|$ *large. We put*

$$v = \sup_{\substack{t \in [\,0,\, T\,],\ x \in R^n \\ |\xi| = 1\ ,\quad j}} |\tau_j(t, x, \xi)|$$

where τ_j *are the zeros with respect to* τ *of* p_m.

For $(t_0, x^0) \in [\,0, T\,] \times R^n$, *we define the backward cone* $C = \{\,(t, x) \mid t \in\,]\,0, t_0\,[\,,\ |x - x_0| < v(t_0 - t)\,\}$; *let* $S_0 = \{\,(0, x) \,\big|\, |x - x_0| < v\,t_0\,\}$ *denote its trace on* t = 0. *Let* $u \in C^m([\,0\ T\,]; \mathcal{D}')$ *be such that* Pu = 0 *in* C *and* $D_t^j\, u_{|t=0} = 0$ *in* S_0 *for* j = 0, ..., m-1. *Then* u = 0 *in* C.

PROOF: First we suppose that $u \in C^m([0, T] \times \mathbb{R}^n)$. For $\epsilon \in]0, t_0[$, we put $t_\epsilon = t_0 - \epsilon$ and we denote by C_ε the backward cone issuing from (t_ε, x_0). Consider, for $0 \leq \lambda < 1$, the part S_γ of the hyperboloid

$$\lambda^3 |x - x_0|^2 - v^2(t - \lambda t_\epsilon)^2 + \lambda^2 t_\epsilon^2 v^2(1 - \lambda) = 0$$

corresponding to $|x - x_0| < v t_\epsilon$ and $0 \leq t \leq \lambda t_\epsilon$.

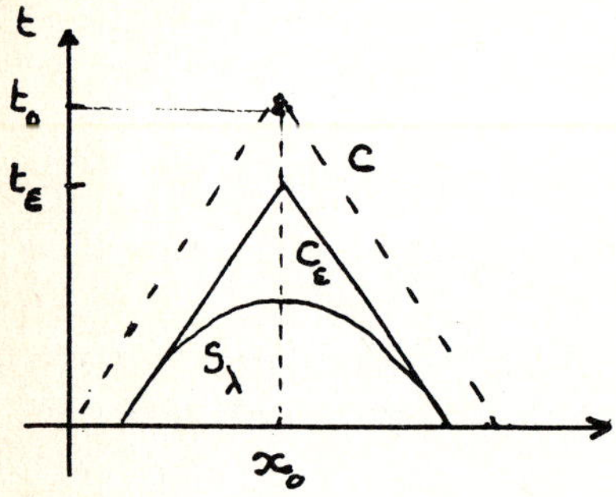

The hypersurfaces S_γ are spacelike for P, and we have

$$C_\epsilon = \bigcup_{0 < \lambda < 1} S_\lambda .$$

We put $\Lambda = \{ \lambda \in [0, 1[\mid D^\alpha u|_{S_\lambda} = 0 \text{ for } |\alpha| \leq m-1 \}$. This is clearly a closed domain containing 0, and Proposition 4.11 shows that it is open.

We deduce from this that $\Lambda = [0, 1[$, and hence u = 0 in C_ε.

Since C is the union of the C_ε, we obtain the conclusion of Theorem 4.13 when $u \in C^m([0, T] \times \mathbb{R}^n)$.

In the general case, we can assume (by truncating u in the neighbourhood of C) that $u \in C^m([0\ T]\ ;\ H^\sigma)$ for some $\sigma \in \mathbb{R}$. We put $Pu = f$, $D_t^{j-1} u|_{t=0} = g_j$ for j = 1 ... m. By hypothesis, we have f = 0 in C and each $g_j = 0$ in S_0. Let

$\rho_k(x)$ be a regularising sequence in $\mathbb{R}^n$; then the solution u_k of the Cauchy problem

$$P u_k = f * \rho_k \quad , \quad D_t^j u_k|_{t=0} = g_j * \rho_k \qquad j = 0,\ldots, m-1$$

is in $C^m([0\ T] \times \mathbb{R}^n)$. We again denote by C_ε the backward cone issuing from $(t_0 - \varepsilon, x^0)$ and by S_ε its trace on $t = 0$. For k sufficiently large, we have $f * \rho_k = 0$ in C_ε and $g_j * \rho_k = 0$ in S_ε; consequently $u_k = 0$ in C_ε. The energy inequality $(4.9.1)_s$ with $s = \sigma - m + 1$ shows that $u_k(t, .)$ tends towards $u(t, .)$ in H^σ when $k \to +\infty$. We thus obtain $u = 0$ in C_ε, which establishes the theorem.

We therefore deduce that the Cauchy problem is well posed for P.

THEOREM 4.14 : *The assumptions on P are the same as those in Theorem 4.13. Then, for* $f \in C^\infty([0\ T] \times \mathbb{R}^n)$, $g_j \in C^\infty(\mathbb{R}^n)$ $j = 0, \ldots, m-1$ *the Cauchy problem*

$$(4.14.1.) \quad Pu = f \quad , \quad D_t^j u|_{t=0} = g_j \qquad j = 0, \ldots, m-1$$

admits a unique solution in $C^\infty([0\ T] \times \mathbb{R}^n)$.

PROOF: The uniqueness follows from Theorem 4.13. We shall now prove the existence. Suppose $\theta(s) \in C_0^\infty(\mathbb{R})$ is identical to unity for $|s| \leq 1$. We put

$$f_k(t, x) = \theta\left(\frac{|x|}{k}\right) f(t, x) \quad \text{and} \quad g_{j,k}(t, x) = \theta\left(\frac{|x|}{k}\right) g_j(x) .$$

Theorems 4.9 and 4.13 show that there exists a unique solution $u_k \in C_0^\infty([0\ T] \times R^n)$ of (4.14.1) for the data f_k and $g_{j,k}$. Theorem 4.13 shows that $u_k(t, x) = u_{k'}(t, x)$ for $k' \geq k$ and $|x| \leq k - vT$. Consequently, the functions u_k admit a limit $u \in C^\infty([0\ T] \times R^n)$ and u satisfies (4.14.1).

REMARK 4.15 : The same method allows us to show that for data $f \in C^0([0\ T]\ ; \mathcal{D}')$, $g_j \in \mathcal{D}'$, there exists a unique $u \in C^m([0\ T], \mathcal{D}')$, the solution of (4.14.1).

REMARK 4.16 : In the case where the operator P is no longer differential, there is in general no notion of a propagation speed (see Exercise 7.1).

5. PARABOLIC CAUCHY PROBLEMS

In this section, we consider operators P of the form $P = D_t - A$ where $A(t, x, D_x)$ is a p.d.o. of degree $m > o$ in x, whose complete symbol $a(t, x, \xi)$ is in $BL^m(\mathbb{R}^{n+1} \times \mathbb{R}^n\ ; \mathcal{L}(\mathbb{C}^N, \mathbb{C}^N))$ and admits a principal part a_m which is homogeneous of degree m in $\xi \in \mathbb{R}^n \backslash 0$.

DEFINITION 5.1 : *We say that P is parabolic for* $t \nearrow$ *if there exists a* C^∞ *mapping:* $(t, x, \xi) \in \mathbb{R}^{n+1} \times (\mathbb{R}^n \backslash 0) \longrightarrow r_o(t, x, \xi)$ *with values in* N × N *hermitian matrices, which is homogeneous of degree* o *in* $\xi \in \mathbb{R}^n \setminus 0$ *and bounded together with all its derivatives for* $|\xi| = 1$, *and which is such that*

$$(5.1.1.)\qquad \begin{aligned} &\operatorname{Im}\,(r_o(t, x, \xi)\, a_m(t, x, \xi) \geq c\,|\xi|^m.\, I \\ &\qquad\qquad r_o(t, x, \xi) \geq c\, I \end{aligned}$$

where c *is a constant*, $c > o$ *independent of* (x, ξ) *and of* t *in a compact subset of* $\mathbb{R}$.

We shall now prove, analogously to Theorem 4.3, the following energy inequality:

THEOREM 5.2 : *Let* $P = D_t - A$ *be a parabolic operator for* $t \uparrow$. *Then, for all* $s \in R$ *and* $T > o$, *there exists* C *such that*

$$(5.2.1)_s\quad \|u(t, .)\|_s^2 + \int_0^T \|u(t', .)\|_{s+m'}^2 \leq C\left[\|u(o, .)\|_s^2 + \int_0^T \|Pu(t', .)\|_{s-m'}^2\, dt'\right]$$

(*with* $m' = \frac{1}{2}m$) *for* $t \in [oT]$, *and* $u \in C^1([oT]\,;\, H^s) \cap C^0([oT]\,,\, H^{s+m'})$.

PROOF: We start by showing that, under the above conditions, there exists C' such that for $t \in [oT]$ we have

$$(5.2.2)_s \quad \|u(t,.)\|_s^2 + \int_0^t \|u(t', .)\|_{s+m'}^2$$
$$\leq C' \left[\|u(o,.)\|_s^2 + \int_0^t (\|u(t',.)\|_s^2 + \|Pu(t',.)\|_{s-m'}^2)dt \right]$$

The inequality obtained by neglecting the second term on the left-hand side allows us, via an argument analogous to that used for Lemma 4.4, to show that in fact we have

$$(5.2.3)_s \; \|u(t, .)\|_s^2 \leq C \left[\|u(o , .)\|_s^2 + \int_0^T \|Pu(t', .)\|_{s-m'}^2 \, dt' \right]$$

We then deduce, by again using $(5.2.2)_s$, that, an upper bound for $\int_0^T \|u(t', .)\|_{s+m'} \, dt'$ is given by the right-hand side of $(5.2.3)_s$. The inequality $(5.2.2)_s$ has the advantage of being preserved if P is perturbed by an operator of degree $m - 1$ in x. In fact, Proposition 2.11 of Chapter II implies that for all $\varepsilon > o$ we have the upper bound

$$\|u(t , .)\|_{s+m'-1} \leq \varepsilon \|u(t , .)\|_{s+m'} + C_\varepsilon \|u(t , .)\|_s \; .$$

As in the proof of Theorem 4.3, we see that it is sufficient to establish $(5.2.2)_o$.

Using notation analogous to that used in the proof of Theorem 4.3, we start from the equality (4.3.3). By virtue of assumption (5.1.1), Gårding's inequality shows that there exist

$c > o$ and C such that

$$\operatorname{Re} i\Big((RA - A^{*}R)\, u \,,\, u\Big) \leq -\, c\|u\|^2_{\frac{m}{2}} + C\,\|u\|^2_o$$

Substituting in (4.3.3), we obtain for $t \in [o\ T]$

$$F(t) \leq -\, c\ \|u(t,.)\|^2_{\frac{m}{2}} + C\left[\|u(t\,,\,.)\|^2_o + \|u(t,.)\|_{\frac{m}{2}}\ \|f(t\,,\,.)\|_{-\frac{m}{2}}\right],$$

from which we deduce that for $t \in [o\ T]$ we have

$$(5.2.4)\ F(t) + \|u(t\,,\,.)\|^2_{\frac{m}{2}} \leq C'\left[\ \|u(t\,,\,.)\|^2_o + \|f(t\,,\,.)\|^2_{-\frac{m}{2}}\ \right]$$

The upper bound $(5.2.2)_o$ follows from (5.2.4) by proceeding as for (4.3.4) and (4.3.5).

Using functional analysis arguments analogous to those used in the proof of Theorem 4.3, and noting that P^* is parabolic for $t \searrow$, we deduce:

THEOREM 5.3 : *Suppose* $P = D_t - A$ *is parabolic for* $t \nearrow$. *Let* $s \in \mathbb{R}$, $T > o$, $f \in L^2([o\ T]\,;\,H^{s-m'})$, $g \in H^s$. *Then there exists a unique* $u \in C^o([o\ T]\,;\,H^s)$ *such that* $Pu = f$ (*in* $\mathcal{D}'(]o\ T[\times\mathbb{R}^n)$) *and* $u(o\,,\,.) = g$. *Furthermore,* $u \in L^2([o\ T]\,;\,H^{s+m'})$ *and satisfies the energy inequality* $(5.2.1)_s$. *If* $f \in C^\infty([o\ T]\,;\,H^{+\infty})$ *and* $g \in H^{+\infty}$, *then* $u \in C^\infty([o\ T]\,;\,H^{+\infty})$.

We shall now show that the classical parabolic operators

are in fact parabolic in the above sense.

DEFINITION AND THEOREM 5.4 : *The operator* $P = D_t - A(t, x, D_x)$ *is said to be parabolic in the sense of Petrovski for* $t \uparrow$ *if there exists* $c > 0$ *such that* $\mathrm{Im}\,\lambda \geq c\,|\xi|^m$ *for any eigenvalue* λ *of* $a_m(t, x, \xi)$. *Assume further that* $a_m(t, x, \xi)$ *is independent of* (t, x) *for* $|t| + |x|$ *sufficiently large; then such an operator* P *is parabolic in the sense of Definition* 5.1.

The proof is based on the following lemma:

LEMMA 5.5 : *Let* M *be an* $N \times N$ *matrix whose eigenvalues have strictly positive imaginary parts. Then there exists a hermitian matrix* H *such that* $H > 0$ *and* $\mathrm{Im}(H.M) > 0$.

PROOF: We know that there exists a Jordan basis $(e_1, \ldots, e_N)$; let T denote the change-of-basis matrix. We have:

$$T M T^{-1} = \begin{bmatrix} \lambda_1 & \alpha_1 & & 0 \\ & \ddots & \ddots & \\ & & \ddots & \alpha_{N-1} \\ 0 & & & \lambda_N \end{bmatrix} \qquad \text{with } \alpha_j = 0 \text{ or } 1, \quad \mathrm{Im}\lambda_j > 0$$

Let $\varepsilon > 0$; replacing the basis $(e_1, \ldots, e_N)$ by the basis

$(e_1, \varepsilon e_2, \ldots, \varepsilon^{N-1} e_N)$, we can assume that $\alpha_j = 0$ or ε.

We obtain

$$T M T^{-1} = D + \varepsilon N \quad \text{with} \quad D = \begin{pmatrix} \lambda_1 & & 0 \\ & \ddots & \\ 0 & & \lambda_N \end{pmatrix} \quad \text{and}$$

$$N = \begin{bmatrix} 0 & \alpha_1 & & 0 \\ & \ddots & \ddots & \\ & & \ddots & \alpha_{N-1} \\ 0 & & & 0 \end{bmatrix} .$$

Furthermore, we can put T in the form $T = U \cdot P$ with U unitary and P strictly positive hermitian. We obtain $M = P^{-1} U^*(D + \varepsilon N) U P$; putting $H = P^2$, we have $H M = P U^*(D + \varepsilon N) U P$. Let $w \in \mathbb{C}^N$; we find that

$$\operatorname{Im} (H M w , w) = \operatorname{Im} (D U P w , U P w) + \varepsilon \operatorname{Im} (N U P w , U P w)$$
$$\geq c \, \|U P w\|^2 - \varepsilon C \, \|U P w\|$$

with $c = \inf_j (\operatorname{Im} \lambda_j) > 0$ and $C = \| N \|$.

Consequently, the matrix H meets our requirements when we take $\varepsilon < \frac{c}{C}$.

This lemma shows that for $(t_0, x^0, \xi^0) \in [0 \; T] \times \mathbb{R}^n \times S_{n-1}$, there exists $H_0 > 0$ such that $\operatorname{Im} (H_0 \, a_m(t_0, x^0, \xi^0)) > 0$; by continuity we also have $\operatorname{Im} (H_0 \, a_m(t, x, \xi)) > 0$ for (t, x, ξ) near (t_0, x^0, ξ^0). By partition of unity, this allows us to construct $r_o(t, x, \xi)$ satisfying (5.1.1) for $|\xi| = 1$, and we next extend r_o into a homogeneous function of degree o in ξ.

EXAMPLE 5.6 : Let $Q(x, D_x)$ be a strongly elliptic scalar differential operator of degree 2m with coefficients independent of x for $|x|$ large. Then the operator $\partial_t + Q(x, D_x)$ is parabolic in the sense of Petrovski for $t\nearrow$; this is the case with the heat operator.

6. SEMIGROUPS OF OPERATORS AND APPLICATIONS

6.1 INTRODUCTION : If we assume in Theorems 4.3 and 5.3 that the operators $A(t, x, D_x)$ are independent of t, then the corresponding Cauchy problems with f = 0 can be written in the form of an abstract differential equation

$$\frac{du}{dt} = Au \quad , \quad u(0) = u_o \tag{6.1.1.}$$

where A is a linear operator with domain $D = H^1(\mathbb{R}^n; \mathbb{C}^N)$ $(D = H^m(\mathbb{R}^n; \mathbb{C}^N)$ for 5.3) which represents $A(x, D_x)$ in the Banach space $E = L^2(\mathbb{R}^n; \mathbb{C}^N)$.

In the special case where A is continuous from E into E, the solution of (6.1.1) is given by $u(t) = G_t u_o$ with

$G_t = \sum_{k \geq 0} \frac{t^k A^k}{k!} = e^{tA}$. In applications, A is not in general continuous; however the solution $\overline{\mathbb{R}}_+ \ni t \longrightarrow u(t)$ of (6.1.1) is continuous for suitable spaces; this leads to:

DEFINITION 6.2 : *We designate as a strongly continuous semigroup (abbreviated to S.G.) defined on a Banach space E, a mapping* $t \in \overline{\mathbb{R}}_+ \longrightarrow G_t \in \mathcal{L}(E, E)$ *such that*

(6.2.1.) . $G_{t_1+t_2} = G_{t_1} \circ G_{t_2}$ *for* $t_1, t_2 \geq 0$, $G_0 = I$

(6.2.2.) . *for all* $x \in E$, *the function*

$$t \in \overline{\mathbb{R}}_+ \longrightarrow G_t\, x \in E \quad \textit{is continuous}.$$

EXAMPLE 6.3 : We take $E = L^2([\,0, +\infty\,[)$, and for $x(.) \in E$ we put $(G_t\, x)(s) = x(s - t)$.

We can define, in a certain sense, the derivative of G_t at $t = 0$.

DEFINITION 6.4 : *We designate as the infinitesimal generator of a S.G.* (G_t) *the operator* A *with domain*

$$D = \{\, x \in E \mid \lim_{s \to 0^+} \frac{1}{s}(G_s\, x - x) \;\; \textit{exists} \,\}$$

defined by $$Ax = \lim_{s \to 0^+} \frac{1}{s}(G_s\, x - x) \quad .$$

Thus, in Example 6.3, we have $D = H^1([\,0, +\infty\,[)$ and $A = \frac{d}{dt}$.

An S.G. has at most exponential growth; that is

PROPOSITION 6.5 : *Let* (G_t) *be an S.G.; then there exist real numbers* ω *and* M *such that*

$$(6.5.1.) \qquad \|G_t\| \leq M e^{\omega t} \quad , \quad \textit{for} \quad t \geq 0 \quad .$$

We can endow E *with an equivalent norm* $|||\ \ |||$ *for which* G_t *satisfies* (6.5.1) *with* M = 1 *and the same* ω.

PROOF: Since the operators G_t are simply bounded for $t \in [0, 1]$, the Banach-Steinhauss theorem implies the existence of M such that

$$\|G_t\| \leq M \qquad \text{for} \qquad t \in [0\ 1] \quad .$$

Any real $t \geq 0$ may be written $t = p + r$ with $p \in \mathbb{N}$ and $r \in [0, 1[$; this allows us to write $G_t = (G_1)^p G_r$. We therefore deduce that

$$\|G_t\| \leq M^{p+1} = M e^{p \operatorname{Lg} M} \leq M e^{t\omega} \text{ with } \omega = \operatorname{Lg} M$$

which proves (6.5.1).

We put $|||x||| = \sup_{t \geq 0} \|e^{-\omega t} G_t x\|$; this is an equivalent norm since the upper bound in (6.5.1) implies $|||x||| \leq M \|x\|$ and we clearly have $\|x\| \leq |||x|||$. Finally,

$$
\begin{aligned}
|||G_t x||| &= \sup_{s \geq 0} \|e^{-\omega s} G_s G_t x\| \\
&= \sup_{s \geq 0} \|e^{-\omega(s+t)} G_{s+t} x\| e^{\omega t} \\
&= \|x\| e^{\omega t} .
\end{aligned}
$$

REMARK 6.6 : If we put $\tilde{G}_t = e^{-\omega t} G_t$, then $(\tilde{G}_t)$ is an S.G. which satisfies $|||\tilde{G}_t||| \leq 1$. We say that it is a contraction S.G.; Proposition 6.5 shows that it is always possible to reduce to this case.

PROPOSITION 6.7 : *The infinitesimal generator* (A, D) *of a semigroup possesses the following properties:*

a/ *For all* $x \in D$, *we have* $G_t x \in D$, $G_t Ax = A G_t x$ *and* $G_t' x = G_t Ax$.

b/ *The domain* D *is dense in* E.

c/ *The operator* A *is closed; in other words its graph* $\{ (x, Ax) \mid x \in D \}$ *is a closed subspace of* $E \times E$.

PROOF OF a/ For $\varepsilon > 0$, we put

$$A_\varepsilon = \frac{1}{\varepsilon}(G_\varepsilon - I) .$$

Then we have the equalities

$$A_\varepsilon G_t x = G_t A_\varepsilon x = \frac{1}{\varepsilon}(G_{t+\varepsilon} x - G_t x) , \quad x \in E .$$

We thus have a/ by taking $x \in D$ and making ε tend to 0.

PROOF OF b/ For $s > 0$, we put

$$M_s x = \frac{1}{s} \int_0^s G_t x \, dt \qquad x \in E \quad .$$

It is clear that $M_s x \xrightarrow[s \to 0^+]{} x$.

It can easily be shown from the definitions that for $x \in E$, we have the equalities

(6.7.1.) $$A_\varepsilon M_s x = A_s M_\varepsilon x = M_s A_\varepsilon x \quad ;$$

making ε tend to 0, the first equality shows that

$$M_s x \in D \text{ and } A M_s x = A_s x \text{ for } x \in E \ ,$$

which proves, in particular, the density of D. Furthermore, if $x \in D$, the second equality in (6.7.1) shows that

(6.7.2.) $$A_s x = M_s Ax \quad , \quad \text{for} \quad x \in D \ .$$

PROOF OF c/ Suppose we have a sequence $x_n \in D$ such that $Ax_n \xrightarrow[n \infty]{} y$ and $x_n \xrightarrow[n \infty]{} x$ in E. From (6.7.2), we have $A_s x_n = M_s Ax_n$; then making n tend towards $+\infty$, we obtain $M_s y = A_s x$; it then follows that $x \in D$ and $Ax = y$.

REMARK 6.8 : By virtue of c/, the norm of the graph $\|x\|_D = \|x\| + \|Ax\|$ endows D with a Banach space structure and, taking account of a/, the S.G. (G_t) induces an S.G. on D.

We now return to the abstract Cauchy problem.

THEOREM 6.9 : *Let* A *be a linear operator with domain* D *in* E. *Suppose that* A *is the infinitesimal generator of an S.G.* (G_t). *Let* $f \in C^0([0\ T]; D)$ *and* $u_o \in D$; *then there exists a unique* $u \in C^1([0\ T]; E) \cap C^0([0\ T]; D)$ *such that*

$$u'(t) = Au(t) + f(t) \tag{6.9.1.}$$

$$\text{for} \quad t \geq 0 \quad , \quad u(0) = u_o \quad .$$

We also have the same conclusion with the assumption

$$f \in C^1([0\ T]; E) \quad .$$

We shall now prove uniqueness. Let u be a solution with $u_o = 0$, $f = 0$. The function $v(s) = G_{t-s}.u(s)$ $(0 \leq s \leq t)$ is differentiable and has the derivative

$$v'(s) = G_{t-s}\, u'(s) - G_{t-s}\, A\, u(s) = 0 \quad .$$

Thus $v(s) = v(0) = 0$, so that $u(t) = v(t) = 0$.

In order to prove existence, we can suppose that $u_o = 0$ since $G_t\, u_o$ is the solution of (6.9.1) with $f = 0$. We shall now show that

$$u(t) = \int_0^t G_{t-s}\, f(s)\, ds$$

is a solution. The mapping $0 \leq s \leq t \longrightarrow G_{t-s}\, f(s) \in D$ is continuous, in view of Remark 6.8; consequently $u \in C^0([0\ T]; D)$. We now calculate the derivative of u; taking $h > 0$, we have:

$$\frac{1}{h}\,(u(t+h)-u(t)) = \int_0^{t+h} G_{t-s}\Big(\frac{G_h - I}{h}\Big)\, f(s)\, ds \;-\; \frac{1}{h}\int_t^{t+h} G_{t-s}\, f(s)\, ds.$$

Since A is a closed operator, the assumption on f shows that for $h \to 0$, the right-hand side possesses a limit equal to $A\,u(t) + f(t)$. This shows that u is a solution of (6.9.1) and is in the space $C^1([\,0\;T\,];\,E)$.

In the case of the assumption $f \in C^1([\,0\;T];\,E)$ we use the same type of argument, taking

$$u(t) = \int_0^t G_s\, f(t-s)\, ds\,, \quad (\text{cf. exercise } 7.3.)\ .$$

In order to apply this theorem, we first need a characterisation of the generators of S.G.'s. In view of Remark 6.6, we shall confine our attention to the case of contraction S.G.'s.

PROPOSITION 6.10 : *Suppose* (G_t) *is a contraction S.G.; then we have the following properties:*

a/ *The mapping* $\lambda \in \mathbb{C} \longrightarrow R_\lambda = \int_0^{+\infty} e^{-\lambda t}\, G_t\, dt \in \mathcal{L}(E, E)$ *is defined and holomorphic for* $\operatorname{Re}\lambda > 0$ *, and satisfies*

$$\|R_\lambda\| \le \frac{1}{\operatorname{Re}\lambda}\ .$$

b/ *The operator* R_λ *is the resolvent of* A *in the open set* $\operatorname{Re}\lambda > 0$; *i.e.*

$$(\lambda - A) \circ R_\lambda = I_E \quad , \quad R_\lambda \circ (\lambda - A) = I_D \quad .$$

The proof of a/ follows immediately from the upper bound for $||G_t||$. In order to prove b/, we start by establishing the following lemma:

LEMMA 6.11 : We have

$$A_\epsilon R_\lambda x \xrightarrow[\epsilon \to 0^+]{} \lambda R_\lambda - x \quad , \quad x \in E \quad .$$

PROOF: We have by definition.

$$A_\epsilon R_\lambda x = \frac{1}{\lambda}\left[\int_0^{+\infty} e^{-\lambda t} G_{t+\epsilon} x - \int_0^{\infty} e^{-\lambda t} G_t x\right]$$

$$= \int_\epsilon^{+\infty} e^{-\lambda t} \left(\frac{e^{\lambda\epsilon} - 1}{\epsilon}\right) G_t x \, dt - \frac{1}{\epsilon} \int_0^\epsilon e^{-ht} G_t x \, dt.$$

It is therefore sufficient to apply the Dominated Convergence Theorem in order to obtain the lemma.

From this lemma, we deduce that $x \in E$ implies $R_\lambda x \in D$ and $A R_\lambda x = \lambda R_\lambda x - x$; in other words $(\lambda - A).R_\lambda = I_E$. Furthermore, A_ε and R_λ commute; consequently, we obtain in the limit $AR_\lambda x = R_\lambda Ax$ for $x \in D$ and this concludes the proof of the proposition.

This proposition has the following important converse:

THEOREM 6.12 (Hille and Yosida) : *Suppose* A *is a linear*

operator defined in a domain D *which is dense in* E. *Suppose that there exists* λ_0 *such that its resolvent* $R_\lambda = (\lambda - A)^{-1}$ *exists and satisfies*

$$\|R_\lambda\| \leq \frac{1}{\lambda} \quad \text{for} \quad \lambda \geq \lambda_0 \quad . \tag{6.12.1.}$$

Then, (A, D) *is the infinitesimal generator of a contraction S.G.*

The proof will be performed in a number of steps.

First step We show that for $x \in E$ we have $\lambda R_\lambda x \xrightarrow[\lambda \to \infty]{} x$. From the definition of R_λ we have $\lambda R_\lambda x = R_\lambda Ax + x$ for $x \in D$, and therefore (6.12.1) implies $\lambda R_\lambda x \to x$ for $x \in D$. Since the operators λR_λ are equicontinuous and D is dense, the convergence holds for all $x \in E$.

Second step Construction of a semigroup G_t. The operator $B_\lambda = \lambda A R_\lambda$ is equal to $\lambda^2 R_\lambda - \lambda$; it is therefore continuous and we can associate with it the group $e^{t B_\lambda}$. This group will serve to approximate (G_t) since

$B_\lambda x = \lambda R_\lambda Ax \xrightarrow[\lambda \to \infty]{} Ax$ when $x \in D$. Since R_λ and R_μ commute, the same is true of B_λ and B_μ; this allows us to write

$$e^{tB_\lambda} - e^{tB_\mu} = e^{tB_\mu}\left[e^{t(B_\lambda - B_\mu)}\right]$$

$$= e^{tB_\mu}\int_0^t e^{s(B_\lambda - B_\mu)}(B_\lambda - B_\mu)\, ds \quad ;$$

using the upper bound

$$\|e^{tB_\lambda}\| = e^{-t\lambda}\,\|e^{t\lambda^2 R_\lambda}\| \leq e^{-t\lambda}\, e^{t\|\lambda^2 R_\lambda\|} \leq 1 ,$$

we then deduce that

$$\|e^{tB_\lambda} - e^{tB_\mu}\| \leq t\,\|B_\lambda - B_\mu\| \quad .$$

This inequality shows that for all $x \in D$ the functions $e^{tB_\lambda}x$ converge uniformly when $\lambda \longrightarrow +\infty$ for t in a compact subset of $\bar{R}_+$. We denote by $G_t x$ the limit function; this is therefore continuous and satisfies $\|G_t x\| \leq 1$ for $x \in D$. Finally, since D is dense, we can extend G_t into a contraction S.G. on the whole of E.

Third step We now show that A is the infinitesimal generator of (G_t). Let $\tilde{A}$ be the generator of (G_t) and let $\tilde{D}$ be its domain; letting $\lambda \longrightarrow +\infty$, the equality

$$e^{tB_\lambda}x - x = \int_0^t e^{sB_\lambda} B_\lambda x\, ds \qquad (x \in D)$$

implies that

$$G_t x - x = \int_0^t G_t Ax\, ds \qquad (x \in D) \quad .$$

We then deduce that $\tilde{A}$ extends A and since $(\lambda_o - A)$ is by hypothesis a bijection of D onto E, we must have $\tilde{D} = D$ and thus $\tilde{A} = A$.

The following corollary is often used in applications:

COROLLARY 6.13 : *Let* A *be a linear operator with domain* D *dense in* E. *Suppose that there exist* λ_o and ω *such that the resolvent* $R_\lambda = (\lambda - A)^{-1}$ *exists and satisfies*

$$(6.13.1.) \qquad \|R_\lambda\| \leq \frac{1}{\lambda - \omega} \qquad \textit{for} \qquad \lambda \geq \lambda_o .$$

Then A *is the infinitesimal generator of an S.G.* (G_t) *such that* $\|G_t\| \leq e^{\omega t}$.

PROOF: We immediately get back to the previous theorem by considering the operator $B = A - \omega$.

REMARK 6.14 : Feller-Miyerdera and Phillips have given a characterisation of S.G. generators without making any assumptions on the growth of $\|G_t\|$. It is necessary to replace (6.13.1) by the following condition: there exist λ_o, ω and M such that

$$\|R_\lambda^n\| \leq \frac{M}{(\lambda - \omega)^n} \qquad \text{for} \qquad \lambda \geq \lambda_o \qquad \text{and} \qquad n \in \mathbb{N} .$$

6.15 APPLICATION TO THE HEAT EQUATION WITH DIRICHLET CONDITIONS

Let Ω be a bounded open subset of $\mathbb{R}^n$ with regular boundary $\partial\Omega$, and let $Q(x, D_x) = \sum_{|\alpha| \leq 2m} q_\alpha(x) D^\alpha$ be a differential operator with coefficients in $C^\infty(\bar{\Omega})$. We suppose that there exists $c > 0$ such that

$$(6.15.1.) \qquad -\operatorname{Re} \sum_{|\alpha| = 2m} q_\alpha(x) \xi^\alpha \geq c |\xi|^{2m} \quad \text{for} \quad (x, \xi) \in \Omega \times \mathbb{R}^n .$$

For given f and u_0, we seek u satisfying

$$(6.15.2) \begin{cases} ((\partial_t - Q(x, D_x)) u(t, x) = f(t, x) & \text{in} \quad \mathbb{R}^+ \times \Omega \\ u(0, x) = u_0(x) & \\ u(t, x)\big|_{\mathbb{R}^+ \times \partial\Omega} = 0 & \text{(Dirichlet condition)} \end{cases}$$

(for $m = 1$, this is an equation of the heat type).

We transform this problem into an abstract differential equation by introducing the linear operator A with domain $D = H^{2m}(\bar{\Omega}) \cap H_0^m(\Omega)$ in the space $E = L^2(\Omega)$, defined by $Au = Qu$. We then have:

THEOREM 6.16 : *The operator* (A, D) *is the generator of a strongly continuous semigroup in* E.

PROOF: The Gårding inequality (8.6.2), Chapter IV, shows that there exist $c > 0$ and λ_0 such that

$$\text{(6.16.1.)} \quad \operatorname{Re}(-Q(x, D)u, u) + \lambda(u, u) \geq c\,\|u\|_m^2$$

$$\text{for} \quad u \in D \quad \text{and} \quad \lambda \geq \lambda_0 .$$

Consequently, the operator $\lambda - A$ is injective from D into E for $\lambda \geq \lambda_0$. Furthermore, we showed in Theorem 5.2, Chapter V, that its index is zero; it is thus bijective. In order to attach an upper bound to the resolvent of A, we minorise the norm

$$\|(\lambda - A)u\|^2 = \lambda^2 \|u\|^2 + \|Au\|^2 - 2\lambda \operatorname{Re}(Au, u) \quad , \quad u \in D .$$

Using (6.16.1) with $\lambda = \lambda_0$, we obtain

$$\|(\lambda - A)u\|^2 \geq (\lambda^2 - 2\lambda\,\lambda_0)\|u\|_0^2 \quad \text{for} \quad u \in D \quad ,$$

from which we deduce

$$\text{(6.16.2.)} \quad \|(\lambda - A)^{-1}\| \leq \frac{1}{\lambda - \lambda_0} \quad \text{for} \quad \lambda \geq \lambda_0 .$$

This inequality and Corollary 6.13 imply Theorem 6.16, since in this case the space D is obviously dense in $L^2(\Omega)$.

From Theorem 6.9 we deduce that for $f \in C^1([0\ T]; L^2(\Omega))$ and $u_0 \in H^{2m}(\bar{\Omega}) \cap H_0^m(\Omega)$, the problem (6.15.2) admits a unique solution

$$u \in C^1([0\ T]; L^2(\Omega)) \cap C^0([0\ T]; H^{2m}(\bar{\Omega}) \cap H_0^m(\Omega)) .$$

We can in fact show that it is sufficient to assume that $f \in C^0([0\ T]; L^2(\Omega))$.

REMARK 6.17 : If we replace Ω by $\mathbb{R}^n$ and if we assume that the coefficients q_α are bounded together with all their derivatives, then Theorem 6.16 remains valid word for word. In fact, the theory of pseudo differential operators in BL^m (see Chapter IV, Section 11.1) allows us to prove the inequality (6.16.1), and the Lax-Milgram theorem (see Section 7.4, Chapter II) shows that $\lambda - A$ is bijective for $\lambda > \lambda_o$.

6.18 APPLICATION TO THE WAVE EQUATION WITH DIRICHLET CONDITIONS

We make the assumptions on Ω and Q which were made in Section 6.15; in addition, we take Q in the form

$\sum_{i,j} \partial_j q_{ij}(x) \partial_i +$ (terms of degree ≤ 1) with real coefficients q_{ij} .

For given f, u_o, u_1, we seek u satisfying

$$(6.18.1.)\begin{cases} (\partial^2_{tt} - Q(x, D_x))\, u(t, x) = f(t, x) & \text{in } \mathbb{R}^+ \times \Omega \\ u(0, x) = u_o(x) & \text{in } \Omega \\ \partial_t u(0, x) = u_1(x) & \text{in } \Omega \\ u(t, x)\big|_{\mathbb{R}^+ \times \partial\Omega} = 0 & \end{cases}$$

(Dirichlet boundary condition) .

In order to reduce to an abstract differential equation, we start by replacing (6.18.1) by an equivalent system of first order. We put $\mathcal{A} = \begin{pmatrix} 0 & 1 \\ Q(x, D_x) & 0 \end{pmatrix}$ then u is a solution of (6.18.1) if and only if $U = \begin{pmatrix} u \\ \partial_t u \end{pmatrix}$ satisfies

$$(6.18.2.)\quad \begin{cases} \partial_t U - \mathcal{A} U = F \\ U(0) = U_0 \\ B U_{|\mathbb{R}^+ \times \partial\Omega} = 0 \end{cases}$$

with $F = \begin{pmatrix} 0 \\ f \end{pmatrix}$, $U_0 = \begin{pmatrix} u_0 \\ u_1 \end{pmatrix}$ and $B = (1,0)$.

We introduce the spaces

$$E = H_0^1(\Omega) \times L^2(\Omega) \quad , \quad D = (H^2(\bar{\Omega}) \cap H_0^1(\Omega)) \times H_0^1(\Omega)$$

and we denote by A the operator defined by $\mathcal{A}$ on D. We then have

THEOREM 6.19 : *The operator* (A, D) *is the generator of a S.G. in the space* E.

PROOF: In order to investigate the existence of the

resolvent of A, we seek $U = \begin{pmatrix} u \\ v \end{pmatrix} \in D$ satisfying the equation $(\lambda - A)\, U = H$ with $H = \begin{pmatrix} h \\ k \end{pmatrix}$ given in E. This equation is equivalent to the system

$$\begin{cases} v = \lambda u - h \\ (\lambda^2 - Q)u = k + \lambda h \end{cases} .$$

However, from the results proved in connection with Theorem 6.16, there exists λ_o such that for $|\lambda| \geq \lambda_o$, the equation

$$(\lambda^2 - Q)u = h$$

admits a unique solution $u \in H^2(\bar{\Omega}) \cap H_0^1(\Omega)$ for any given h in $L^2(\Omega)$. This implies that the resolvent R_λ of A exists when $|\lambda| \geq \lambda_0$. In order to attach an upper bound to its norm, it is convenient to endow the Hilbert space E with a new equivalent Hilbert norm. We define a new inner product ((,)) by putting

$$((U, U')) = \sum_{i,j=1}^{n} (q_{ij}\, \partial_i u, \partial_j u') + C(u, u') + (v, v')$$

with $U = \begin{pmatrix} u \\ v \end{pmatrix}$, $U' = \begin{pmatrix} u' \\ v' \end{pmatrix}$, where C is such that we have the Gårding inequality

$$\sum_{i,j=1}^{n} (q_{ij}\,\partial_i u, \partial_j u) + C(u, u) \geq c\,\|u\|_1^2 , \quad \text{for} \quad u \in H_0^1(\Omega) .$$

Thus the associated norm $|||\cdot|||$ is equivalent to the old norm on E.

The upper bound for the resolvent will follow from

LEMMA 6.20 : *There exists* C' *such that*

(6.20.1.) $$|\mathrm{Re}((AU, U))| \leq C'\,|||U|||^2 \quad \text{for} \quad U \in D .$$

PROOF: Putting $U = \begin{pmatrix} u \\ v \end{pmatrix} \in D$, we have by definition

(6.20.2.) $$((AU, U)) = \sum_{i,j=1}^{n} (q_{ij}\,\partial_i v, \partial_j u) + C(v, u) - (Qu, v).$$

Since $C_o^\infty(\Omega)$ is dense in $H_o^1(\Omega)$, we have the equality $(-Qu, v) = \sum_{i,j=1}^{n} (q_{ij}\,\partial_i u, \partial_j v) + (Lu, v)$ for all $u \in H^2(\bar{\Omega})$ and $v \in H_o^1(\Omega)$, and L is a differential operator of first order. Substituting this equality into (6.20.2), we obtain

$$\mathrm{Re}((AU, U)) = C\,\mathrm{Re}(v, u) + \mathrm{Re}(Lu, v) \quad ;$$

we thus have the upper bounds

$$\begin{aligned} |\mathrm{Re}((AU, U))| &\leq C\,(\|u\|_0\,\|v\|_0 + \|u\|_1\,\|v\|_0) \\ &\leq C\,\|u\|_1^2 + \|v\|_0^2 = C\,\|U\|^2 \\ &\leq C'\,|||U|||^2 \end{aligned}$$

This proves the lemma.

Finally, proceeding exactly as for (6.16.2), we deduce from (6.20.1) that

$$|||(\lambda - A)^{-1}||| \leq \frac{1}{|\lambda| - \lambda_0} \quad \text{for} \quad |\lambda| \geq \lambda_0 \quad ,$$

and this proves the theorem.

(Using Exercise 7.2 below, it can be shown that A is the generator of a strongly continuous group).

As an application, we deduce that for

$$f \in C^1[0\ T]; L^2(\Omega)) \quad , u_1 \in H_0^1(\Omega) \ , \quad u_0 \in H_0^1(\Omega) \cap H^2(\overline{\Omega}) \quad ,$$

problem (6.18.1) admits a unique solution

$$u \in C^2([0\ T]; L^2(\Omega)) \cap C^1([0\ T]; H_0^1(\Omega)) \cap C^0([0\ T]; H^2(\Omega)) .$$

REMARK 6.21 : If we replace Ω by $\mathbb{R}^n$, we have the analogue of Remark 6.17.

REMARK 6.22 : In (6.18.1), we can replace the Dirichlet condition by the Neumann condition $\frac{\partial u}{\partial n}\Big|_{R^+ \times \partial\Omega} = 0$; we then obtain analogous results (see Exercise 7.5 below).

7. ADDITIONAL NOTES AND EXERCISES

7.1 COUNTER-EXAMPLE TO THE PROPERTY OF FINITE PROPAGATION SPEED IN THE NON-DIFFERENTIAL CASE

Let $P = D_t - A(D_x)$, where $A \in L^1(\mathbb{R})$ is the pseudo-differential operator with symbol $|\xi|$. The operator P is thus strictly hyperbolic with constant coefficients and of degree one. Show that the solution of the Cauchy problem

$$\begin{cases} Pu = 0 \\ u|_{t=0} = \delta \end{cases}$$

is given by $\hat{u}(t, \xi) = e^{-it|\xi|}$.

Show that for $t \neq 0$, the inverse Fourier transform of $\xi \longmapsto e^{-it|\xi|}$ does not have compact support (note that $\xi \longmapsto e^{-it|\xi|}$ is not C^∞).

7.2 GENERATOR OF A STRONGLY CONTINUOUS GROUP

Let E be a Banach space; we have the notion of a strongly continuous group if we replace $\bar{\mathbb{R}}^+$ by $\mathbb{R}$ in Definition 6.2.

a) Let $t \in \mathbb{R} \longrightarrow G_t \in \mathcal{L}(E, E)$ be a strongly continuous group. We denote by (A , D) the generator of the semigroup $(G_t)_{t \geq 0}$; show that (-A, D) generates a semigroup.

b) Conversely, let (A, D) be an operator such that (A, D) and (-A, D) generate semigroups; then show that the semigroup generated by A extends on $\mathbb{R}$ into a group.

[Let $G_1(t)$ and $G_2(t)$ be the semigroups generated by A and $-A$; we define $G(t)$ by $G_1(t)$ for $t \geq 0$ and by $G_2(t)$ for $t \leq 0$. Show that for $x \in D$ the function $u(t) = G(t)x$ is in $C^1(\mathbb{R},E) \cap C^0(\mathbb{R},D)$ and satisfies $u'(t) = A\,u(t)$ for $t \in \mathbb{R}$; deduce from this that $G(t)$ is a group].

7.3 FURTHER DISCUSSION OF THEOREM 6.9

Let E be a Banach space and suppose (A, D) generates a semigroup (G_t) in E. Consider the differential equation

$$(7.3.1) \quad u'(t) = Au(t) + f(t) \quad \text{for} \quad t \geq 0 \text{ and } u(0) = u_0 \in D .$$

a) Take $f \in C^0(\bar{\mathbb{R}}^+, D)$ and $u_0 \in D$ and let u be the solution of (7.3.1). Show that for all T there exists C such that

$$(7.3.2) \qquad \| u(t) \| \leq C(\| u_0 \| + \int_0^T \| f(s) \| ds) \quad \text{for } t \in [0,T].$$

If in addition $f \in C^1(\bar{\mathbb{R}}_+, E)$, we have

$$(7.3.3) \quad \| u'(t) \| \leq C(\| Au_0 + f(0) \| + \int_0^T \| f'(s) \| ds) \quad \text{for } t \in [0,T]$$

[use the expression $u(t) = G_t u_0 + \int_0^t G_s \cdot f(t-s)ds$].

b) Let $f \in C^1(\bar{\mathbb{R}}_+, E)$; we define for $\varepsilon > 0$ the function f_ε by $f_\varepsilon(t) = M_\varepsilon f(t)$ where M_ε is as defined in the proof of

Proposition 6.7, Show that $f_\varepsilon \in C^1(\overline{R}_+ ; E) \cap C^0(\overline{R}_+ ; D)$ and that f_ε converges to f in $C^1(\overline{R}_+ ; E)$.

c) Let $f \in C^1(\overline{R}_+ ; E)$ and $u_0 \in D$; prove that (7.3.1) admits a unique solution $u \in C^1(\overline{R}_+ ; E) \cap C^0(\overline{R}_+ ; D)$ and that this solution satisfies the inequalities (7.3.2) and (7.3.3).

[Let u_ε be the solution corresponding to f_ε; use the inequalities (7.3.2) and (7.3.3) to show that u_ε is a Cauchy sequence].

d) Let $k \in \mathbb{N}$; we define by recurrence the spaces D_k by $D_k = \{u \in D_{k-1} \mid A u \in D_{k-1}\}$ and we put $D_0 = E$. Show that D_k is stable under G_t and is dense in E, and show that it is a Banach space for the norm $\| u \|_{D_k} = \sum_{j=0}^{k} \| A^j u \|$.

Let $k \geq 2$; take $f \in C^k(E) \cap C^{k-2}(D_1) \cap \ldots \cap C^0(D_{k-1})$ and $u_0 \in D_k$; show that the solution u of (7.3.1) is in the space $C^k(E) \cap C^{k-1}(D_1) \cap \ldots \cap C^0(D_k)$.

7.4 ENERGY INEQUALITY FOR THE WAVE EQUATION

We adopt the conditions of Theorem 6.19. Let $T > 0$; show that there exists C such that if u is the solution of (6.18.1) corresponding to data $f \in C^1([0,T]; L^2(\Omega))$, $u_0 \in H^2(\overline{\Omega}) \cap H_0^1(\Omega)$, $u_1 \in H_0^1(\Omega)$ we have

$$\| u(t,\cdot)\|_1 + \| \partial_t u(t,\cdot)\|_0 \leq C \left[\|u(0,\cdot)\|_1 + \|\partial_t u(0,\cdot)\|_0 + \int_0^T \| f(s)\| ds\right]$$

[use the inequality (7.3.2)].

7.5 WAVE EQUATION WITH NEUMANN BOUNDARY CONDITIONS.

We adopt the assumptions of problem (6 .18.1) and we assume that the coefficients of the operator $Q(x, D_x)$ are real. In what follows we shall consider only real-valued functions. In (6.18.1) we replace the Dirichlet conditions by the Neumann condition:

$$(7.5.1) \qquad \left(\frac{d}{dn} u + \rho u\right)\Big|_{R^+ \times \partial\Omega} = 0$$

where $\frac{d}{dn}$ denotes differentiation in the direction conormal to Q,

i.e. $\frac{du}{dn}(x) = \sum q_{ij}(x)\nu_i(x) \frac{\partial}{\partial x_j} u(x)$ and $\rho \in C^\infty(\partial\Omega ; R)$

We wish to extend the proof of Theorem 6.19 to this situation.

a) We define on $E = H^1(\bar{\Omega}) \times L^2(\Omega)$ the symmetric bilinear form

$$((U , U')) = \sum_{i,\ j=1}^{n} (q_{ij}(x)\ \partial_i u, \partial_j u') + c(u,u') + (v,v') + \int_{\partial\Omega} \rho\ uu' d\sigma$$

where $U = (u,v)$ and $U' = (u',v')$ are in E. Show that it is possible to choose c such that the norm $\| \cdot \|$ associated with ((,)) defines the topology of the product space E.

b) Let $H^2_N(\Omega)$ be the closed subspace of $H^2(\Omega)$ constituted by the set of the $u \in H^2(\Omega)$ which satisfy (7.5.1). We define the product space $D = H^2_N(\Omega) \times H^1(\Omega)$ and the operator

$A = \begin{pmatrix} 0 & , & I \\ Q & , & 0 \end{pmatrix}$ from D into E. Show that there exists C such that

$$|\mathrm{Re}\ ((AU\ ,\ U))| \leq C|||\ U\ |||^2 \quad \text{for all} \quad U \in D\ .$$

c) Prove that the operator $(A - \lambda): D \to E$ is bijective for $|\lambda| > C$; deduce from this that (A, D) generates a strongly continuous group in E.

d) Suppose $T > 0$; take $f \in C^1([0\ T]; L^2(\Omega))$, $u_0 \in H^2_N(\Omega)$, $u_1 \in H^1(\bar{\Omega})$; show that there exists

$u \in C^2([0\ T]; L^2(\Omega)) \cap C^1([0\ T]; H^1(\bar{\Omega})) \cap C^0([0\ T]; H^2_N(\Omega))$

the unique solution of

$$\begin{cases} (\partial^2_{tt} - Q(x,D_x))u = f(t,x) & \text{in} \quad [0\ T] \times \Omega \\ u(0,x) = u_0(x)\ , \quad \dfrac{\partial u}{\partial t}(0,x) = u_1(x) & \text{in} \quad \Omega \\ \left(\dfrac{d}{dn} u(t,x) + p(x)u(t,x)\right)\Big|_{[0\ T] \times \partial\Omega} = 0\ . \end{cases}$$

e) Use the inequality (7.3.3) in order to prove that the solution of d) satisfies the following energy inequality:

$$\| u(t) \|_2 + \| \partial_t u(t) \|_1 + \| \partial_t^2 u(t) \|_0$$

$$\leq C[\| u_0 \|_2 + \| u_1 \|_1 + \| f(0) \|_0 + \int_0^t \| f'(s) \|_0 \, ds]$$

for $t \in [0,T]$, with C independent of the solution u. (It should be pointed out that the method of semigroups can also be applied to mixed problems of order greater than 2; see MIZHOHATA [1]).

7.6 EQUATIONS OF THE SCHRÖDINGER TYPE

Suppose $Q(x, D_x)$ is a formally self-adjoint differential operator of degree 2m in $\mathbb{R}^n$. Suppose that its coefficients are bounded, together with all their derivatives, and that there exists $c_0 > 0$ such that

$$|Q_{2m}(x,\xi)| \geq c_0 |\xi|^{2m} \quad \text{for} \quad (x,\xi) \in R^n \times R^n$$

a) Let $\lambda \in \mathbb{R} \setminus 0$ and $w \in L^2$; prove that there exists a unique $u \in H^{2m}$ such that $(\lambda - iQ)u = w$. [Use the results of Exercise 11.1 of Chapter IV, and the Lax-Milgram theorem (see Section 7.4, Chapter II)].

b) Deduce from this that the operator Q, considered as an unbounded operator in L^2 with domain $D = H^{2m}$, generates a strongly continuous group of unitary transformations.

c) Apply this result to the study of the equation of Schrödinger type: $(\partial_t - i\,Q)u = f,\ u(0) = u_o$.

CHAPTER 7

MIXED HYPERBOLIC PROBLEMS

1. INTRODUCTION

Suppose $P(t, x, D_t, D_x)$ is a differential operator with C^∞ coefficients in $\mathbb{R}^{n+1}$ ($t \in \mathbb{R}$, $x \in \mathbb{R}^n$), which is strictly hyperbolic with respect to the "time variable" t (see Definition 4.9, Chapter VI).

Let Ω be a regular open subset of $\mathbb{R}^n$ and $T > 0$; purely with a view to simplifying the notation in what follows, we shall assume $\partial\Omega$ to be connected. The mixed Cauchy-Dirichlet problem for the wave equation (see Section 6.18, Chapter VI) can be generalised into a mixed problem of the following form: given f, g, $h = (h_o, \ldots, h_{m-1})$, find u satisfying

$$(1.1)\begin{cases} (Pu)(t, x) = f(t, x) & \text{in }]0,T[\times \Omega \\ (Bu)(t, x') = g(t, x') & \text{in }]0,T[\times \partial\Omega \\ & \text{(boundary conditions)} \\ D_t^k u(0, x) = h_k(x) \quad \text{in } \Omega & \text{for } k = 0, \ldots, m-1 \\ & \text{(initial conditions)} \end{cases}$$

More precisely, we shall, in this chapter, distinguish two types of problem (1.1): the systems case and the scalar case. In the systems case, P is of degree $m = 1$, but its coefficients are $N \times N$ matrices; $B(t, x')$ is a given $d \times N$ matrix of class C^∞ for $(t, x') \in \mathbb{R} \times \partial\Omega$. In the scalar case, P is of degree m with scalar coefficients, and $Bu = (B_j u)_{j=1,\dots,d}$, with

$$B_j u = \sum_{k=1}^{m} B_{j,k}(t,x',D_t,D_{x'})\, \gamma_{k-1} u \;, \qquad \text{where each } B_{j,k}$$

is a given differential operator of degree $r_j - k+1$ with coefficients which are scalar and C^∞ in $\mathbb{R} \times \partial\Omega$; we have put

$$\gamma_k u = [(\tfrac{1}{i}\, \nu)^k u]_{|\partial\Omega}, \qquad \text{where } \nu \text{ is a given field of}$$

real-valued vectors transversal to $\partial\Omega$.

Throughout the subsequent discussion we shall use the following assumptions:

(1.2.) $\mathbb{R} \times \partial\Omega$ is non-characteristic for P.

(1.3.) $\partial\Omega$ is compact, or Ω is a half space $\mathbb{R}^n_+$

(1.4.) P and B have constant coefficients outside a compact set.

Then, under a supplementary condition known as the *uniform Lopatinski condition*, we shall give existence and uniqueness results for the solutions of (1.1). These results are based essentially on an a priori estimate for the problem without initial conditions

$$\text{(1.5)} \qquad \begin{cases} Pu = f & \text{in } \mathbb{R} \times \Omega \\ Bu = g & \text{in } \mathbb{R} \times \partial\Omega \end{cases}$$

in the spaces $e^{\gamma t} L^2$ when the parameter γ is sufficiently large. This a priori estimate was established by KREISS [1] in the case of systems and by SAKAMOTO [1] in the scalar case. Here we shall adopt the method of Kreiss, and we shall show how this also allows the a priori estimate to be obtained for the scalar case when $\Omega = R^n_+$, $\gamma = \frac{\partial}{\partial x_n}$ by using a reduction to first order with the aid of suitable pseudo-differential operators in tangential variables t, x' (see AGRANOVITCH [1]).

We shall see in particular that the uniform Lopatinski condition is satisfied in the case of a scalar operator of order m with Dirichlet boundary data; this forms a generalisation of problem (6.18.1) of Chapter VI. On the other hand, this condition is not satisfied in the case of Neumann boundary data (see Remark 6.22 in Chapter VI, and Exercise (9.4).

2. OPERATORS AND SPACES USED

2.1 Background and notation.

If Ω is a regular open subset of $\mathbb{R}^n$ and if $\gamma \in \mathbb{R}$, we denote by $e^{\gamma t}L^2(R \times \Omega)$ the Hilbert space of functions u in $\mathbb{R} \times \Omega$ such that $e^{-\gamma t}u \in L^2(R \times \Omega)$, endowed with the inner product

$$((u,v))_{0;\gamma} = ((e^{-\gamma t}u , e^{-\gamma t}v)) = \iint_{R\times\Omega} e^{-2\gamma t} u\bar{v} \, dt \, dx ,$$

and with the norm $\|u\|_{0;\gamma}$ defined by

$$\|u\|^2_{0;\gamma} = \|e^{-\gamma t}u\|^2 = \iint_{R\times\Omega} e^{-2\gamma t}|u|^2 dt\, dx \quad .$$

We define in similar fashion the Hilbert space $e^{\gamma t}L^2(R \times \partial\Omega)$, endowed with the inner product:

$$(u,v)_{0;\gamma} = (e^{-\gamma t}u , e^{-\gamma t}v) = \int_{R\times\partial\Omega} e^{-2\gamma t}u\bar{v}\, dt\, dx' \quad ,$$

and with the norm $|u|_{0;\gamma}$ defined by

$$|u|^2_{0;\gamma} = |e^{-\gamma t}u| = \int_{R\times\partial\Omega} e^{-2\gamma t}\, |u|^2\, dt\, dx' \quad ,$$

where dx' denotes the measure induced by the Riemannian structure of $\partial\Omega$.

We shall describe the operators which will be used in the case where Ω is the half space $\mathbb{R}^n_+$. First we consider

$$R \times \partial\Omega = R^n = \{(t, x') \mid t \in R , x' \in R^{n-1}\} \quad ,$$

and we denote by σ, ξ' the dual variables of t, x'. We refer to Section 11.2, Chapter IV, for the definition of the symbols $A \in PS^m(R^n \times R^n \times [\bar{\gamma}_0, +\infty[\,)$ which depend on a parameter $\gamma \in [\bar{\gamma}_0, +\infty[$, as well as for the properties of the pseudo-differential operators $A^{(\gamma)} = A(t, x' ; D_t, D_{x'}, \gamma)$.
We define the operator A_γ by $A_\gamma u = e^{\gamma t} A^{(\gamma)} e^{-\gamma t} u$.
Thus, for $u \in C_0^\infty(\mathbb{R}^n)$ and by putting $\tau = \sigma - i\gamma$, we have:

$$(A_\gamma u)(t, x') = \int e^{it\tau + ix'\xi'} A(t, x' ; \sigma, \xi', \gamma)\, \hat{u}(\tau, \xi')\, d\!\!\!/\sigma\, d\!\!\!/\xi'.$$

In particular, we have $e^{\gamma t}D_t e^{-\gamma t} = D_t + i\gamma$, and it is convenient to use the notation

$$A_\gamma = A(t, x' ; D_t + i\gamma , D_{x'}, \gamma) \quad .$$

When $A(t, x' ; \sigma, \xi', \gamma) = P(t, x', \tau, \xi')$ is a polynomial in τ, ξ' with $\tau = \sigma - i\gamma$, we see that $A_\gamma = P(t, x', D_t, D_{x'})$ for all γ. We clearly have for the operators A_γ properties analogous to those of the operators $A^{(\gamma)}$. In particular, if A is of degree 0, there exists a constant C such that

$|A_\gamma u|_{0;\gamma} \leq C |u|_{0;\gamma}$ for all $\gamma \geq \bar{\gamma}_0$ and for all $u \in e^{\gamma t} L^2(\mathbf{R}^n)$. In the following, we shall use in particular the symbols $\Lambda^s = (\gamma^2 + \sigma^2 + |\xi'|^2)^{s/2}$ and the operators Λ^s_γ. We note that if A is of degree -1, there exists a constant C such that

$$\text{(2.1.1)} \qquad |A_\gamma u|_{0;\gamma} \leq \frac{C}{\gamma} |u|_{0;\gamma} \quad \text{for all} \quad \gamma \geq \overline{\gamma_0}$$
$$\text{and for all} \quad u \in e^{\gamma t} L^2(\mathbf{R}^n) \quad .$$

In fact, since the operator $\Lambda_\gamma A_\gamma$ is of degree 0, we have

$$|A_\gamma u|_{0;\gamma} = |A_\gamma \Lambda_\gamma \Lambda_\gamma^{-1} u|_{0;\gamma} \leq C|\Lambda_\gamma^{-1} u|_{0;\gamma}$$

which gives (2.1.1) since we have

$$|\Lambda_\gamma^{-1} u|^2_{0;\gamma} = \int (\gamma^2 + \sigma^2 + |\xi'|^2)^{-1} |\widehat{e^{-\gamma t} u}(\sigma,\xi')|^2 \, d\!\!\!/\sigma \, d\!\!\!/\xi' \leq \frac{1}{\gamma^2} |u|^2_{0;\gamma}$$

for all $u \in C_0^\infty(\mathbf{R}^n)$.

We now consider the half space

$$\overline{\mathbf{R}}^{n+1}_+ = \{(t, x', x_n) \mid t \in \mathbf{R} , x' \in \mathbf{R}^{n-1} , x_n \geq 0 \}$$
$$= \mathbf{R}^n \times \overline{\mathbf{R}}^+ \quad .$$

We here use tangential operators which depend on the

normal variable x_n: if $A = A(t, x ; \sigma, \xi', \gamma)$ is a symbol in

$$PS^m(\bar{R}_+^{n+1} \times R^n \times [\bar{\gamma}_0, +\infty[) ,$$

we associate with it, for $\gamma \geq \bar{\gamma}_0$, the operator
$A_\gamma : C_0^\infty(\bar{R}_+^{n+1}) \to C^\infty(\bar{R}_+^{n+1})$ defined by

$$(A_\gamma u)(t, x', x_n) = A(t, x', x_n ; D_t + i\gamma , D_{x'}, \gamma)\, u(\cdot, \cdot, x_n) .$$

If A is of degree 0, we have an estimate of the form

$$\|A_\gamma u\|_{0;\gamma} \leq C \|u\|_{0;\gamma}$$

for all $\gamma \geq \bar{\gamma}_0$ and for all $u \in e^{\gamma t} L^2(\bar{R}_+^{n+1})$.

Reduction 2.2

We return to problem (1.5) in the case where P is scalar of order m, $\Omega = R_+^n$, $\nu = \frac{\partial}{\partial x_n}$.

We also assume that the coefficient of $D_{x_n}^m$ in P does not vanish, so that we can reduce to the case where P is of the form

$$P = D_{x_n}^m - \sum_{j=0}^{m-1} P_{m-j}(t, x, D_t, D_{x'})D_{x_n}^j ,$$

with P_j of degree j.

As in 4.8, Chapter VI, we put $u_j = \Lambda_\gamma^{m-j} D_{x_n}^{j-1} u$, and $U = (u_1, \ldots, u_m)$. Then (1.5) can be written

$$
(2.2.1) \qquad \begin{cases} D_{x_n} U - A_\gamma U = F & (x_n > 0) \\ \mathcal{B}_\gamma U = G & (x_n = 0) \end{cases}
$$

where

$$
A_\gamma = \begin{pmatrix} 0 & \Lambda_\gamma & 0 \cdots\cdots & 0 \\ 0 & 0 & \Lambda_\gamma \cdots\cdots & 0 \\ & & \ddots & \\ 0 & 0 & \cdots\cdots & \Lambda_\gamma \\ A_{1\gamma} & \cdots\cdots & \cdots\cdots & A_{m\gamma} \end{pmatrix}
$$

$$
A_{j\gamma} = P_{m-j+1} \Lambda_\gamma^{-m+j} , \qquad F = (0, \ldots, 0, f)
$$

$$
\mathcal{B}_\gamma = (\mathcal{B}_{j,k,\gamma})_{\substack{j=1,\ldots,d \\ k=1,\ldots,m}}
$$

$$
\mathcal{B}_{j,k,\gamma} = \Lambda_\gamma^{m-1-r_j} B_{j,k} \Lambda_\gamma^{-m+k}
$$

$$
G = (G_1, \ldots, G_d) , \quad G_j = \Lambda_\gamma^{m-1-r_j} g_j .
$$

We see that A is of degree 1, $\mathcal{B}$ is of degree 0, and that the principal part (homogeneous of degree 1) of A has the characteristic polynomial $\xi_n \mapsto p(t, x ; \sigma - i\gamma , \xi)$, where p is the principal symbol of P. We thus come down to a situation analogous to that of problem (1.5) in the case of systems, but with pseudo-differential operators in tangential variables.

In order to encompass both types of problem (when $\Omega = \mathbb{R}^n_+$, $\nu = \frac{\partial}{\partial x_n}$), we are led to consider the problem in $\bar{\mathbb{R}}^{n+1}_+$:

Problem 2.3

$$\begin{cases} D_{x_n}u - A_\gamma u = f & (x_n > 0) \\ B_\gamma u = g & (x_n = 0) \end{cases}$$

where $A \in PS^1(\bar{\mathbb{R}}^{n+1}_+ \times \mathbb{R}^n \times [\gamma_0 , +\infty[)$ is of N × N matrix form with principal part $a(t, x; \sigma, \xi' ; \gamma)$ homogeneous of degree 1, and $B \in PS^0(\mathbb{R}^n \times \mathbb{R}^n \times [\bar{\gamma}_0 , +\infty[)$ is of d × N matrix form with principal part $b(t, x' ; \sigma, \xi', \gamma)$ homogeneous of degree 0.

We put $X = (t, x ; \sigma , \xi', \gamma)$ and $\Delta(X, \xi_n) = \det(\xi_n I - a(X))$. We adopt the following "strict hyperbolicity" assumption:

(2.3.1) When we put $\tau = \sigma - i\gamma$, then $\Delta(X, \xi_n)$ is a polynomial of degree N in (τ, ξ_n) which possesses N zeros in τ, which are real and simple when $\xi \in \mathbb{R}^n \setminus 0$. Furthermore, we have $\Delta \neq 0$ when $\xi = 0$, $\gamma = 0$, $\sigma \in \mathbb{R} \setminus 0$.

We note that assumption (2.3.1) is satisfied in the case of

problem (1.1) in view of the strict hyperbolicity of P. We also note that assumption (2.3.1) implies

(2.3.2) For $\gamma > 0$, $a(X)$ has no real eigenvalues.

(2.3.3) For $\gamma = 0$, the eigen-subspaces associated with the real eigenvalues of $a(X)$ are of dimension 1.

To prove (2.3.3), we fix t, x, σ, ξ' and put $a(X) = a(\gamma)$, $\Delta(X, \xi_n) = \Delta(\gamma, \xi_n)$. Let λ be a real eigenvalue of $a(0)$ and $e_1, \ldots, e_N$ a basis of $\mathbf{C}^N$ such that e_1 is an eigenvector of $a(0)$ associated with λ. We denote by $C_1(\gamma), \ldots, C_N(\gamma)$ the columns of the matrix $\lambda I - a(\gamma)$ in this basis. We thus have $C_1(0) = 0$, and assumption (2.3.1) implies that

$$\frac{\partial \Delta}{\partial \gamma}(0, \lambda) = \det (C_1'(0), C_2(0), \ldots, C_N(0)) \neq 0 .$$

Thus the columns $C_2(0), \ldots, C_N(0)$ are independent, and this proves (2.3.3).

We finally make the following assumption which, in the case of problem (1.1), follows from (1.4) after using a local chart of the boundary.

(2.3.4) The symbols a and b are constant for $|x|$ sufficiently large and are independent of t for $|t|$ sufficiently large.

We conclude this section with two results which will be

used in Section 6. These are in fact very simple special cases of general properties of partial hypoellipticity at the boundary (HORMANDER [2]). For $s \in \mathbb{R}$, we put

$H^{0,s}(\bar{\mathbb{R}}^{n+1}_+) = L^2(\bar{\mathbb{R}}^+, H^s(\mathbb{R}^n))$. For $k \in \mathbb{N}$, $H^{0,k}(\bar{\mathbb{R}}^{n+1}_+)$ is therefore the space of the $u \in L^2(\mathbb{R}^{n+1}_+)$ such that

$D^{\alpha'}_{t,x'} u \in L^2(\mathbb{R}^{n+1}_+)$ for $|\alpha'| \leq k$. We denote by $BL^m(\bar{\mathbb{R}}^{n+1}_+)$ the space of the operators $A : C^\infty_0(\bar{\mathbb{R}}^{n+1}_+) \to C^\infty(\bar{\mathbb{R}}^{n+1}_+)$ defined by $(Au)(t, x', x_n) = A(t, x', x_n, D_t, D_{x'})u(\cdot,\cdot,x_n)$ for $u \in C^\infty_0(\bar{\mathbb{R}}^{n+1}_+)$, where A is a symbol from the space

$BS^m(\bar{\mathbb{R}}^{n+1}_+ \times \mathbb{R}^n)$ defined in 11.1, Chapter IV. Such an operator is clearly continuous from $H^{0,s}(\bar{\mathbb{R}}^{n+1}_+)$ into $H^{0,s-m}(\bar{\mathbb{R}}^{n+1}_+)$.

PROPOSITION 2.4: *Let* $A \in BL^1(\bar{\mathbb{R}}^{n+1}_+)$. *We put* $P = D_{x_n} - A$. *For* k *an integer* ≥ 1, *let* $u \in H^{0,k}(\bar{\mathbb{R}}^{n+1}_+)$ *be such that* $Pu \in H^{k-1}(\bar{\mathbb{R}}^{n+1}_+)$. *Then* $u \in H^k(\bar{\mathbb{R}}^{n+1}_+)$.

PROOF: We require to prove that $D^\alpha u \in L^2(\mathbb{R}^{n+1}_+)$ for $\alpha = (\alpha', \alpha_n)$, $|\alpha| \leq k$. This is true by hypothesis when $\alpha_n = 0$. We also have $D^{\alpha'} D^{\alpha_{n+1}} u = D^{\alpha'} D^{\alpha_n} Au + D^{\alpha'} D^{\alpha_n} Pu$: this gives the required conclusion by recurrence over α_n since

$$Au \in H^{0,k-1}(\bar{\mathbb{R}}^{n+1}_+), \quad Pu \in H^{k-1}(\bar{\mathbb{R}}^{n+1}_+).$$

THEOREM 2.5 : *Let* X *be a regular open subset of* $\mathbb{R}^n$ *and let* $P(x, D_x)$ *be an* $N \times N$ *differential system of degree* 1 *with coefficients in* $C^\infty(\mathbb{R}^n)$. *Suppose that* ∂X *is non-characteristic for* P. *Let* $E = \{u \in L^2_{loc}(\bar{X}) \mid Pu \in L^2_{loc}(\bar{X})\}$, *endowed with the seminorms of* (u, Pu) *in* $L^2_{loc}(\bar{X}) \times L^2_{loc}(\bar{X})$.

Then $C_0^\infty(\bar{X})$ *is dense in* E *and the trace operator*

$$\gamma : C_0^\infty(\bar{X}) \longrightarrow C_0^\infty(\partial X)$$
$$u \longmapsto u|_{\partial X}$$

extends uniquely into a continuous operator (also denoted by γ*) from* E *into* $H^{-1/2}_{loc}(\partial X)$.

PROOF: When $C_0^\infty(\bar{X})$ is endowed with the topology induced by E, the continuity of $\gamma : C_0^\infty(\bar{X}) \to H^{-1/2}_{loc}(\partial X)$ is an immediate consequence, after using local charts of $\bar{X}$ at the boundary, of the following lemma:

(2.5.1) LEMMA *We adopt the conditions of Theorem 2.5, assuming that* $X = \mathbb{R}^n_+$, *that the coefficients of* P *are constant outside a compact domain and that the coefficient of* D_{x_n} *is the identity. Then there exists a constant* C *such that*

$$\|\gamma u\|_{-\frac{1}{2}} \leq C(\|Pu\|_0 + \|u\|_0) \quad \textit{for all} \quad u \in C_0^\infty(\bar{X}) .$$

In order to establish this lemma, we consider the extension u^0 of u by 0 for $x_n < 0$. The jump formula

$P(u^0) = (Pu)^0 + \frac{1}{i}\,\gamma u \otimes \delta(x_n)$ shows that

$$\|\gamma u \otimes \delta(x_n)\|_{-1} \le \|P(u^0)\|_{-1} + \|(Pu)^0\|_{-1}$$

$$\le c\|u^0\|_0 + \|(Pu)^0\|_0 = c\|u\|_0 + \|Pu\|_0 ,$$

which gives Lemma (2.5.1) since, in view of Proposition 4.6, Chapter II, there exists a constant $c > 0$ such that

$$\|\gamma u \otimes \delta\|_{-1} = c\|\gamma u\|_{-\frac{1}{2}} .$$

In order to conclude the proof of Theorem 2.5, it now remains to prove that $C_0^\infty(\overline{X})$ is dense in E. By truncation and partition of unity, we see that it is sufficient to approximate by elements u_j from $C_0^\infty(\overline{X})$ a given element $u \in E$ with compact support in the case $X = \mathbb{R}^n_+$; let ρ_j be a regularising sequence (see (1.1.1), Chapter I) with $\operatorname{supp} \rho_j \subset \{x_n \le 0\}$; we put $u_j = (u^0 * \rho_j)|_X$; we know that $u_j \to u$ in $L^2(X)$. Furthermore, we have

$$Pu_j = (\rho_j * P(u^0))|_X + ([P, \rho_j *] u^0)|_X$$

where $\rho_j *$ is the operator of convolution with ρ_j. Now $P(u^0) - (Pu)^0$ is supported by $\{x_n = 0\}$; thus $(\rho_j * P(u^0))|_X = (\rho_j * (Pu)^0)|_X$ which tends towards Pu in $L^2(X)$. Finally, Friedrich's lemma (see Section 10, Chapter IV) shows that $[P, \rho_j *]u^0 \to 0$ in $L^2(\mathbb{R}^n)$. We conclude from this that $u_j \to u$ in E.

REMARK 2.6 : Under the conditions of Lemma 2.5.1, the above argument shows that Theorem 2.5 remains valid when we replace $L^2_{loc}(\bar{X})$, $H^{-\frac{1}{2}}_{loc}(\partial X)$ by $L^2(X)$, $H^{-\frac{1}{2}}(\partial X)$.

REMARK 2.7 : Under the conditions of Theorem 2.5, we see immediately (by continuity and density of $C^\infty_0(\bar{X})$ in E) that:

(2.7.1) $\gamma(\phi u) = \gamma\phi . \gamma u$ for $u \in E$ and $\phi \in C^\infty(\bar{X})$.

This allows us to localise in the neighbourhood of a point of ∂X.

(2.7.2) The trace commutes with transport by diffeomorphism (global or local).

(2.7.3) In the case $X = \mathbb{R}^n_+$, the jump formula $P(u^0) = (Pu)^0 + \frac{1}{i} \gamma u \otimes \delta$ remains valid for $u \in E$.

3. THE UNIFORM LOPATINSKI CONDITION

3.1 Introduction

We consider problem (1.5) in the systems case, with $\Omega = \mathbb{R}^n_+$, P homogeneous with constant coefficients, B constant, f = 0:

$$\text{(3.1.1)} \quad \begin{cases} D_{x_n} u - A(D_t, D_{x'})u = 0 & (x_n > 0) \\ Bu(t, x', 0) = g(t, x') & \end{cases}$$

By partial Fourier-Laplace transformation relative to the tangential variables (t, x'), we formally obtain the ordinary differential system with initial conditions, and parameters (τ, ξ'), where $\xi' \in \mathbb{R}^{n-1}$, $\tau = \sigma - i\gamma$, $\sigma \in \mathbb{R}$, $\gamma > 0$:

$$(3.1.2)\quad \begin{cases} D_{x_n} \hat{u}(\tau, \xi', x_n) - A(\tau, \xi')\hat{u}(\tau, \xi', x_n) = 0 & (x_n > 0) \\ B\hat{u}(\tau, \xi', 0) = \hat{g}(\tau, \xi') & . \end{cases}$$

Before studying (3.1.2) it will be useful to give some properties of the solutions v of the differential system

$$(3.1.3)\qquad D_{x_n} v - A(\tau, \xi')v = 0 \qquad (x_n \in \mathbb{R}) \quad .$$

The solution v of (3.1.3) such that $v(0) = w$ is $v(x_n) = e^{ix_n A(\tau, \xi')} w$. We denote by λ_j (of order m_j ; $j = 1,\dots,\ell$) the eigenvalues of the matrix $A(\tau, \xi')$, i.e. the zeros at ξ_n of the equation $\det P(\tau, \xi', \xi_n) = 0$, where $P(D_t, D_{x'}, D_{x_n}) = D_{x_n} - A(D_t, D_{x'})$. The hyperbolicity of P shows that $\lambda_j \notin \mathbb{R}$ when $\xi' \in \mathbb{R}^{n-1}$ and $\operatorname{Im} \tau = \gamma > 0$. We consider the decomposition

$$\mathbb{C}^N = \bigoplus_{j=1}^{\ell} \operatorname{Ker}\left[(\lambda_j I - A(\tau, \xi'))^{m_j}\right] \quad \text{of } \mathbb{C}^N$$

into a direct sum of the generalised eigen-subspaces of $A(\tau, \xi')$.

Denoting by $w = \sum_{j=1}^{\ell} w_j$ the corresponding decomposition of w, we obtain

$$v(x_n) = \sum_{j=1}^{\ell} e^{i\lambda_j x_n} \sum_{p=0}^{m_j - 1} \frac{x_n^p}{p!} (A(\tau, \xi') - \lambda_j I)^p w_j$$

and we then readily deduce that v remains bounded for $x_n \geq 0$ if and only if $w_j = 0$ when $\operatorname{Im} \lambda_j < 0$. As in the case of elliptic operators (see Section 3 of Chapter V), the space of solutions v of (3.1.3) is therefore written $S^+(\tau, \xi') \oplus S^-(\tau, \xi')$, where $S^\pm(\tau, \xi')$ is the subspace of the solutions v of (3.1.3) which remain bounded for $\pm x_n \geq 0$. We denote by $E^\pm(\sigma, \xi', \gamma)$ the subspace of $\mathbf{C}^N$ defined by $E^\pm(\sigma, \xi', \gamma) = \{v(0) \mid v \in S^\pm(\tau, \xi')\}$. We thus have

$$E^\pm(\sigma, \xi', \gamma) = \bigoplus_{\pm \operatorname{Im} \lambda_j > 0} \operatorname{Ker} [(\lambda_j I - A(\tau, \xi'))^{m_j}] ,$$

$\mathbf{C}^N = E^+(\sigma,\xi',\gamma) \oplus E^-(\sigma,\xi',\gamma)$ and the dimension $\mu = \sum_{\operatorname{Im} \lambda_j > 0} m_j$ of $E^+(\sigma, \xi', \gamma)$ is independent of σ, ξ', γ (for $\xi' \in \mathbf{R}^{n-1}$, $\sigma \in \mathbf{R}$, $\gamma > 0$).

We denote by C a contour of $\mathbf{C}$ enclosing the eigenvalues $\lambda_j (j = 1,\ldots,\ell)$ of $A(\tau, \xi')$, and by $C^\pm$ a contour of the half-plane $\{\pm \operatorname{Im} \xi_n > 0\}$ enclosing $\{\lambda_j \mid \pm \operatorname{Im} \lambda_j > 0\}$.

We know that

$$e^{ix_n A(\tau, \xi')} = \frac{1}{2i\pi} \int_C e^{ix_n \xi_n} [\xi_n I - A(\tau, \xi')]^{-1} d\xi_n ,$$

and therefore the solution v of (3.1.3) such that $v(0) = w$ is given by

$$v(x_n) = \frac{1}{2i\pi} \int_C e^{ix_n \xi_n} [\xi_n I - A(\tau, \xi')]^{-1} w \, d\xi_n = v^+(x_n) + v^-(x_n)$$

where $$v^{\pm}(x_n) = \frac{1}{2i\pi}\int_{C^{\pm}} e^{ix_n\xi_n}[\xi_n I - A(\tau, \xi')]^{-1} w \, d\xi_n .$$

It is obvious that $v^{\pm} \in S^{\pm}(\tau, \xi')$, so that

$$v(0) = w = P^{+}w + P^{-}w \quad \text{with} \quad P^{\pm} = \frac{1}{2i\pi}\int_{C^{\pm}} [\xi_n I - A(\tau, \xi')]^{-1} d\xi_n ,$$

and $P^{\pm}w \in E^{\pm}(\sigma, \xi', \gamma)$. Thus P^{+} is the projector from $\mathbb{C}^N$ onto $E^{+}(\sigma, \xi', \gamma)$ parallel to $E^{-}(\sigma, \xi', \gamma)$, and in particular we have $E^{+}(\sigma, \xi', \gamma) = \text{Image } P^{+}$.

We now return to problem (3.1.2). It is clear that the problem

$$\begin{cases} D_{x_n} v - A(\tau, \xi')v = 0 & \text{for} \quad x_n > 0 \\ B\, v(0) = w \end{cases}$$

has, for all $w \in \mathbb{C}^d$, a unique bounded solution v for $x_n \geq 0$ if and only if the restriction $B^{+}(\sigma, \xi', \gamma)$ of B to $E^{+}(\sigma, \xi', \gamma)$ is an isomorphism of $E^{+}(\sigma, \xi', \gamma)$ onto $\mathbb{C}^d$. By homogeneity in (σ, ξ', γ), we see that this condition is realised for all γ sufficiently large, for all $\sigma \in \mathbb{R}$, and for all $\xi' \in \mathbb{R}^{n-1}$ if and only if it is realised for all σ, ξ', γ such that $\sigma^2 + |\xi'|^2 + \gamma^2 = 1$, $\gamma > 0$. We then say that the Lopatinski condition is satisfied. This allows us to give some existence and uniqueness results for (3.1.1) (see HERSH [1], CHAZARAIN-PIRIOU [1], SAKAMOTO [1]). However, this condition is unstable and therefore not suitable for general problems with variable coefficients. In fact, we shall show later that the subspace $E^+(\sigma, \xi', \gamma)$ has a limit when $\gamma \to 0_+$ with $(\sigma, \xi',) \neq 0$. The uniform Lopatinski condition for (3.1.1) consists of requiring $B^+(\sigma, \xi', \gamma)$ to be an isomorphism not only for $\gamma > 0$, but also in the limit for $\gamma = 0_+$, $(\sigma, \xi') \neq 0$.

We now return to the general problem (1.1) in the case of systems. Suppose $(t_o, x'_o, \sigma_o, \xi'_o) \in T^*(R \times \partial\Omega)$ and $\gamma_o > 0$; we shall now give a natural definition of a subspace $E^+(t_o, x'_o, \sigma_o, \xi'_o, \gamma_o)$: we consider a local chart χ of $\bar{\Omega}$ at x'_o, and the corresponding local charts of $T^*(R^n)$, $\partial\Omega$, $T^*(\partial\Omega)$; we use the notation $\sim$ to signify transport by these charts, and we denote by p the principal symbol of P. We have

PROPOSITION 3.2 : *The subspace* $E^+(\sigma_o, \tilde{\xi}'_o, \gamma_o)$ *corresponding to the operator* $\tilde{p}(t_o, \tilde{x}'_o ; D_t, D_{\tilde{x}})$ *is independent of the local chart* χ *used. We denote this by*

$E^+(t_o, x'_o ; \sigma_o, \xi'_o, \gamma_o)$.

PROOF: We can obviously assume that Ω is already of the form $\mathbb{R}^n_+$. We consider a coordinate change $\tilde{x} = \chi(x)$, and put

$$\begin{cases} \tilde{x}' = \tilde{x}'(x', x_n) \\ \tilde{x}_n = \tilde{x}_n(x', x_n) \end{cases} .$$

We have $p(t_o, x'_o, \tau, \xi) = \tilde{p}(t_o, \tilde{x}_o, \tau, \tilde{\xi})$ with

$$\begin{cases} \xi' = \dfrac{\partial \tilde{x}'}{\partial x'}(x'_o)\tilde{\xi}' \\ \xi_n = \dfrac{\partial \tilde{x}'}{\partial x_n}(x'_o).\tilde{\xi} + \dfrac{\partial \tilde{x}_n}{\partial x_n}(x'_o)\tilde{\xi}_n \end{cases}$$

With the notation used in the Introduction 3.1,

$E^+(\sigma_o, \xi'_o, \gamma_o)$ corresponds to the equation

$(D_{x_n} - A(t_o, x'_o, \tau_o, \xi'_o)) \quad v(x_n) = 0$, whereas

$\tilde{E}^+(\sigma_o, \tilde{\xi}'_o, \gamma_o)$ corresponds to the equation

$(a D_{\tilde{x}_n} + bI - A(t_o, x'_o, \tau_o, \xi'_o)) \tilde{v}(\tilde{x}_n) = 0$

where $a = \dfrac{\partial \tilde{x}_n}{\partial x_n}(x'_o) > 0$, $b = \dfrac{\partial \tilde{x}'}{\partial x_n}(x'_o) . \tilde{\xi}'_o \in \mathbb{R}$.

Now we can reduce this equation to the preceding equation by successively putting $\tilde{v}(\tilde{x}_n) = w\left(\dfrac{\tilde{x}_n}{a}\right)$ and $w(x_n) = e^{-ibx_n} v(x_n)$; since both these transformations keep the solutions bounded in $\overline{\mathbb{R}}^+$, together with their initial values, (3.2) is proved. It should be pointed out that the properties of invariance of the subspaces E^+ will be investigated more systematically in Section 8.

In order to bring together the case of systems and the scalar case for problem (1.1), we give:

DEFINITION 3.3 : *We consider the situation of Problem*

2.3, *and we put* $\mathcal{X} = \{X = (t, x ; \sigma, \xi', \gamma) \mid t \in R, x \in \bar{R}^n_+, \sigma \in R, \xi' \in R^{n-1}, \gamma \geq 0, (\sigma, \xi', \gamma) \neq 0\}$.
For $\gamma > 0$, *we put* $E^+(X) = \bigoplus_{\operatorname{Im} \lambda_j > 0} [(\operatorname{Ker} \lambda_j I - a(X))^{m_j}]$,

where the λ_j *are the eigenvalues (of order* m_j ; $j = 1, \ldots, \ell$*)* *of* $a(X)$. *We denote by* $\mu = \sum_{\operatorname{Im} \lambda_j > 0} m_j$ *the dimension (independent of* X*) of* $E^+(X)$.

We obviously have properties for the subspaces $E^+(X)$ analogous to those given in Section 3.1 for the subspaces $E^+(\sigma, \xi', \gamma)$. In particular, we again have $E^+(X) =$ Image $P^+(X)$, where $P^+(X) = \frac{1}{2i\pi} \int_{C^+} (\xi_n I - a(X))^{-1} d\xi_n$; since $P^+(X)$ is C^∞ with respect to X and remains of constant rank μ, the subspaces $E^+(X)$ are organised into a sub-bundle E^+ of class C^∞ of the trivial bundle $\mathbb{C}^N$ over $\mathcal{X} \cap \{\gamma > 0\}$. We shall now investigate the behaviour of $E^+(X)$ when $\gamma \to 0$. For $\gamma > 0$, we put

$$\Delta(X, \xi_n) = \det(\xi_n I - a(X)) = \prod_{j=1}^{\ell} (\xi_n - \lambda_j)^{m_j},$$

$$\Delta^{\pm}(X, \xi_n) = \prod_{\pm \operatorname{Im} \lambda_j > 0} (\xi_n - \lambda_j)^{m_j}$$

such that $\Delta = \Delta^+ \Delta^-$. We consider the matrices

$$Q^{\pm}(X) = \Delta^{\mp}(X, a(X)) = \prod_{\mp \operatorname{Im} \lambda_j > 0} (a(X) - \lambda_j I)^{m_j} . \tag{3.3.1}$$

The decomposition $\mathbb{C}^N = \bigoplus_{j=1}^{\ell} \operatorname{Ker} [(a(X) - \lambda_j I)^{m_j}]$

shows that $E^{\pm}(X) = \mathrm{Ker}\; Q^{\mp}(X)$, Rank $Q^{\pm}(X) = \dim E^{\pm}(X)$. Furthermore, the Cayley-Hamilton theorem shows that $Q^{+}(X)\, Q^{-}(X) = \Delta(X, a(X)) = 0$, so that $\mathrm{Image}\; Q^{+}(X) \subset \mathrm{Ker}\; Q^{-}(X) = E^{+}(X)$, and consequently we obtain

$$E^{+}(X) = \mathrm{Image}\; Q^{+}(X) \quad \text{for} \quad \gamma > 0 \quad . \tag{3.3.2}$$

The coefficients of the polynomials (in ξ_n) $\Delta^{\pm}(X, \xi_n)$ have C^{∞} dependence on X for $\gamma > 0$, since $\Delta^{\pm}(X, \xi_n)$ is the characteristic polynomial of the restriction of $a(X)$ to $E^{\pm}(X)$. We shall now show that the coefficients of $\Delta^{\pm}(X, \xi_n)$ extend continuously to $\mathcal{X}$: suppose $X_0 \in \mathcal{X} \cap \{\gamma = 0\}$, and denote by λ_j (of order m_j ; $j = 1,\ldots,\ell$) the eigenvalues of $a(X_0)$. Let $D_1,\ldots,D_\ell$ be discs which are mutually disjoint centred respectively at $\lambda_1,\ldots,\lambda_\ell$. We know that there exists a neighbourhood V of X_0 in $\mathcal{X}$ such that, for $X \in V$, the eigenvalues (counted according to their order) of $a(X)$ are distributed as m_1 eigenvalues in $D_1,\ldots, m_\ell$ in D_ℓ. Let μ_j denote the number of eigenvalues of $a(X)$ in D_j which have positive imaginary part when $X \in V \cap \{\gamma > 0\}$. We put

$$\Delta^{+}(X_0, \xi_n) = \prod_{j=1}^{\ell} (\xi_n - \lambda_j)^{\mu_j} \quad , \quad \Delta^{-}(X_0, \xi_n) = \prod_{j=1}^{\ell} (\xi_n - \lambda_j)^{m_j-\mu_j}$$

and it can readily be shown, using the relationships between coefficients and roots, that the coefficients of the polynomials $\Delta^{\pm}(X, \xi_n)$ thus extended to $\mathcal{X}$ are continuous in $\mathcal{X}$. The formulas (3.3.1) and (3.3.2) lead us to extend the subspace $E^{+}(X)$ to $\mathcal{X}$ by putting $Q^{+}(X) = \Delta^{-}(X, a(X))$, and

$E^+(X) = \text{Image } Q^+(X)$ for all $x \in \mathcal{X}$. We know that $Q^+(X)$ is of rank μ for $\gamma > 0$, and we shall show that the same applies for $\gamma = 0$.

First, we shall give a reduced form of $a(X)$ in the neighbourhood of an arbitrary point X_o of $\mathcal{X}$:

PROPOSITION 3.4 : *For* $X_o \in \mathcal{X}$, *we denote by* $\{\lambda_1,\dots,\lambda_k\}$ *the (possibly empty) set of the real eigenvalues of* $a(X_o)$, *and by* m_j *the multiplicity of* $\lambda_j (j=1,\dots,k)$. *We denote by* $m^\pm$ *the number (possibly zero) of eigenvalues (counted according to their multiplicity) of* $a(X_o)$ *located in the half plane* $\{\pm \operatorname{Im} \lambda > 0\}$. *Then, there exists a change of basis of* $\mathbf{C}^N$, *with* C^∞ *dependence on* X *in the neighbourhood of* X_o *in* $\mathcal{X}$, *which puts* $a(X)$ *into block-diagonal form, having blocks of dimension* $m_1,\dots m_k$, m^+, m^- :

$$\begin{pmatrix} a_1(X) & & & & 0 \\ & \ddots & & & \\ & & a_k(X) & & \\ & & & a^+(X) & \\ 0 & & & & a^-(X) \end{pmatrix} \tag{3.4.1}$$

where the eigenvalues of $a^\pm(X)$ *are in the half plane* $\{\pm \operatorname{Im} \lambda > 0\}$ *and where, for* $j = 1,\dots k$, *we have*

$$(3.4.2) \qquad a_j(X_o) = \begin{pmatrix} \lambda_j & 1 & & 0 \\ & \ddots & \ddots & \\ & & \ddots & 1 \\ 0 & & & \lambda_j \end{pmatrix} .$$

PROOF: Let $C_j (j = 1,\dots,k)$ be a small contour of $\mathbb{C}$ enclosing only λ_j. Let $C_\pm$ be a contour of the half plane $\{\pm \operatorname{Im} \lambda > 0\}$ enclosing only the eigenvalues of $a(X_o)$ which are located in this half plane. For $X \in \mathcal{X}$ sufficiently near X_o, consider the projectors

$$P_j(X) = \frac{1}{2i\pi} \int_{C_j} (\xi_n I - a(X))^{-1} d\xi_n \quad , \qquad j = 1,\dots,k$$

$$P_\pm(X) = \frac{1}{2i\pi} \int_{C_\pm} (\xi_n I - a(X))^{-1} d\xi_n \quad .$$

These have C^∞ dependence on X, and we know that:

$$\text{Rank } P_j(X) = m_j \quad (j = 1,\dots,k) \quad , \qquad \text{Rank } P_\pm(X) = m^\pm ,$$

$$\mathbb{C}^N = \bigoplus_{j=1}^{k} \text{Image } P_j(X) \oplus \text{Image } P_+(X) \oplus \text{Image } P_-(X) \quad .$$

The image under $P_j(X)$ of a basis of Image $P_j(X_o)$ is for X sufficiently near X_o, a basis of Image $P_j(X)$ having C^∞ dependence on X. We construct similarly a basis of Image $P_\pm(X)$, which gives a basis of $\mathbf{C}^N$ putting $a(X)$ into a form (3.4.1) in which the eigenvalues of $a^{\pm}(X)$ are in the half plane $\{\pm \operatorname{Im} \lambda > 0\}$ and, for $j = 1,\ldots,k$, the matrix $a_j(X_o)$ has only the eigenvalue λ_j (of order m_j). In each subspace $\mathbf{C}^{m_j}$ ($j = 1,\ldots,k$), we apply a change of basis which puts $a_j(X_o)$ in Jordan form; we thus obtain (3.4.2) with ones everywhere on the diagonal immediately above the principal diagonal, in view of (2.3.3). It may be noted, from (2.3.2) that the blocks $a_j(X_o)$ do not appear in (3.4.1) when $X_o \in \mathcal{X} \cap \{\gamma > 0\}$.

THEOREM 3.5 : *The subspaces* $E^+(X)$ *are organised into a sub-bundle of class* C^∞ *of the trivial sub-bundle* $\mathbf{C}^N$ *over* $\mathcal{X} \cap \{\gamma > 0\}$, *which extends into a continuous sub-bundle* E^+*(of rank* μ*) of the trivial bundle* $\mathbf{C}^N$ *over* $\mathcal{X}$.

PROOF: In view of the arguments put forward following the statement of Definition 3.3, we require to prove that Rank $Q^+(X_o) = \mu$ for $X_o \in \mathcal{X} \cap \{\gamma = 0\}$. By Proposition 3.4, we may assume that $a(X_o)$ is of the form (3.4.1) with $X = X_o$, and that (3.4.2) is satisfied. We then obtain

$$Q^+(X_o) = \begin{pmatrix} Q_1 & & & & 0 \\ & \ddots & & & \\ & & Q_k & & \\ & & & Q_+ & \\ 0 & & & & Q_- \end{pmatrix}$$

with $Q_j = \Delta^-(X_o, a_j(X_o))$ for $j = 1,\dots,k$, and $Q_\pm = \Delta^-(X_o, a^\pm(X_o))$. Now

$$\Delta^-(X_o, \xi_n) = \prod_{j=1}^{k} (\xi_n - \lambda_j)^{m_j-\mu_j} \Delta_-(\xi_n)\ ,$$ where Δ_- is the characteristic polynomial of $a^-(X_o)$. We thus obtain $Q_- = 0$, Q_+ invertible, and $Q_j = (a_j(X_o) - \lambda_j I)^{m_j-\mu_j} T_j$ with T_j invertible for $j = 1,\dots,k$. Now, from (3.4.2), we see that the image of $(a_j(X_o) - \lambda_j I)^{m_j-\mu_j}$ is generated by the first μ_j canonical basis vectors of $\mathbf{C}^{m_j}$, so that we have

$$\text{Rank } Q^+(X_o) = \sum_{j=1}^{k} \mu_j + m^+ = \mu \quad .$$

REMARK 3.6 : The conclusion of the above proof shows that, for $X_o \in \mathfrak{X} \cap \{\gamma = 0\}$, and following a change of basis of $\mathbf{C}^N$ of the type indicated in Proposition 3.4, $E^+(X_o)$ is identified with the subspace $\bigoplus_{j=1}^{k} \mathbf{C}^{\mu_j} \oplus \mathbf{C}^{m^+}$ of $\bigoplus_{j=1}^{k} \mathbf{C}^{m_j} \oplus \mathbf{C}^{m^+} \oplus \mathbf{C}^{m^-}$, where $\mathbf{C}^{\mu_j}$ is the subspace generated by the first μ_j canonical basis vectors of $\mathbf{C}^{m_j}$.

Looking ahead to Section 5, it will be useful at this point to determine the number μ_j.

PROPOSITION 3.7 : *Using the notation of Proposition* 3.4 *in the case where* $X_o \in \mathfrak{X} \cap \{\gamma = 0\}$, *we put*

$$D_j(X) = \frac{1}{i} \frac{\partial a_j}{\partial \gamma}(X)\ ,$$ *and denote by* α_j *the lower left element*

of the matrix $D_j(X_o)$. *Then* α_j *is real and non-zero, and* μ_j *is given by:*

$$\mu_j = m_j/2 \quad \text{when } m_j \text{ is even}$$

$$\mu_j = \frac{m_j \pm 1}{2} \quad \text{when } m_j \text{ is odd and } \pm\alpha_j > 0 .$$

PROOF: We put $X_o = (t_o, x_o ; \sigma_o, \xi'_o, 0)$, $\tau = \sigma - i\gamma$. We consider the characteristic polynomials

$$\Delta(\tau, \xi_n) = \det(\xi_n I - a(t_o, x_o ; \sigma, \xi'_o, \gamma)) ,$$

$$\Delta_j(\tau, \xi_n) = \det(\xi_n I - a_j(t_o, x_o; \sigma, \xi'_o, \gamma)) \quad \text{for } j = 1,\dots,k ,$$

$$\widetilde{\Delta}(\tau, \xi_n) = \det(\xi_n I - a^+(t_o, x_o ; \sigma, \xi'_o, \gamma))$$

$$\times \det(\xi_n I - a^-(t_o, x_o ; \sigma, \xi'_o, \gamma)) .$$

For $\xi_n \in \mathbb{R}\setminus 0$, assumption (2.3.1) shows that

$$\Delta(\tau, \xi_n) = c \prod_{\ell=1}^{N} (\tau - \tau_\ell(\xi_n)) , \text{ with } \tau_\ell(\xi_n) \in \mathbb{R} \text{ and } c \neq 0 . \tag{3.7.1}$$

Furthermore, $\Delta(\tau, \xi_n)$ is of the form

$$\Delta(\tau, \xi_n) = \xi_n^N + c_1(\tau)\xi_n^{N-1} + \dots + c_N(\tau) . \tag{3.7.2}$$

By equating the terms of highest degree in ξ_n in (3.7.1) and (3.7.2) we obtain $c \in \mathbb{R}$ and we then deduce from (3.7.1) that $\Delta(\tau, \xi_n)$ and $\frac{1}{i}\frac{\partial\Delta}{\partial\gamma}(\tau, \xi_n)$ are real for $\gamma = 0$ and $\xi_n \in \mathbb{R}$. It then follows that the zeros ξ_n of $\Delta(\tau, \xi_n)$ form imaginary conjugate pairs when $\gamma = 0$; the same therefore applies for the zeros ξ_n of $\Delta_j(\tau, \xi_n)$ and of $\widetilde{\Delta}(\tau, \xi_n)$; thus $\Delta_j(\tau, \xi_n)$ and

$\widetilde{\Delta}(\tau, \xi_n)$ are real for $\gamma = 0$ and $\xi_n \in \mathbb{R}$. Since $\Delta = \Delta_1 \ldots \Delta_k \widetilde{\Delta}$, we then deduce that $\frac{1}{i} \frac{\partial \Delta_j}{\partial \gamma}(\sigma_0, \lambda_j)$ is real; since $\frac{\partial \Delta}{\partial \gamma}(\sigma_0, \lambda_j)$ is non-zero by virtue of assumption (2.3.1), we similarly see that $\frac{\partial \Delta_j}{\partial \gamma}(\sigma_0 ; \lambda_j) \neq 0$. We have thus proved that $\frac{1}{i} \frac{\partial \Delta_j}{\partial \gamma}(\sigma_0, \lambda_j)$ is real and non-zero. Furthermore a straightforward calculation gives

$$\frac{1}{i} \frac{\partial \Delta_j}{\partial \gamma}(\sigma_0, \lambda_j) = \frac{1}{i} \frac{\partial}{\partial \gamma}(\lambda_j I - a_j(X))\Big|_{X = X_0} = -\alpha_j$$

since

$$\lambda_j I - a_j(X_0) = - \begin{pmatrix} 0 & \boxed{\mathrm{Id}} \\ \vdots & \\ 0 & 0 \cdots 0 \end{pmatrix}, \qquad \text{hence } \alpha_j \in \mathbb{R} \setminus 0 .$$

Finally, we consider the polynomial equation $\Delta(\sigma_0 - i\gamma, \xi_n) = 0$. For $\gamma = 0$, the value $\xi_n = \lambda_j$ is a root of order m_j; for $\xi_n = \lambda_j$, the value $\gamma = 0$ is a simple root. Thus, when $\gamma \to 0$, the roots ξ_n near λ_j (i.e. the roots ξ_n of the equation $\Delta_j(\sigma_0 - i\gamma, \xi_n) = 0$) are, in a certain sense, near to the roots ξ_n of the equation $(\xi_m - \lambda_j)^{m_j} - \rho\gamma = 0$, with $\rho = -\frac{\partial \Delta_j}{\partial \gamma}(\sigma_0, \lambda_j) = i\alpha_j$. More precisely, the theory of Puiseux expansions (see LEFCHETZ [1]) shows that these can be written, for $\gamma \geq 0$ sufficiently small, in the form:

$$\lambda_j^{(r)}(\gamma) = \lambda_j + \gamma^{1/m_j} (i\alpha_j)^{1/m_j} + O(\gamma^{2/m_j}) \qquad (r = 1, \ldots, m_j)$$

where the $(i\alpha_j)_r^{1/m_j}$ denote the m_j possible evaluations of

$(i\alpha_j)^{1/m_j}$. It then follows that μ_j is equal to the number of these evaluations with positive imaginary part; this then allows us to conclude by means of an elementary calculation.

We can now give the following definition

DEFINITION 3.8 : *We say that problem* (2.3) *satisfies the uniform Lopatinski condition if the restriction* $b^+(t, x' ; \sigma, \xi', \gamma)$ *of* $b(t, x' ; \sigma, \xi', \gamma)$ *to* $E^+(t, x' ; \sigma, \xi', \gamma)$ *is an isomorphism for all* $t \in \mathbb{R}$, $x' \in R^{n-1}$, $\sigma \in R$, $\xi' \in R^{n-1}$, $\gamma \geq 0$ *such that* $(\sigma, \xi', \gamma) \neq 0$.

We refer to Exercise 9.5 for the study of an example.

The uniform Lopatinski condition implies that the number d of scalar boundary conditions is equal to the rank μ of E^+. Proposition 3.2 allows us to give a natural definition of the uniform Lopatinski condition for problem (1.1) in the case of systems. In the scalar version of problem (1.1), and under the conditions of the reduction 2.2, we reduced problem (1.5) to problem (2.2.1). We shall now give an interpretation of the uniform Lopatinski condition for this problem. Remembering that $p = p(t, x, \tau, \xi)$ is the principal symbol (of degree m) of P we define, for $X \in \mathcal{X} \cap \{\gamma > 0\}$, a subspace $D^+(X)$

analogous to the subspace $D^+(x', \xi')$ appearing in Proposition 3.3, Chapter V : this is the space of the $w \in (w_1,\ldots,w_m) \in \mathbb{C}^m$ of the form $w_j = D^{j-1}v(0)$ for $j = 1,\ldots,m$, where v is a solution, bounded for $x_n \geq 0$, of the differential equation

$$p(t, x', \sigma - i\gamma, \xi', D_{x_n}) v(x_n) = 0 \quad . \tag{3.8.1}$$

It can easily be shown that the subspace $E^+(X)$ corresponding to the reduced problem (2.2.1) is given, for $\gamma > 0$, by:

$$E^+(X) = \left\{\left((\gamma^2 + \sigma^2 + |\xi'|^2)^{\frac{m-j}{2}} w_j\right)_{j=1,\ldots,m} \Big| \; w \in D^+(X)\right\}. \tag{3.8.2}$$

This equality shows, in particular, that the subspaces $D^+(X)$ are organised into a bundle of rank μ over $\mathcal{X}$ which has properties analogous to those of the bundle E^+ (see Theorem 3.5). Returning to (2.2.1), we note that the principal part of $\mathbb{B}$ is the matrix

$$\left((\gamma^2 + \sigma^2 + |\xi'|^2)^{\frac{k-1-r_j}{2}} b_{j,k} \right)_{\substack{j=1,\ldots,d \\ k=1,\ldots,m}} ,$$

where $b_{j,k}$ denotes the principal symbol of degree $r_j - k + 1$ of $B_{j,k}$, or more precisely, its value at $(t, x', \sigma - i\gamma, \xi')$. From (3.5.2), we therefore see that the uniform Lopatinski condition for the reduced problem (2.2.1) is expressed by the condition

(3.8.3.) *If we put* $b(X) = (b_{j,k}(t, x', \sigma - i\gamma, \xi'))_{\substack{j=1,\dots,d\\ k=1,\dots,m}}$

then the restriction $b^+(X)$ *of* $b(X)$ *to* $D^+(X)$ *is an isomorphism for all* $X \in \mathcal{X}_{|x_n=0}$.

In the general case of the scalar problem (1.1), by local charts of the boundary transporting the field ν into $\frac{\partial}{\partial x_n}$ and by arguing as for Proposition 3.2, we define the bundle D^+ over

$$\{(t, x', \sigma, \xi', \gamma) | (t, x', \sigma, \xi') \in T^*(\mathbb{R} \times \partial\Omega), \gamma \geq 0, (\sigma, \xi', \gamma) \neq 0\}$$

with the uniform Lopatinski condition defined by taking (3.8.3).

REMARK 3.9 : Lopatinski determinant in the scalar case.

We return to problem (1.1) in the scalar case with $\Omega = \mathbb{R}^n_+$, $\nu = \frac{\partial}{\partial x_n}$ and $d = \mu$. We denote by p the principal symbol of P, and we put $p(t, x', \sigma - i\gamma, \xi', \xi_n) = p(X, \xi_n)$. As in Section 4.5, Chapter V, we consider, for $\gamma > 0$, a factorisation $p(X, \xi_n) = p^+(X, \xi_n)\, p^-(X, \xi_n)$, where $p^{\pm}(X, \xi_n)$ is a polynomial in ξ_n whose zeros are in the half plane $\{\pm \operatorname{Im} \xi_n > 0\}$. The solutions, bounded for $x_n \geq 0$, of equation (3.8.1) are the solutions of the differential equation $p^+(X, D_{x_n}) v(x_n) = 0$. By successive differentiations, we therefore deduce that we can parametrise the bundle D^+, over $\mathcal{X} \cap \{\gamma > 0\}$, by

$$(3.9.1) \qquad \begin{pmatrix} w_{\mu+1} \\ \vdots \\ w_m \end{pmatrix} = M(X) \begin{pmatrix} w_1 \\ \vdots \\ w_\mu \end{pmatrix}$$

where the elements of the matrix $M(X)$ are expressed in terms of the coefficients of $p^+(X, \xi_n)$ by means of fixed polynomials. Now we saw previously that the coefficients of $p^+(X, \xi_n)$ extend continuously to $\mathcal{X}$; the same therefore applies to the matrix M, and (3.9.1) thus furnishes a global frame for the bundle D^+ over $\mathcal{X}$. We denote by

$$b_j(X, \xi_n) = \sum_{k=1}^{m} b_{j,k}(t, x', \sigma - i\gamma, \xi')\xi_n^{k-1}$$

the principal symbol of the operator B_j, and we denote by

$$b_j^+(X, \xi_n) = \sum_{k=1}^{\mu} b_{j,k}^+(X)\xi_n^{k-1}$$

the remainder of the Euclidian division of $b_j(X, \xi_n)$ by $p^+(X, \xi_n)$. Then we have

$$(3.9.2) \qquad \sum_{k=1}^{m} b_{j,k}(t, x', \sigma - i\gamma, \xi')w_k = \sum_{k=1}^{\mu} b_{j,k}^+(X)w_k$$

for all $w = (w_1, \ldots, w_m) \in D^+(X)$.

In fact it is sufficient, by continuity, to show that this equality holds for $\gamma > 0$. Then we have $w_k = D_{x_n}^{k-1}v(0)$ for $k = 1, \ldots, m$, with $p^+(X, D_{x_n})v(x_n) = 0$. The left-hand side of (3.9.2) is thus given by

$$(b_j(X, D_{x_n})v(x_n))_{|x_n=0} = (b_j^+(X, D_{x_n})\, v(x_n))_{|x_n=0}$$

since $b_j(X, \xi_n) \equiv b_j^+(X, \xi_n)$ modulo $p^+(X, \xi_n)$, which gives (3.9.2).

If we parametrise D^+ by $w_1, \ldots, w_\mu$ and use (3.9.2), we are led to consider the (Lopatinski) determinant

$$\mathrm{Lop}_{P,B}(X) = \det\, (b^+_{j,k}(X))_{\substack{j=1,\ldots,\mu \\ k=1,\ldots,\mu}}$$

and to express the uniform Lopatinski condition (3.8.3) by

$$\mathrm{Lop}_{P,B}(X) \neq 0 \quad \text{for all} \quad X \in \mathcal{X}\ . \tag{3.9.3}$$

EXAMPLE 3.10 :- Cauchy-Dirichlet problem for a scalar operator of order m. We are here concerned with problem (1.1) in the scalar case with $d = \mu$, $B_j u = \gamma_{j-1} u$ $(j = 1,\ldots,\mu)$. Using the notation of Remark 3.9, we have $b_j = b_j^+ = \xi_n^{j-1}$ for $j = 1,\ldots,\mu$, and $b^+_{j,k} = \delta_{j,k}$ for $j, k = 1,\ldots,\mu$. Therefore $\mathrm{Lop}_{P,B} = 1$, and the uniform Lopatinski condition (3.9.3) is satisfied.

COUNTER-EXAMPLE 3.11 : Cauchy-Neumann problem.

We take as an example the scalar problem (1.1) in the case

$$\Omega = \mathbb{R}^n_+ \ , \quad P = D_t^2 - \sum_{j=1}^{n} D^2_{x_j} \ , \quad Bu = (D_{x_n} u)_{|x_n} = 0 \ .$$

Using the notation of Remark 3.9, we have $d = \mu = 1$, $b_1 = \xi_n$,

$$p^+(X, \xi_n) = \xi_n - \sqrt[+]{(\sigma - i\gamma)^2 - |\xi'|^2} \ ,$$ where $\sqrt[+]{\ }$ denotes

the root with positive imaginary part when $\gamma > 0$, or its limit when $\gamma \to 0_+$. We thus have $b_1^+(X, \xi_n) = {}^+\sqrt{(\sigma - i\gamma)^2 - |\xi'|^2}$,

hence $\mathrm{Lop}_{P,B}(X) = {}^+\sqrt{(\sigma - i\gamma)^2 - |\xi'|^2}$,

this quantity vanishing for $\gamma = 0$, $\sigma^2 = |\xi'|^2$. The uniform Lopatinski condition (3.9.3) is therefore not satisfied.

4. ENERGY INEQUALITIES

We start by establishing an a priori estimate for the symmetrisable operators in the unrestricted space $\mathbb{R}^{n+1}$. For $u \in C^\infty(\mathbb{R}^{n+1})$ and $\gamma > 0$, we define the norm

$$\|u\|_{0;\gamma} \quad \text{by} \quad \|u\|^2_{0;\gamma} = \iint_{\mathbb{R}^{n+1}} e^{-2\gamma t} |u(t,x)|^2 dt\, dx .$$

THEOREM 4.1 : *Suppose* $P = D_t - A(t, x, D_x)$ *is a first-order operator which is symmetrisable in the sense of Definition 4.2, Chapter VI. We also assume that the constant* $c > 0$ *appearing in* (4.2.2), *Chapter VI, can be chosen independently of* (t, x, ξ). *Then there exist positive constants* C *and* γ_0 *such that*

$$\gamma\|u\|_{0;\gamma} \leq C\, \|Pu\|_{0;\gamma} \qquad \textit{for all} \qquad \gamma \geq \gamma_0 \tag{4.1.1}$$

and for all $u \in C_0^\infty(\mathbb{R}^{n+1})$.

PROOF: This is analogous to the proof of Theorem 4.3, Chapter VI, the notation of which we use here. We calculate

$G(t) = \frac{d}{dt} e^{-2\gamma t}(Ru, u)$. Putting $Pu = f$, we obtain

$$\frac{1}{2} G(t) = - \gamma e^{-2\gamma t}(Ru, u) - e^{-2\gamma t} \operatorname{Im}(RAu, u) + e^{-2\gamma t} \operatorname{Im}(Ru, f) + \frac{1}{2} e^{-2\gamma t}(R_t' u, u) \quad . \tag{4.1.2}$$

Integrating both sides of this equality with respect to t from $-\infty$ to $+\infty$, we obtain

$$\gamma \int (R\, e^{-\gamma t}u , e^{-\gamma t}u)dt = - \int \operatorname{Im}(RA\, e^{-\gamma t}u , e^{-\gamma t}u)\, dt + \int (R\, e^{-\gamma t}u , e^{-\gamma t}f)dt + \frac{1}{2} \int (R_t'\, e^{-\gamma t}u, e^{-\gamma t}u)dt \quad .$$

As in the proof of Theorem 4.3, Chapter VI, we then deduce by using the Gårding and Schwarz inequalities that

$$c'\gamma \|u\|^2_{0;\gamma} \le C' \|u\|^2_{0;\gamma} \|f\|^2_{0;\gamma} + C' \|u\|^2_{0;\gamma} \quad ,$$

with $c' > 0$.

We thus have, for $\varepsilon > 0$:

$$c'\gamma \|u\|^2_{0;\gamma} \le \epsilon\gamma \|u\|^2_{0;\gamma} + \frac{C'^2}{4\epsilon\gamma} \|f\|^2_{0;\gamma} + C' \|u\|^2_{0;\gamma}$$

from which we deduce (4.1.1) by choosing $\varepsilon < c'$ and γ_0 sufficiently large.

We shall now give a priori estimates for problems (1.5) and 2.3 with boundary conditions. We refer to Section 2.1 for the definition of the spaces $e^{\gamma t}L^2$ and the corresponding norms. For $u \in C_0^\infty(\mathbb{R} \times \overline{\Omega})$, we put $|u|_{0;\gamma} = |u_{|\mathbb{R}\times\partial\Omega}|_{0;\gamma}$.

THEOREM 4.2 : *We consider problem* 2.3, *assuming the uniform Lopatinski condition (see 3.8) to be satisfied , and we put* $P_\gamma = D_{x_n} - A_\gamma$. *Then there exist positive constants* C *and* γ_o *such that*

$$\gamma\|u\|^2_{o;\gamma} + |u|^2_{o;\gamma} \le C\left(\frac{1}{\gamma}\|P_\gamma u\|^2_{o;\gamma} + |B_\gamma u|^2_{o;\gamma}\right) \tag{4.2.1}$$

for all $\gamma \ge \gamma_o$ *and for all* $u \in C_o^\infty(\overline{\mathbb{R}}_+^{n+1})$.

This theorem forms one of the fundamental results of this chapter, and it will be established later on. For the moment, we shall show how we can deduce from it energy inequalities for problem (1.1).

COROLLARY 4.3 : *We consider problem* (1.1) *in the case of systems, assuming the uniform Lopatinski condition to be satisfied. Then there exist positive constants* C *and* γ_o *such that*

$$\gamma\|u\|^2_{o;\gamma} + |u|^2_{o;\gamma} \le C\left(\frac{1}{\gamma}\|Pu\|^2_{o;\gamma} + |Bu|^2_{o;\gamma}\right) \tag{4.3.1}$$

for all $\gamma \ge \gamma_o$ *and for all* $u \in C_o^\infty(\mathbb{R} \times \overline{\Omega})$.

PROOF: Since P and B are differential, we have $P_\gamma = P$ and $B_\gamma = B$ for all γ, and thus (4.3.1) follows from (4.2.1) in the case where $\Omega = \mathbb{R}^n_+$. In the case where Ω is a regular open subset of $\mathbb{R}^n$ with $\partial\Omega$ compact, we consider a finite covering of $\partial\Omega$ by open coordinate patches $U_j (1 \leq j \leq J)$ of $\bar{\Omega}$ to the boundary; suppose we have functions $\varphi_j \in C_0^\infty(U_j)$ and $\varphi_0 \in C_0^\infty(\Omega)$ such that $\sum_{j=0}^{J} \varphi_j = 1$ in $\bar{\Omega}$. From Theorem 4.1, we have

$$\gamma\|\varphi_0 u\|^2_{0;\gamma} \leq \frac{C}{\gamma}\|P(\varphi_0 u\|^2_{0;\gamma} \quad \text{for} \quad \gamma \geq \gamma_0 \quad \text{and} \quad u \in C_0^\infty(\mathbb{R} \times \bar{\Omega}) .$$

Now $P(\varphi_0 u) = \varphi_0 Pu + [P, \varphi_0]u$, with $[P, \varphi_0]$ of degree 0. We thus deduce an upper bound of the form

$$\gamma\|\varphi_0 u\|^2_{0;\gamma} \leq \frac{C'}{\gamma}\|Pu\|^2_{0;\gamma} + \frac{C''}{\gamma}\|u\|^2_{0;\gamma} \quad \text{for} \quad \gamma \geq \gamma_0 .$$

Similarly from (4.2.1) it follows that for $j = 1,\ldots,J$ and γ sufficiently large:

$$\gamma\|\varphi_j u\|^2_{0;\gamma} + |\varphi_j u|^2_{0;\gamma} \leq C\left(\frac{1}{\gamma}\|P\,\varphi_j u\|^2_{0;\gamma} + |B\,\varphi_j u|^2_{0;\gamma}\right)$$

$$\leq C'\left(\frac{1}{\gamma}\|Pu\|^2_{0;\gamma} + |Bu|^2_{0;\gamma}\right) + \frac{C''}{\gamma}\|u\|^2_{0;\gamma} .$$

Adding together the inequalities obtained, we have

$$\gamma\|u\|^2_{0;\gamma} + |u|^2_{0;\gamma} \leq \tilde{C}'\left(\frac{1}{\gamma}\|Pu\|^2_{0;\gamma} + |Bu|^2_{0;\gamma}\right) + \frac{\tilde{C}''}{\gamma}\|u\|^2_{0;\gamma}$$

which gives (4.3.1) for γ sufficiently large.

By reducing the scalar case to the case of systems of first order (see Example 4.8 of Chapter VI, and 2.2), Theorems 4.1 and 4.2 allow an energy inequality to be stated for problem (1.5) in the scalar case. For $s \in \mathbb{R}$ and $\gamma \in \mathbb{R} \setminus 0$, we define the Hilbert space $H^s_\gamma(\mathbb{R} \times \overline{\Omega})$ by restriction of the elements of $H^s_\gamma(\mathbb{R}^{n+1})$ to $\mathbb{R} \times \Omega$ (cf. Exercise 9.1). For $u \in e^{\gamma t} H^s_\gamma(\mathbb{R} \times \overline{\Omega})$, we denote by $\|u\|_{s;\gamma}$ the norm of $e^{-\gamma t}u$ in $H^s_\gamma(\mathbb{R} \times \overline{\Omega})$. In particular, it can be shown (see Exercise 9.1 later on in this chapter) that, for $k \in \mathbb{N}$, $e^{\gamma t} H^k_\gamma(\mathbb{R} \times \overline{\Omega})$ is the space of the $u \in e^{\gamma t} L^2(\mathbb{R} \times \Omega)$ such that $e^{-\gamma t} D^\alpha u \in L^2(\mathbb{R} \times \Omega)$ for any multi-index α (of differentiation with respect to t, x) of length k, and that we have the uniform norm equivalence with respect to γ:

$$\|u\|_{k;\gamma} \sim \sum_{|\alpha| = k} \|e^{-\gamma t} D^\alpha u\|_0 \quad .$$

We define the Hilbert space $H^s_\gamma(\mathbb{R} \times \partial\Omega)$ from $H^s_\gamma(\mathbb{R}^n)$, by fixing once and for all a finite atlas of $\partial\Omega$ and a partition of unity. For $v \in e^{\gamma t} H^s_\gamma(\mathbb{R} \times \partial\Omega)$, we denote by $|v|_{s;\gamma}$ the norm of $e^{-\gamma t}v$ in $H^s_\gamma(\mathbb{R} \times \partial\Omega)$. We then obtain (see Exercise 9.2) the following theorem:

THEOREM 4.4 : *We consider problem* (1.1) *in the scalar case* (P *of order* m), *assuming the uniform Lopatinski condition to be satisfied. Then there exist* C, $\gamma_0 > 0$ *such that:*

$$\gamma \|u\|^2_{m-1;\gamma} + \sum_{j=0}^{m-1} |\gamma_j u|^2_{m-1-j;\gamma} \le C \left(\frac{1}{\gamma} \|Pu\|^2_{0;\gamma} + \sum_{j=1}^{d} |B_j u|^2_{m-1-r_j;\gamma} \right)$$

for all $\gamma \ge \gamma_0$ *and for all* $u \in C^\infty_0(\mathbb{R} \times \overline{\Omega})$.

We now return to the proof of the fundamental theorem 4.2. We shall achieve this using a procedure analogous to that which allowed Theorem 4.1 to be proved by symmetrisation using the operator R. In fact, in the next section we shall establish the following fundamental result:

THEOREM 4.5 : *We put* $\Sigma = \mathcal{X} \cap \{\gamma^2 + \sigma^2 + |\xi'|^2 = 1\}$ *and we consider problem* 2.3, *assuming the uniform Lopatinski condition to be satisfied. Then, for all* $X_0 = (t_0, x_0; \sigma_0, \xi'_0, \gamma_0) \in \Sigma$, *there exist matrices* $r(X)$, $T(X)$ *of class* C^∞ *in a neighbourhood* V *of* X_0 *in* Σ *such that:* $r(X)$ *is hermitian,* $T(X)$ *is invertible and, putting* $T^{-1} a T = \tilde{a}$, $bT = \tilde{b}$, *there exist constants* C *and* c *with* $c > 0$ *and:*

$$\operatorname{Im} r(X)\, \tilde{a}(X) = \left(\begin{array}{c|c} h(X) & 0 \\ \hline 0 & e(X) \end{array}\right)$$

is a block diagonal matrix (possibly with just a single block) which satisfies:

(4.5.1) $\frac{h(X)}{\gamma}$ *is* C^∞ *up to the boundary, for* $\gamma \geq 0$, $\frac{h(X)}{\gamma} \geq cI$

(4.5.2) $e(X) \geq cI$

(4.5.3) $-r(X) + C\, \tilde{b}^*(X)\, \tilde{b}(X) \geq cI$ *for* $x_n = 0$.

We shall now show how Theorem 4.2 is deduced from Theorem 4.5. By pseudo-differential partition of unity, we see that it is sufficient to establish the energy inequality (4.2.1) with u replaced by $u_\gamma = \Phi_\gamma u$, where the symbol Φ is the extension to a homogeneous function of degree 0 (with respect to σ, γ, ξ') of a function $\varphi \in C_0^\infty(V)$. In fact it can be seen, as in the proof of Corollary 4.3, that the errors due to the commutators $[P_\gamma, \Phi_\gamma]$ and $[B_\gamma, \Phi_\gamma]$ are absorbed by the terms $\gamma\|u\|^2_{0;\gamma}$ and $|u|^2_{0;\gamma}$ when γ is sufficiently large (see in particular (2.1.1)). Suppose $\psi \in C_0^\infty(V)$ is equal to unity in the neighbourhood of supp φ , and let Ψ denote the extension of ψ to a homogeneous function of degree 0. Extending T in similar fashion, we see that it is sufficient to establish the energy inequality (for u_γ) with A_γ replaced by $(\Psi T^{-1})_\gamma A_\gamma(\Psi T)_\gamma$ and B_γ replaced by $B_\gamma(\Psi T)_\gamma$. In fact, when we put $g = (\Psi T^{-1})_\gamma f$, we clearly have the estimate $|g|_{0;\gamma} \le C|f|_{0;\gamma}$. Furthermore, we have $(\Psi T)_\gamma(\gamma T^{-1})_\gamma\Phi_\gamma \equiv \Phi_\gamma$ modulo an operator of degree -1, and this allows us to obtain estimates in the other direction. Finally, using an argument based on extension of the symbols (more precisely, their restriction to Σ) to symbols which are constant outside a compact neighbourhood of supp φ (cf. the proof of Lemma 4.12, Chapter VI), we reduce to the case where, in the statement of Theorem 4.5, r exists globally and $\tilde{a} = a$, $\tilde{b} = b$. We then consider a symbol R with principal part r, and we put $K(x_n) = \frac{d}{dx_n}(R_\gamma u, u)_{0;\gamma}$, where the inner product is that of $e^{\gamma t}L^2(\mathbb{R}^n)$ for $x_n \ge 0$ fixed.

Putting $D_{x_n} u_\gamma - A_\gamma u_\gamma = f_\gamma$, and suppressing the index γ for the operators and the functions, we obtain (cf. (4.1.2)):

$$K(x_n) = -2\,\mathrm{Im}(RAu, u)_{0;\gamma} + 2\,\mathrm{Im}(Ru, f)_{0;\gamma} + (R'_{x_n} u, u)_{0;\gamma} .$$

Integrating both sides of this equality, with respect to x_n from 0 to $+\infty$, we obtain

$$(4.5.4) \qquad ((Ru, u))_{0;\gamma} = 2\int_0^{+\infty} \mathrm{Im}(RAu, u)_{0;\gamma}\, dx_n$$
$$- 2\,\mathrm{Im}\int_0^{+\infty} (Ru, f)_{0;\gamma}\, dx_n - \int_0^{+\infty} (R'_{x_n} u, u)_{0;\gamma}\, dx_n .$$

Now we have, with a constant $c' > 0$:

$$(4.5.5) \qquad \mathrm{Im}(RAu, u)_{0;\gamma} \geq c'\gamma\, |u|^2_{0;\gamma}$$

for γ sufficiently large .

In fact, this follows from (4.5.1), (4.5.2): for the contribution of the symbol $h(X)$ we apply Gårding's inequality, considering $\frac{h(X)}{\gamma}$ as a symbol of degree 0. For the contribution of the symbol $e(X)$ we apply Gårding's inequality, considering $\Lambda^{-\frac{1}{2}} e(X) \Lambda^{-\frac{1}{2}}$ as a symbol of degree 0, and conclude by means of the obvious lower bound $|\Lambda_\gamma^{\frac{1}{2}} u|^2_{0;\gamma} \geq \gamma |u|^2_{0;\gamma}$. From (4.5.3) and Gårding's inequality, we obtain

$$(4.5.6) \qquad (Ru, u)_{0;\gamma} \leq C|Bu|^2_{0;\gamma} - c'|u|^2_{0;\gamma}$$

for γ sufficiently large .

Using (4.5.4), (4.5.5), (4.5.6), we obtain:

$$c'\gamma\|u\|^2_{0;\gamma} + c'|u|^2_{0;\gamma} \le C'\|u\|^2_{0;\gamma}\,\|f\|^2_{0;\gamma} + C'\|u\|^2_{0;\gamma} + C|Bu|^2_{0;\gamma},$$

and arguing as at the end of Theorem 4.1, we then deduce the inequality (4.2.1) for γ sufficiently large.

5. CONSTRUCTION OF THE SYMMETRISER

This section is devoted to the proof of Theorem 4.5. We thus consider the neighbourhood of a point $x_0 \in \Sigma$. We start with a situation analogous to that of elliptic boundary-value problems: this corresponds to the case where the matrix $a(X_0)$ has no real eigenvalues. We then say that X_0 is an elliptic point. From (2.3.2), a point X_0 such that $\gamma_0 > 0$ is, in particular, elliptic.

5.1 Construction of r in the neighbourhood of an elliptic point X_0.

In view of Proposition 3.4, a change of basis of $\mathbf{C}^N$ of class C^∞ in the neighbourhood of X_0 allows us to reduce the problem to the case where $a(X) = \begin{pmatrix} a^+(X) & 0 \\ 0 & a^-(X) \end{pmatrix}$ where the eigenvalues of $a^+(X)$ (resp. of $a^-(X)$) have positive imaginary part (resp. negative). Let $\mathbf{C}^N = \mathbf{C}^{m_+} \oplus \mathbf{C}^{m_-}$ denote the corresponding decomposition of $\mathbf{C}^N$. The subspace $E^+(X)$

corresponds to the subspace $\mathbf{C}^{m_+}$ through the change of basis. Using a procedure analogous to that used for Lemma 5.5, Chapter VI, changes of basis in $\mathbf{C}^{m_+}$ and $\mathbf{C}^{m_-}$ allow us to reduce the problem to the case where $\pm \operatorname{Im} a^{\pm}(X) \geq c'I$ for X near X_0, where c' is a positive constant. We seek r(X) in the form

$$r(X) = \begin{pmatrix} I & 0 \\ 0 & -\rho I \end{pmatrix},$$ where ρ is a positive constant. We then have Im r(X)a(X) = e(X), with

$$e(X) = \begin{pmatrix} \operatorname{Im} a^+(X) & 0 \\ 0 & -\rho \operatorname{Im} a^-(X) \end{pmatrix} \geq \begin{pmatrix} c'I & 0 \\ 0 & \rho c'I \end{pmatrix},$$

so that (4.5.2) is satisfied (here there is no condition (4.5.1) to satisfy). We shall show that (4.5.3) is satisfied when we choose ρ sufficiently large.

For $w \in \mathbf{C}^N = \mathbf{C}^{m_+} \oplus \mathbf{C}^{m_-}$, we put $w = (w^+, w^-)$, $\tilde{b}(X_0)w = b^+w^+ + b^-w^-$. The uniform Lopatinski condition shows that b^+ is an isomorphism. There therefore exists a constant C' such that

$$(5.1.1) \quad |w^+|^2 \leq C'(|w^-|^2 + |\tilde{b}(X_0)w|^2) \qquad \text{for all} \quad w \in \mathbf{C}^N$$

Now we have

$$-(r(X_0)w, w) = -|w^+|^2 + \rho|w^-|^2 = |w^+|^2 + \rho|w^-|^2 - 2|w^+|^2$$

and therefore, using (5.1.1):

$$-(r(X_0)w,\ w) \geq |w^+|^2 + \rho|w^-|^2 - 2C'|w^-|^2 - 2C'|\tilde{b}(X_0)w|^2$$
$$= |w^+|^2 + (\rho - 2C')|w^-|^2 - 2C'|\tilde{b}(X_0)w|^2$$

and (4.5.3) is then deduced by choosing ρ such that $\rho - 2C' > 0$.

We now proceed to the fundamental part of this section; this consists of the construction of r in the neighbourhood of a non-elliptic point X_0 (which is such that $\gamma_0 = 0$). This construction will prove to be relatively complicated, except in the case where the real eigenvalues of $a(X_0)$ are all simple. We start by discussing this case, and we in fact assume in order to simplify the notation, that $a(X_0)$ has no eigenvalues which are not real (the contribution of such eigenvalues would in fact give rise to arguments analogous to those for the elliptic case 5.1, and, furthermore, will be treated fully in the general case). We then say that X_0 is a strictly hyperbolic point.

5.2 Construction of r in the neighbourhood of a strictly hyperbolic point X_0.

In accordance with the notation used for Proposition 3.4, we denote by $\lambda_1, \ldots, \lambda_N$ the eigenvalues of $a(X_0)$, and by $a_1(X), \ldots, a_N(X)$ the eigenvalues of $a(X)$ in the neighbourhood of X_0 (with $a_j(X_0) = \lambda_j$). The $a_j(X)$ are real for $\gamma = 0$, and a change of basis of $\mathbf{C}^N$ of class C^∞ in the neighbourhood of X_0

allows a reduction to the diagonal form

$$\tilde{a}(X) = \begin{pmatrix} a_1(X) & & 0 \\ & \ddots & \\ 0 & & a_N(X) \end{pmatrix} .$$

In accordance with the notation introduced in Proposition 3.7, we put $\alpha_j = \frac{1}{i} \frac{\partial a_j}{\partial \gamma} (X_0)$, and put $\mu_j = 1$ or $\mu_j = 0$ depending on whether $\operatorname{Im} a_j(X) > 0$ or $\operatorname{Im} a_j(X) < 0$ for $\gamma > 0$ sufficiently small. From Proposition 3.7, or preferably by a straightforward direct verification, we can see that

$\alpha_j \in \mathbf{R} \setminus 0$, $\mu_j = 1$ or 0 depending on whether $\alpha_j > 0$ or $\alpha_j < 0$, and $E^+(X_0)$ corresponds in the new basis to

$\tilde{E}^+(X_0) = \{w \in \mathbf{C}^N \mid w_j = 0 \text{ if } \mu_j = 0\}$. For

$X = (t, x ; \sigma ; \xi', \gamma)$, we put

$\tilde{X} = (t, x ; \sigma, \xi', 0)$; Taylor's formula with respect to the variable γ shows that

$$a_j(X) = a_j(\tilde{X}) + i\gamma D_j(\tilde{X}) + \gamma^2 H_j(X) \tag{5.2.1}$$

where $D_j = \frac{1}{i} \frac{\partial a_j}{\partial \gamma}$ and where H_j is C^∞.

We seek $r(X)$ in the diagonal form

$$r(X) = \begin{pmatrix} r_1 & & \\ & \ddots & \\ & & r_N \end{pmatrix}$$

where the r_j are real constants. Since $a_j(\tilde{X})$ is real, (5.2.1) shows that $\mathrm{Im}\ r_j a_j(X) = \gamma r_j\ \mathrm{Re}\ D_j(\tilde{X}) + \gamma^2 r_j\ \mathrm{Im}\ H_j(X)$ and therefore that conditions of the type (4.5.1) are satisfied when

$$r_j \alpha_j > 0 \quad \text{for} \quad j = 1, \ldots, N \quad .$$

Suppose we choose, for example $r_j = 1$ for $\alpha_j > 0$, and $r_j = -\rho$ for $\alpha_j < 0$, where ρ is a positive constant. For $w \in C^N$, we denote by w^+ its projection onto $\tilde{E}^+(X_o)$, and we put $w = w^+ + w^-$. Then $-(r(X_o)w, w) = -|w^+|^2 + \rho|w^-|^2$, and the uniform Lopatinski condition then allows us to deduce, exactly as for the elliptic case 5.1, that condition (4.8.3) is satisfied when ρ is sufficiently large.

We now go on to the general construction of r in the neighbourhood of a point X_o such that $\gamma_o = 0$. Proposition 3.4 and the arguments employed in the elliptic case 5.1 allow us to reduce to the case where $\tilde{a}(X)$ has, in a neighbourhood of X_o, the block diagonal form (3.4.1) with, furthermore, $\pm\ \mathrm{Im}\ a^{\pm}(X)$ positive definite. We now introduce, or recall, the following notation:

NOTATION 5.3

For $X = (t, x\ ;\ \sigma, \xi', \gamma)$, we put $\tilde{X} = (t, x\ ;\ \sigma, \xi', 0)$.

For $j = 1, \ldots, k$, consider the matrix with dimensions m_j

$$A_j = \begin{pmatrix} 0 & 1 & & Q \\ & \ddots & \ddots & \\ & & \ddots & 1 \\ 0 & & & 0 \end{pmatrix} .$$

We put
$$B_j(\tilde{X}) = a_j(\tilde{X}) - a_j(X_0) \quad ,$$
$$D_j(\tilde{X}) = \frac{1}{i} \frac{\partial a_j}{\partial \gamma} (\tilde{X}) \quad .$$

We thus have

(5.3.1) $$a_j(X) = \lambda_j I + A_j + B_j(\tilde{X}) + i\gamma D_j(\tilde{X}) + \gamma^2 H_j(X)$$

where the matrices B_j, D_j, H_j are C^∞, $B_j(X_o) = 0$, and where the left lower element α_j of $D_j(X_o)$ is real and non-zero (see Proposition 3.7). Furthermore, we can assume that $B_j(\tilde{X})$ has real coefficients in view of the following lemma, which is due to RALSTON [1].

LEMMA 5.4 : *In Proposition* 3.4, *it is possible to choose the change of basis such that the matrices* $B_j(\tilde{X})$ *have real coefficients* $(j = 1, \ldots, k)$.

PROOF: In the situation of Proposition 3.4, we shall show that there exists, for $j = 1, \ldots, k$ an isomorphism $U_j(\tilde{X})$ of $\mathbf{C}^{m_j}$, with C^∞ dependence on $\tilde{X}$ in the neighbourhood of X_o such that

(5.4.1) $$U_j^{-1}(\tilde{X})\ (A_j + B_j(\tilde{X}))\ U_j(\tilde{X}) = A_j + B'_j(\tilde{X})$$

with a matrix $B'_j(\tilde{X})$ which has real coefficients. Suppressing the index j in order to shorten the notation, we denote by $e_1, \ldots, e_m$ the canonical basis of $\mathbb{C}^m = \mathbb{C}^{m_j}$. Since $(A_j^{\ell})e_m = e_{m-\ell}$ for $\ell = 0, \ldots, m-1$ and since $B_j(X_o) = 0$, the vectors $(A_j + B_j(\tilde{X}))^{\ell} e_m$ constitute, a basis for $\mathbb{C}^m$ for $\ell = 0, \ldots, m-1$ and $\tilde{X}$ sufficiently near X_o. We define the isomorphism $U_j(\tilde{X})$ by $U_j(\tilde{X})\, A_j^{\ell}\, e_m = (A_j + B_j(\tilde{X}))^{\ell}\, e_m$, $\ell = 0, \ldots, m-1$. We then have (4.5.1) with a matrix $B'_j(\tilde{X})$ of the form

$$\begin{pmatrix} b_1(\tilde{X}) & 0 & \cdots & 0 \\ \vdots & \vdots & & \vdots \\ b_m(\tilde{X}) & 0 & \cdots & 0 \end{pmatrix} .$$

Now it is clear that, for $i = 1, \ldots, m$, $b_i(\tilde{X})$ is, apart from the sign, equal to the coefficient of ξ_n^{m-1} in $\det(A_j + B'_j(\tilde{X}) - \xi_n I) = \det(a_j(\tilde{X}) - \xi_n I)$, and we saw during the proof of Proposition 3.7 that these coefficients are actually real since in this case $\gamma = 0$. Taking account of this lemma, we seek $r(X)$ in block diagonal form

$$r(X) = \begin{pmatrix} r_1(X) & & & & \\ & \ddots & & & \\ & & r_k(X) & & \\ & & & I & \\ & & & & -\rho I \end{pmatrix} \tag{5.4.2}$$

with $r_j(X)$ hermitian $(j = 1, \ldots, k)$ and $\rho > 0$. Conditions of the type (4.5.2) are then clearly satisfied by the last two blocks. We shall construct the $r_j(X)$ such that conditions of the type (4.5.1) are satisfied by the first k blocks. To do this, we seek $r_j(X)$ in the form

$$(5.4.3) \qquad r_j(X) = E_j + F_j(\tilde{X}) + i\gamma \; G_j$$

where E_j, $F_j(\tilde{X})$ are real and symmetric, $F_j \in C^\infty$, $F_j(X_o) = 0$, and where G_j is real and anti-symmetric. Under these conditions the matrix $r_j(X)$ is hermitian and in view of (5.3.1) we have:

$$\text{Im } r_j a_j = \text{Im } [(E_j + F_j)\;(A_j + B_j)] + \gamma \text{ Re}(E_j D_j + G_j A_j) + \gamma M ,$$

where $M = M(X)$ is C^∞ and of the form $O(\gamma) + O(|X-\tilde{X}|)$.

In order to meet the conditions (4.5.1), we therefore impose:

$$(5.4.4) \qquad (E_j + F_j(\tilde{X}))\,(A_j + B_j(\tilde{X})) \quad \text{symmetric}$$

$$(5.4.5) \qquad \text{Re }(E_j D_j(X_o) + G_j A_j) \quad \text{positive definite} .$$

We note that condition (5.4.4) reduces to $E_j A_j$ symmetric for $\tilde{X} = X_o$. Now the following lemma can be proved in elementary fashion:

LEMMA 5.5 : *The symmetric matrices* E_j *such that* $E_j A_j$ *is symmetric, are the matrices of the form*

$$E_j = \begin{pmatrix} 0 & & & e_1 \\ & & \iddots & e_2 \\ & \iddots & \iddots & \vdots \\ e_1 & e_2 & \cdots & e_{m_j} \end{pmatrix} \tag{5.5.1}$$

We therefore take E_j in the form (5.5.1). Then condition (5.4.4) reduces to

$$F_j(\tilde{X})(A_j + B_j(\tilde{X})) + E_j B_j(\tilde{X}) \quad \text{symmetric} \; . \tag{5.5.2}$$

This property is embodied in the coefficients of $F_j(\tilde{X})$ through a linear system of $m_j(m_j - 1)/2$ equations. We seek $F_j(\tilde{X})$ in the form

$$F_j(\tilde{X}) = \begin{pmatrix} \Phi_j(\tilde{X}) & 0 \\ 0 & 0 \end{pmatrix} ,$$

where $\Phi_j(\tilde{X})$ is a symmetric matrix with dimensions $m_j - 1$. Then the linear system is square, and the associated homogeneous system is equivalent to $F_j(X_o)A_j$ symmetric, for $\tilde{X} = X_o$, or, in view of Lemma 5.5, to $F_j(X_o) = 0$. We thus see that (5.5.2) is expressed in terms of a Cramer system for $\tilde{X}$ near X_o, and therefore admits a unique solution $F_j(\tilde{X})$ which is of class C^∞ and which is equal to zero for $\tilde{X} = X_o$. With this choice of $F_j(\tilde{X})$ we have condition (5.4.4). We shall see that condition (5.4.5) is satisfied with a suitable choice of G_j; first, we recall that α_j denotes the lower left element of $D_j(X_o)$; there therefore exists a constant C which depends only upon the coefficients of E_j (see 5.5.1) such that

$$\operatorname{Re}(E_j D_j(X_o)w, w) \geq e_1\alpha_j |w_1|^2 - (|w_1|^2 + C|w'|^2$$

for all $w \in \mathbb{C}^{m_j}$, with $w' = (w_2, \ldots, w_{m_j})$.

We impose upon E_j the condition

$$e_1\alpha_j \geq 3 \qquad (5.5.3)$$

Then (5.4.5) is an obvious consequence of

LEMMA 5.7 : *For any constant* C, *there exists a real, antisymmetric matrix* G_j *such that*

$$\operatorname{Re}(G_j A_j w, w) \geq -|w_1|^2 + C|w'|^2 \quad \textit{for all} \quad w \in \mathbb{C}^{m_j}. \qquad (5.7.1)$$

PROOF: We take

$$G_j = \begin{pmatrix} 0 & \ddots & \ddots & 0 \\ \ddots & \ddots & \ddots & \ddots \\ 0 & & \ddots & 0 \end{pmatrix} = (g_{i,\ell})$$

a tridiagonal antisymmetric matrix. Then

$$\operatorname{Re}(G_j A_j w, w) = \sum_{i=2}^{m_j} g_{i,i-1}|w_i|^2 - \operatorname{Re}\sum_{i=1}^{m_j-2} g_{i+1,i}\, w_{i+2}\bar{w}_i$$

and we leave it to the reader to verify that (5.7.1) is

satisfied when the coefficients $g_{i,i-1}$ are chosen successively to be sufficiently large.

To summarise, we have shown that if the matrix E_j is chosen to be of the form (5.5.1), and satisfies (5.5.3), then we can construct $r_j(X)$ in the form (5.4.3) such that conditions of the type (4.5.1) are satisfied for $h(x) = r_j(X)\ a_j(X)$. To conclude the proof of Theorem 4.5, it now remains to satisfy condition (4.5.3). By continuity, it is sufficient to satisfy this condition at the point X_o, i.e.

$$(5.7.2)\quad -(r(X_o)w, w) \geq c|w|^2 - C|\tilde{b}(X_o)w|^2 \quad \text{for} \quad w \in \mathbb{C}^m$$

where we recall that

$$(5.7.3)\qquad r(X_o) = \begin{pmatrix} E_1 & & & & 0 \\ & \ddots & & & \\ & & E_k & & \\ & & & I & \\ 0 & & & & -\rho I \end{pmatrix}.$$

We also recall (see Remark 3.6) that the subspace $E^+(X_o)$ corresponds, in the new basis, to $\tilde{E}^+(X_o) = \bigoplus_{j=1}^{k} \mathbb{C}^{\mu_j} \oplus \mathbb{C}^{m_+}$.

For $w \in \mathbb{C}^N$, we put $w = w^+ + w^-$, where w^+ is the projection of w on $\tilde{E}^+(X_o)$. Using the uniform Lopatinski condition and arguing exactly as at the end of the construction 5.1, we see that in order to establish (5.7.2) and to conclude the proof, it is sufficient, to apply the following lemma:

LEMMA 5.8 : *For any number* $\rho > 0$, *there exist matrices* E_j *of the form* (5.5.1), *satisfying* (5.5.3), *such that the matrix* $r(X_o)$ *defined by* (5.7.3) *satisfies the following property: there exists a constant* C'' *with*

$$(5.8.1)\quad -(r(X_o)w,\ w) \geq C''\left(-|w^+|^2 + \rho|w^-|^2\right) \quad \textit{for all } w \in \mathbb{C}^N.$$

PROOF: It is obviously sufficient to satisfy the inequality (5.8.1) for all $w \in \mathbb{C}^{m_j}$ when $r(X_o)$ is replaced by E_j, and w^+ (resp. w^-) is replaced by the projection of w in $\mathbb{C}^{\mu_j}$ (resp. in $\mathbb{C}^{m_j-\mu_j}$), i.e.

$$(5.8.2)\quad -(E_j w,\ w) \geq C''\left(-|w_+|^2 + \rho|w_-|^2\right) \quad \text{for all } w \in \mathbb{C}^{m_j}$$

where

$$w_+ = (w_1, \ldots, w_{\mu_j}),\quad w_- = (w_{\mu_j+1}, \ldots, w_{m_j}) .$$

We have, putting $e_\ell = -\sigma_\ell$ for $\ell = 1,\ldots,m_j$ (cf. (5.5.1)):

$$-(E_j w,\ w) = \sigma_1 \sum_{i+\ell = m_j+1} w_\ell \bar{w}_i + \sigma_2 \sum_{i+\ell = m_j+2} w_\ell \bar{w}_i + \cdots\cdots$$

Let us first assume that m_j is even; from Proposition 3.7 we then have $\mu_j = m_j/2$; we fix $\sigma_1 \in \mathbb{R}$ satisfying condition (5.5.3), $\sigma_1\alpha_j \leq -3$, and it can easily be seen that (5.8.2) is satisfied when we take $\sigma_3 = \sigma_5 = \ldots = 0$, and when the positive

σ_2, σ_4, ... are chosen in turn to be sufficiently large. In the case where m_j is odd and $\alpha_j > 0$, we have $\mu_j = \frac{m_j+1}{2}$; we fix $\sigma_1 < 0$ such that $\sigma_1\alpha_j \leq -3$, and it can be shown that (5.8.2) is satisfied when we take $\sigma_2 = \sigma_4 = \cdots = 0$ and when the positive numbers σ_3, σ_5 ... are chosen in turn to be sufficiently large. Finally, when m_j is odd and $\alpha_j < 0$, we have $\mu_j = \frac{m_j - 1}{2}$; we fix $\sigma_1 > 0$ such that $\sigma_1\alpha_j \leq -3$, and we continue as in the preceding case.

6. SOLUTION OF THE PROBLEM WITHOUT INITIAL CONDITIONS

Before solving the mixed problem proper, we shall first investigate in the spaces $e^{\gamma t} H^k$ the boundary-value problem (without initial conditions) $Pu = f$, $Bu = g$. We shall establish existence and uniqueness results using functional analysis methods, starting from the energy inequality (4.3.1) and the analogous inequality for an adjoint problem. In order to shorten the description, we consider the case of first-order systems. We start by defining the concept of an adjoint problem.

6.1 NOTATION AND ASSUMPTIONS

We consider in $\mathbb{R}^{n+1}$ an $N \times N$ matrix differential operator of first-order

$$P = D_t + \sum_{j=1} M_j D_{x_j} + C ,$$

an open subset Ω of $\mathbb{R}^n$, and, for $(t,x') \in \mathbb{R} \times \partial\Omega$, a surjective matrix $B(t,x') : \mathbb{C}^N \longrightarrow \mathbb{C}^d$. We adopt the following assumptions:

(6.1.1.) The coefficients of P and B have C^∞ dependence on $(t,x) \in \mathbb{R}^{n+1}$ and $(t,x') \in \mathbb{R} \times \partial\Omega$, are constant for $|x|$ sufficiently large, and are independent of t for $|t|$ sufficiently large.

(6.1.2.) The hypersurface $\mathbb{R} \times \partial\Omega$ is non-characteristic for P.

We endow $\mathbb{R}^n$ with its canonical Euclidian structure and $\mathbb{R} \times \partial\Omega$ with the corresponding Riemannian density. Stokes' formula shows that

$$(6.1.3)\quad ((P\varphi, \psi)) = ((\varphi, P^* \psi)) + i(M\varphi, \psi) \text{ for } \varphi, \psi \in C_0^\infty(\mathbb{R} \times \overline{\Omega}; \mathbb{C}^N)$$

where ((,)) and (,) respectively denote the inner product of $L^2(\mathbb{R} \times \Omega; \mathbb{C}^N)$ and of $L^2(\mathbb{R} \times \partial\Omega; \mathbb{C}^N)$ and where we have put

$$M(t,x') = \sum_{j=1} \nu_j(x')\, M_j(t,x'),$$

$\nu = (\nu_1, \ldots, \nu_n)$ denoting the inward unit normal to $\partial\Omega$. The assumption (6.1.2) signified that $M(t, x')$ is an invertible $N \times N$ matrix for all $(t,x') \in \mathbb{R} \times \partial\Omega$.

We now propose to transform the boundary term $(M\varphi, \psi)$ in the equality (6.1.3) so that it contains $B\varphi$. This leads us to pose the following question.

<u>Question 6.2</u>; Find matrices $\tilde{B}(t,x')$, $A(t,x')$ in $\mathcal{L}(\mathbb{C}^N, \mathbb{C}^{N-d})$ and $\tilde{A}(t,x') \in \mathcal{L}(\mathbb{C}^N, \mathbb{C}^d)$ depending on $(t,x') \in \mathbb{R}$ as in (6.1.1), with, at each point

$(t,x') \in \mathbb{R} \times \partial\Omega$:

(6.2.1) $\quad <M\alpha, \beta>_{\mathbb{C}^N} = <B\alpha, \tilde{A}\beta>_{\mathbb{C}^d} + <A\alpha, \tilde{B}\beta>_{\mathbb{C}^{N-d}}$

for all $\alpha, \beta \in \mathbb{C}^N$, where the brackets denote respectively the canonical hermitian structures of $\mathbb{C}^N$, $\mathbb{C}^d$, $\mathbb{C}^{N-d}$.

Let us assume for the moment that this question has been resolved. If $\alpha \in \operatorname{Ker} A \cap \operatorname{Ker} B$, we have $<M\alpha, \beta> = 0$ for all $\beta \in \mathbb{C}^N$ and thus $\alpha = 0$; this shows that $\operatorname{Ker} A \cap \operatorname{Ker} B = \{0\}$ and consequently $A_2 = A_{|\operatorname{Ker} B}: \operatorname{Ker} B \to \mathbb{C}^{N-d}$ is an isomorphism since $\dim \operatorname{Ker} B = N-d$ in view of the surjectivity of B. We then deduce that A is surjective and that $\mathbb{C}^N = \operatorname{Ker} A \oplus \operatorname{Ker} B$, We thus have canonical isomorphisms $\mathbb{C}^{N-d} \approx (\mathbb{C}^N/\operatorname{Ker} A) \approx \operatorname{Ker} B$. Consequently, Ker B, considered as a vector bundle over $\mathbb{R} \times \partial\Omega$, must be trivialisable since it is isomorphic to the trivial bundle $(\mathbb{R} \times \partial\Omega) \times \mathbb{C}^{N-d}$. In order to resolve Question 6.2, we are therefore led to make the following assumption:

(6.2.2.) Assume that there exists a basis

$e_1(t,x'), \ldots, e_{N-d}(t,x')$ of Ker B(t, x'), depending as in (6.1.1) on $(t,x') \in \mathbb{R} \times \partial\Omega$.

Under these conditions, we have

PROPOSITION 6.3 : *We adopt the assumption* (6.2.2). *Then:*

(6.3.1.) *There exists* $A(t,x')$ *depending on* $(t,x') \in \mathbb{R} \times \partial\Omega$ *as in* (6.1.1), *such that* $\mathbb{C}^N = \operatorname{Ker} A \oplus \operatorname{Ker} B$.

(6.3.2.) *Having chosen such an* A, *there exist unique* $\tilde{A}$, $\tilde{B}$ *such that* $\tilde{B}$, A, $\tilde{A}$ *satisfy Question* 6.2.

(6.3.3.) *Moreover,* $\tilde{B}$ *is surjective, and we have*

$$\operatorname{Ker} \tilde{B} = (M.\operatorname{Ker} B)^{\perp}.$$

PROOF: In order to establish (6.3.1), we start by extending $e_1,\dots,e_{N-d}$ into a basis $e_1,\dots,e_N$ for $\mathbb{C}^N$, again satisfying conditions of the type (6.1.1): for example, we note that $\mathbb{C}^N = \operatorname{Im}(B^*) \oplus \operatorname{Ker} B$, and we take for the vectors $e_{N-d+1},\dots,e_N$ the column vectors of B^*. Then, it is sufficient to take for $A(t, x')$ the matrix which associates with an element of $\mathbb{C}^N$ its first N-d components in the basis $e_1(t,x'),\dots,e_N(t,x')$.

To establish (6.3.2), we decompose an element $\alpha \in \mathbb{C}^N = \operatorname{Ker} A \oplus \operatorname{Ker} B$ into $\alpha = \alpha_1 + \alpha_2$ and we see that (6.2.1) is equivalent to

$$(6.3.4) \quad \langle M\alpha_1, \beta\rangle_{\mathbb{C}^N} = \langle B\alpha_1, \tilde{A}\beta\rangle_{\mathbb{C}^d} \text{ for all } \alpha_1 \in \operatorname{Ker} A, \beta \in \mathbb{C}^N$$

$$(6.3.5) \quad \langle M\alpha_2, \beta\rangle_{\mathbb{C}^N} = \langle A\alpha_2, \tilde{B}\beta\rangle_{\mathbb{C}^{N-d}} \text{ for all } \alpha_2 \in \operatorname{Ker} B, \beta \in \mathbb{C}^N.$$

We put $M_2 = M_{|\operatorname{Ker} B} : \operatorname{Ker} B \to \mathbb{C}^N$. Since $A_2 = A_{|\operatorname{Ker} B}$ is an isomorphism, (6.3.5) is equivalent to

$$\langle M_2 A_2^{-1}\gamma, \beta\rangle_{\mathbb{C}^N} = \langle \gamma, \tilde{B}\beta\rangle_{\mathbb{C}^{N-d}} \quad \text{for all } \gamma \in \mathbb{C}^{N-d}, \beta \in \mathbb{C}^N,$$

i.e.

(6.3.6) $$\tilde{B} = \left(M_2 A_2^{-1}\right)^* .$$

Similarly, (6.3.4) is equivalent to $\tilde{A} = \left(M_1 B_1^{-1}\right)^*$ where M_1 and B_1 denote the restrictions of M and B to Ker A.

Definition (6.3.6) shows that

$$\text{rank } \tilde{B} = \text{rank } (M_2 A_2^{-1}) = \dim \operatorname{Im}\left(M|_{\text{Ker } B}\right) = \dim \text{Ker } B = N - d ,$$

hence the surjectivity of B.

Finally, it can be shown that we have the equalities

$$\text{Ker } \tilde{B} = \text{Ker}\left(M_2 A_2^{-1}\right)^* = \left(\operatorname{Im} M_2 A_2^{-1}\right)^{\perp} = \left(\operatorname{Im} M|_{\text{Ker } B}\right)^{\perp} = \left(M \cdot \text{ker } B\right)^{\perp}.$$

DEFINITION 6.4 : *Let* Ω, P, B *be as in* 6.1, *and suppose the triviality assumption* (6.2.2) *applies for* Ker B. *Consider the associated boundary-value problem:*

(6.4.1) $$Pu = f \quad \textit{in} \quad \mathbb{R} \times \Omega \quad \textit{and} \quad Bu = g \quad \textit{in} \quad \mathbb{R} \times \partial\Omega .$$

Suppose $\tilde{B}$, A, $\tilde{A}$ *satisfy Question* 6.2. *The boundary-value problem associated with* Ω, P*, $\tilde{B}$ *is called an adjoint problem of* (6.4.1).

We note that we do not have uniqueness of $\tilde{B}$, but that (6.3.3) still applies. From (6.1.3) and (6.2.1), we have

(6.4.2) $$((P\varphi, \psi)) = ((\varphi, P^*\psi)) + i(B\varphi, \tilde{A}\psi) + i(A\varphi, \tilde{B}\psi)$$

$$\text{for } \varphi, \psi \text{ in } C_0^\infty(\mathbb{R} \times \overline{\Omega}; \mathbb{C}^N) .$$

We note that problem (6.4.1) is an adjoint of the problem defined by Ω, P^*, $\tilde{B}$ and that the arguments presented after the statement of Question 6.2 show that

$$(6.4.3) \quad \{\tilde{A}\psi \mid \psi \in C_0^\infty(\mathbb{R} \times \partial\Omega ; \mathbb{C}^N),\ \tilde{B}\psi = 0\} = C_0^\infty(\mathbb{R} \times \partial\Omega ; \mathbb{C}^d) .$$

REMARK 6.5 : In the case where $\Omega = \mathbb{R}^n_+$, the triviality assumption (6.2.2) is satisfied automatically since $\mathbb{R} \times \partial\Omega = \mathbb{R}^n$ is contractile. In the general case, this supplementary assumption has to be made for B in order to have the notion of an adjoint problem with a boundary condition of the same type as that of the original problem. This supplementary assumption can be avoided by generalising the boundary condition of (6.4.1) as follows: we replace $\mathbb{C}^d$ by a complex vector bundle $\mathcal{G}$ of rank d on $\mathbb{R} \times \partial\Omega$, and B by a surjective isomorphism of $\mathbb{C}^N$ onto $\mathcal{G}$. We make for $\mathcal{G}$ and for this isomorphism assumptions analogous to those appearing in (6.1.1) for $|x|$ or $|t|$ large. The whole of the above argument adapts readily to this situation, which is locally isomorphic to the previous situation.

PROPOSITION 6.6 : *We consider the situation* 6.1, *assuming that* P *is strictly hyperbolic with respect to* t *and that problem* (6.4.1) *satisfies the uniform Lopatinski condition. Then, any problem which is adjoint to* (6.4.1) *satisfies the backward uniform Lopatinski condition (i.e. the condition by changing* $\gamma \geq o$ *into* $\gamma \leq o$ *in the definitions in Section 3).*

PROOF: Throughout the following, we omit the variable $(t,x') \in \mathbb{R} \times \partial\Omega$ and we indicate by the sign $\sim$ the notation relating to a problem which is adjoint to (6.4.1).

We start by showing that it is sufficient to establish that

$$(6.6.1) \qquad \tilde{E}^+(\sigma, \xi', -\gamma) = \left(M E^+(\sigma, \xi', \gamma)\right)^\perp \text{ for } \gamma \geq o, \quad (\sigma, \xi', \gamma) \neq o;$$

In fact, taking account of (6.3.3), we then have

$$\operatorname{Ker} \tilde{B} \cap \tilde{E}^+(\sigma, , \xi', -\gamma) = (M \ker B)^\perp \cap \left(M E^+(\sigma, \xi', \gamma)\right)^\perp$$

$$= \left(M \operatorname{Ker} B + M E^+(\sigma, \xi', \gamma)\right)^\perp = (M\mathbb{C}^N)^\perp = \{o\}$$

which proves Proposition 6.6 since we have $\tilde{\mu} = N - \mu = N - d$.

In order to establish (6.6.1), we can, by continuity, confine our attention to $\gamma > o$. By local charts, we reduce the problem to $\Omega = \mathbb{R}^n_+$. The principal symbol of P is $p(\tau, \xi) = \tau \, Id + \sum_1^n \xi_j M_j$ and that of P^* is $\tilde{p}(\tau, \xi) = \tau \, Id + \sum_1^n \xi_j M_j^*$. We note that for $\xi' \in \mathbb{R}^{n-1}, \xi_n \in \mathbb{C}, \tau = \sigma - i\gamma, \sigma \in \mathbb{R}, \gamma > o$, we have $\tilde{p}(\bar{\tau}, \xi', \xi_n) = p^*(\tau, \xi', \xi_n)$.

Since M_n is invertible, it can easily be seen that it is sufficient to establish (6.6.1) in the case where $M_n = Id$. We saw in Section 3 that the subspaces $E^{\pm}(\tau, \xi')$ are the images of the projectors $P^{\pm} = \frac{1}{2\pi i} \int_{C^{\pm}} p(\tau, \xi', \xi_n)^{-1} \, d\xi_n$

where $C^{\pm}$ encloses the zeros at ξ_n of $\det p(\tau, \xi', \xi_n)$ such that $\pm \operatorname{Im} \xi_n > 0$. Similarly, the subspace $\tilde{E}^{\pm}(\sigma, \xi', -\gamma)$ is the image of the projector

$$\tilde{p}^{+} = \frac{1}{2\pi i} \int_{\overline{C^{-}}} \tilde{p}(\bar{\tau}, \xi', \xi_n)^{-1} \, d\xi_n$$

$$= -\frac{1}{2\pi i} \int_{C^{-}} p^*(\tau, \xi', \xi_n)^{-1} \, \overline{d\xi_n} = (P^{-})^*.$$

Thus finally we have

$$E^{+}(\sigma, \xi', -\gamma) = \operatorname{Im}(P^{-})^* = (\operatorname{Ker} P^{-})^{\perp} = (\operatorname{Im} P^{+})^{\perp} = E^{+}(\sigma, \xi', \gamma)^{\perp}.$$

The concept of an adjoint problem allows us to give an existence result. In Proposition 6.7 and 6.8 below, we consider problem (1.1) in the case of systems, assuming the uniform Lopatinski condition to be satisfied.

PROPOSITION 6.7 : *There exists* $\gamma_0 > 0$ *such that for* $\gamma \geq \gamma_0$, $f \in e^{\gamma t} L^2(\mathbb{R} \times \Omega)$ *and* $g \in e^{\gamma t} L^2(\mathbb{R} \times \partial\Omega)$, *there exists* $u \in e^{\gamma t} L^2(\mathbb{R} \times \Omega)$ *such that* $Pu = f$ *and* $Bu = g$.

PROOF: First, we note that the boundary condition $Bu = g$ is meaningful, since Theorem 2.5 shows that if u and Pu are in $e^{\gamma t} L^2(\mathbb{R} \times \Omega)$, then u admits a trace $u|_{\mathbb{R} \times \partial\Omega} \in H^{-\frac{1}{2}}_{loc}(\mathbb{R} \times \partial\Omega)$.

To prove the existence of u, we proceed as in the proof of Theorem 4.5, Chapter VI. Suppose $(P^*,\tilde{B})$ is an adjoint problem; we define the space $E = \left\{\psi \in C_0^\infty\left(\mathbb{R}\times\overline{\Omega}; \mathbb{C}^N\right);\ \tilde{B}\psi = 0\right\}$; we shall show that it is possible to define a linear form ℓ on P^*E by putting

$$\ell(P^*\psi) = ((f,\psi)) - i(g,\tilde{A}\psi).$$

We have the obvious upper bounds

$$|((f,\psi))| \le \|f\|_{0;\gamma}\cdot\|\psi\|_{0;-\gamma}$$

and

$$|(g,\tilde{A}\psi)| \le C|g|_{0;\gamma}\ |\psi|_{0;-\gamma}\ .$$

We then deduce from the energy inequality (4.3.1) applied to the backward adjoint problem (for $\gamma \ge \gamma_0$):

$$|\ell(P^*\psi)| \le C\|P^*\psi\|_{0;-\gamma}\left(\frac{1}{\gamma}\|f\|_{0;\gamma} + \frac{1}{\sqrt{\gamma}}\ |g|_{0;\gamma}\right).$$

This shows that, for $\gamma \ge \gamma_0$, ℓ is well defined and is, furthermore, continuous for the topology induced on P^*E by $e^{-\gamma t}\ L^2(\mathbb{R}\times\Omega)$. Consequently, the Hahn-Banach theorem implies the existence of an element $u \in e^{\gamma t}\ L^2(\mathbb{R}\times\Omega)$ such that

(6.7.1) $((u,P^*\psi)) = ((f,\psi)) - i(g,\tilde{A}\psi)$ for all $\psi \in E$.

Taking ψ in $C_0^\infty(\mathbb{R}\times\Omega;\mathbb{C}^N)$, (6.7.1) shows that

(6.7.2) $Pu = f$.

Theorem 2.5 and the equality (6.4.2) show that we again have (by density and continuity):

$$(6.7.3)\quad ((Pu,\psi)) = ((u,P^*\psi)) + i(Bu,\tilde{A}\psi) + i(Au,\tilde{B}\psi)$$

for all $\psi \in C_0^\infty(\mathbb{R}\times\overline{\Omega};\mathbb{C}^N)$.

Combining this equality with (6.7.1) and (6.7.2), we obtain

$$(Bu-g,\tilde{A}\psi) = 0 \qquad \text{for all} \quad \psi \in E,$$

and consequently Bu = g, in view of (6.4.3).

In order to investigate the regularity of the solution, we shall now establish

PROPOSITION 6.8 : *We put* $\mathbb{R}\times\overline{\Omega} = \Sigma$. *We take an integer* $k \geq -1$; *then there exists* $\gamma_k > 0$ *such that if* $u \in e^{\gamma t} H^k(\Sigma) \cap e^{\gamma t} L^2(\Sigma)$, $Pu \in e^{\gamma t} H^{k+1}(\Sigma)$, $u|_{\partial\Sigma} \in e^{\gamma t} H^k(\partial\Sigma)$ *and* $Bu \in e^{\gamma t} H^{k+1}(\partial\Sigma)$ *then we have* $u \in e^{\gamma t} H^{k+1}(\Sigma)$ *and* $u|_{\partial\Sigma} \in e^{\gamma t} H^{k+1}(\partial\Sigma)$.

PROOF: Since M_0 is invertible, we can reduce to the case where M_0 = Id. If we put $v = e^{-\gamma t} u$, then $e^{-\gamma t} Pu = P_{(\gamma)}v$ with $P_{(\gamma)} = P - i\gamma$ and the energy inequality (4.3.1) is written

$$(6.8.1)\quad \gamma\|\varphi\|_0^2 + |\varphi|_0^2 \leq C\left[\frac{1}{\gamma}\|P_{(\gamma)}\varphi\|_0^2 + |B\varphi|_0^2\right]$$

for $\gamma \geq \gamma_0$ and $\varphi \in H^1(\Sigma)$.

This means that we have to show that if $v \in L^2(\Sigma) \cap H^k(\Sigma)$, $P_{(\gamma)} v \in H^{k+1}(\Sigma)$, $v|_{\partial\Sigma} \in H^k(\partial\Sigma)$ and $Bv \in H^{k+1}(\partial\Sigma)$ then $v \in H^{k+1}(\Sigma)$ and $v|_{\partial\Sigma} \in H^{k+1}(\partial\Sigma)$. We note that for $k = -1$, it is sufficient to show that $v|_{\partial\Sigma} \in L^2(\partial\Sigma)$.

Since $\partial\Omega$ is bounded (we shall confine our attention to this case; the method given below may be adapted to the case $\Omega = \mathbb{R}^n_+$ and does in fact then become simpler) we can cover $\bar{\Omega}$ by a finite number of open coordinate patches $U_j \xrightarrow{\chi_j} \tilde{U}_j$ $j = 0, \dots, J$, such that for $j \neq 0$ U_j is bounded and $\tilde{U}_j \subset \bar{\mathbb{R}}^n_+$ and for $j = 0$ we have $U_0 \subset \Omega$ and $\chi_0 =$ Identity. Let $(\varphi_j)_{j=0,\dots,J}$ be a partition of unity subordinate to this covering with $\varphi_j \in C_0^\infty(U_j)$ for $j \neq 0$. We require to show that we have:

$$\varphi_j v \in H^{k+1}(\Sigma) \text{ and } \varphi_j v|_{\partial\Sigma} \in H^{k+1}(\partial\Sigma) \quad \text{for all } j ,$$

that is

$$\chi_{j*}(\varphi_j v) \in H^{k+1}(\bar{\mathbb{R}}^{n+1}_+) \text{ and } \chi_{j*}(\varphi_j v) \in H^{k+1}(\mathbb{R}^n)$$

$$\text{for all } j \neq 0 \quad \text{and } \chi_0 {*}(\varphi_0 v) \in H^{k+1}(\mathbb{R}^{n+1}).$$

In order to avoid having to distinguish the case $j = 0$, in the following we shall read $\mathbb{R}^{n+1}$ instead of $\bar{\mathbb{R}}^{n+1}_+$ in the statements concerned with this case.

We put $\psi_j = \chi_{j*}\varphi_j$, $w_j = \chi_{j*}\left(v|_{U_j}\right)$, $Q_j = \chi_{j*} \circ P(\gamma) \circ \chi_j^*$, and $B_j = \chi_{j*} \circ B \circ \chi_j^*$. Since Q_j is a differential operator of order 1, the commutator which appears in the equality

$$Q_j(\psi_j w_j) = \psi_j(Q_j w_j) + [Q_j, \psi_j] w_j$$

is an operator of multiplication by a function $\theta_j \in C_0^\infty(\tilde{U}_j)$ for $j \neq 0$ and θ_0 is bounded together with all its derivatives;

the assumptions on v and Pv imply that for all j

$$Q_j(\psi_j w_j) \in H^k(\overline{\mathbb{R}}_+^{n+1}) \cap L^2(\overline{\mathbb{R}}_+^{n+1}) \quad .$$

Consequently, Proposition 2.4, shows that it is sufficient to prove that for all j : $\psi_j w_j \in H^{o,k+1}(\overline{\mathbb{R}}_+^{n+1})$, i.e.

(6.8.2.) $\int_o^{+\infty} |\psi_j w_j(\cdot,x_n)|_{k+1}^2 \, dx_n < +\infty$, (with our convention, this integral is taken between $-\infty$ and $+\infty$ for j = o)

(6.8.3) $$\psi_j w_j|_{x_n=o} \in H^{k+1}(\mathbb{R}^n) .$$

For this, we shall use the technique introduced in Section 6, Chapter II. For $\varphi \in H^k(\mathbb{R}^n)$, we put

$$n_\delta(\varphi) = \int_o^1 |\rho_\varepsilon * \varphi|_o^2 \, \varepsilon^{-2(k+1)} \left(1 + \frac{\delta^2}{\varepsilon^2}\right)^{-1} \frac{d\varepsilon}{\varepsilon}$$

where ρ_ε satisfies the assumptions of Theorem 6.4, Chapter II, with h > k + 1. With this notation, condition (6.8.3) is equivalent to the existence of a constant C such that

(6.8.4) $n_\delta\left(\psi_j w_j|_{x_n=o}\right) \le C$ for all $o < \delta \le 1$ and all j

since we know that $\psi_j w_j|_{x_n=o} \in H^k(\mathbb{R}^n)$. Similarly for $\psi \in H^k(\overline{\mathbb{R}}_+^{n+1}) \cap L^2(\overline{\mathbb{R}}_+^{n+1})$, we put $n_\delta(\psi) = \int_o^{+\infty} n_\delta(\psi(\cdot,x_n)) \, dx_n$, in this case condition (6.8.2) is equivalent to the existence of a constant C such that

(6.8.5) $N_\delta(\psi_j w_j) \le C$ for all $o<\delta\le 1$ and all j.

To obtain these upper bounds, we apply the inequality (6.8.1) transported by χ to $\rho_\varepsilon *' \psi_j w_j$ ($*'$ denotes convolution with respect to (t, x')); we obtain for $\gamma \geq \gamma_0$,

$$\gamma \|\rho_\varepsilon *' \psi_j w_j\|_0^2 + |\rho_\varepsilon *' \psi_j w_j|_0^2 \leq \frac{C}{\gamma}\Big[\|\rho_\varepsilon *' \psi_j Q_j w_j\|_0^2 + \|[\rho_\varepsilon *', Q_j]\psi_j w_j\|_0^2 + \|\rho_\varepsilon *' \theta_j w_j\|_0' \Big] + C\Big[|\rho_\varepsilon *' \psi_j B_j w_j|_0^2 + |[\rho_\varepsilon *', B_j]\psi_j w_j|_0^2\Big].$$

We multiply this inequality by $\varepsilon^{-2(k+1)}\left(1+\frac{\delta^2}{\varepsilon^2}\right)^{-1}\frac{1}{\varepsilon}$, and then we integrate with respect to $\varepsilon \in]0,1]$; after changing the order of integration with respect to ε and x_n and after using Theorems 6.4 and 6.5 of Chapter II and Remark 6.6 of Chapter II (with x_n as parameter); we obtain

$$(6.8.6)_j \quad \gamma\, 'N_\delta(\psi_j w_j) + \eta_\delta(\psi_j w_j) \leq \frac{C'}{\gamma}\Big[\int_0^{+\infty} |\psi_j Q_j w_j(\cdot, x_n)|_{k,\delta}^2\, dx_n + \int_0^{+\infty} |\psi_j w_j(\cdot, x_n)|_{k,\delta}^2\, dx_n + \int_0^{+\infty} |\theta_j w_j(\cdot, x_n)|_{k,\delta}^2\, dx_n\Big] + C'\Big[|\psi_j B_j w_j|_{k,\delta}^2 + |\psi_j w_j|_{k-1,\delta}^2\Big].$$

We have immediately the following upper bounds:

$$\int_0^{+\infty} |\psi_j Q_j w_j(\cdot, x_n)|_{k,\delta}^2\, dx_n \leq C\|P_{(\gamma)} v\|_{k+1}^2 ,$$

$$|\psi_j B_j w_j|_{k,\delta}^2 \leq C|B v|_{k+1}^2 ,$$

$$|\psi_j w_j|_{k-1,\delta}^2 \leq |\psi_j w_j|_k^2 \leq C|v|_k^2 .$$

$$(6.8.7)\qquad \int_0^\infty |\psi_j w_j(\cdot,x_n)|^2_{k,\delta}\, dx_n \le C\, N_\delta(\psi_j w_j) + C\|v\|_k^2$$

and by appropriately augmenting γ, we can absorb the term $C\, N_\delta(\psi_j w_j)$ into the left-hand side of $(6.8.6)_j$. In order to have an analogous treatment for the term in $\theta_j w_j$, we use an idea due to TARTAKOFF [1]. This consists of specifying the choice of the charts χ_j for $j \neq o$. We again denote by ν a vector field defined in the neighbourhood of $\partial\Omega$ and transversal to $\partial\Omega$, and we take for χ_j charts adapted to $(\bar{\Omega}, \nu)$ in the sense of Chapter V, Section 1. With this choice, the coordinate transitions are of the form

$$x = (x', x_n) \in \tilde{U}_j \xrightarrow{\ \chi\ } \chi(x) = (y', y_n) \in \tilde{U}_i$$

with $y' = y'(x')$ and $y_n = x_n$, and Proposition 6.3 of Chapter II shows that there exists C such that

$$(6.8.8)\qquad \int_0^{+\infty} |\chi^*\psi(\cdot,x_n)|^2_{k,\delta}\, dx_n \le C\int_0^{+\infty} |\psi(\cdot,x_n)|^2_{k,\delta}\, dx_n \quad \text{for } 0<\delta\le 1$$

and for all $\psi \le H^k(\tilde{U}_i) \cap L^2(\tilde{U}_i)$ with support in a fixed compact domain.

We now return to the term in $\theta_j w_j$ in $(6.8.6)_j$; since $\Sigma \varphi_j = 1$, we have

$$(6.8.9)\qquad \theta_j w_j = \sum_{i=0}^{J} (\chi_{j*}\varphi_i)\theta_j w_j$$

and from (6.8.8), we deduce

$$\int_0^{+\infty} |((\chi_j * \varphi_i)\theta_j w_j)(\cdot,x_n)|^2_{k,\delta}\, dx_n \le C\int_0^{+\infty} |((\chi_{i*}\theta_i)\psi_i w_i)(\cdot,y_n)|^2_{k,\delta}\, dy_n$$

Adding, and taking account of Proposition 6.3 of

Chapter II, we obtain:

$$\int_0^{+\infty} |\theta_j w_j(\cdot, x_n)|^2_{k,\delta}\, dx_n \leq C \sum_{i=0}^{J} \int_0^{+\infty} |\psi_i w_i(\cdot, y_n)|^2_{k,\delta}\, dy_n$$

$$\leq C' \sum_{i=0}^{J} N_\delta(\psi_i w_i) + C' \|v\|^2_k$$

and this allows this term to be treated like (6.8.7) after adding together the inequalities $(6.8.6)_j$ for $j = o,\ldots,J$.

Finally, for $\gamma \geq \gamma_k$ sufficiently large, we obtain

$$\sum_{j=0}^{J} \left[\gamma N_\delta(\psi_j w_j) + n_\delta(\psi_j w_j)\right] \leq \frac{C}{\gamma}\left[\|P_\gamma v\|^2_{k+1} + \|v\|^2_k\right] + C\left[|Bv|^2_{k+1} + |v|^2_k\right]$$

which proves (6.8.4) and (6.8.5), and therefore Proposition 6.8.

COROLLARY 6.9 : *Consider an integer* $k \geq o$; *then there exists* $\gamma_k > o$, *such that if* $\gamma \geq \gamma_k$, *and if* u *satisfies:* $u \in e^{\gamma t} L^2(\Sigma)$, $Pu \in e^{\gamma t} H^k(\Sigma)$, $Bu \in e^{\gamma t} H^k(\partial\Sigma)$, *then* $u \in e^{\gamma t} H^k(\Sigma)$ *and* $u_{|\partial\Sigma} \in e^{\gamma t} H^k(\partial\Sigma)$.

PROOF: We start by proving that $u_{|\partial\Sigma} \in e^{\gamma t} H^o(\partial\Sigma)$. To do this we apply Proposition 6.8 with $k = -1$, noting that Remark 2.6 ensures that $u_{|\partial\Sigma} \in e^{\gamma t} H^{-1}(\partial\Sigma)$. Then, repeated application of Proposition 6.8 shows that $u \in e^{\gamma t} H^j(\Sigma)$ and that $u_{|\partial\Sigma} \in e^{\gamma t} H^j(\partial\Sigma)$ for $j = 1,\ldots,k$.

Finally, we have

THEOREM 6.10 (Existence, uniqueness and regularity)

We consider problem (1.1) *in the case of systems, assuming that the uniform Lopatinski condition is satisfied. Then there exist* γ_0 *and* C *such that: for* $\gamma \geq \gamma_0$, $f \in e^{\gamma t} L^2(\mathbb{R} \times \Omega)$ $g \in e^{\gamma t} L^2(\mathbb{R} \times \partial\Omega)$ *there exists a unique* $u \in e^{\gamma t} L^2(\mathbb{R} \times \Omega)$ *which satisfies*

$$(6.10.1) \qquad Pu = f \quad \text{in} \quad \mathbb{R} \times \Omega \quad , \; Bu = g \quad \text{in} \quad \mathbb{R} \times \partial\Omega \; .$$

Furthermore, we have

$$(6.10.2) \qquad u|_{\mathbb{R}\times\partial\Omega} \in e^{\gamma t} L^2(\mathbb{R} \times \partial\Omega) \; ,$$

$$(6.10.3) \qquad \gamma\|u\|^2_{0;\gamma} + |u|^2_{0;\gamma} \leq C\left(\frac{1}{\gamma}\|f\|^2_{0;\gamma} + |g|^2_{0;\gamma}\right) .$$

For any integer $k \geq 0$, *there exists* γ_k *(with* $\gamma_k \geq \gamma_0$*) such that: for* $\gamma \geq \gamma_k$, *for* $f \in e^{\gamma t} H^k(\mathbb{R} \times \overline{\Omega})$, $g \in e^{\gamma t} H^k(\mathbb{R} \times \partial\Omega)$, *the solution* u *of* (6.10.1) *satisfies:*

$$u \in e^{\gamma t} H^k(\mathbb{R} \times \overline{\Omega}), \; u|_{\mathbb{R}\times\partial\Omega} \in e^{\gamma t} H^k(\mathbb{R} \times \partial\Omega) \; .$$

PROOF: In view of Proposition 6.7 and Corollary 6.9, it only remains to establish the uniqueness of u, together with (6.10.2) and (6.10.3). First, using Corollary 4.3 and from the density of $C_0^\infty(\mathbb{R} \times \bar{\Omega})$ in $e^{\gamma t} H^1(\mathbb{R} \times \overline{\Omega})$ it can easily be seen that

(6.10.4) $\gamma\|u\|^2_{o;\gamma} + |u|^2_{o;\gamma} \leq C\left(\frac{1}{\gamma}\|Pu\|^2_{o;\gamma} + |Bu|^2_{o;\gamma}\right)$

for all $\gamma \geq \gamma_o$ and all $u \in e^{\gamma t} H^1(\mathbb{R}\times\bar{\Omega})$.

In particular, Corollary 6.9 shows that the inequality (6.10.4) holds for all $u \in e^{\gamma t} L^2(\mathbb{R}\times\Omega)$ satisfying $Pu = o$, $Bu = o$ when γ is sufficiently large; this gives the uniqueness of u.

In order to prove (6.10.2) and (6.10.3), we consider sequences $f_j \in C^\infty_o(\mathbb{R}\times\bar{\Omega})$, $g_j \in C^\infty_o(\mathbb{R}\times\partial\Omega)$ converging respectively to f, g in $e^{\gamma t}L^2$. For γ sufficiently large, we know that there exists $u_j \in e^{\gamma t} H^1(\mathbb{R}\times\bar{\Omega})$ such that $Pu_j = f_j$, $Bu_j = g_j$. The inequality (6.10.4) applied to the differences $u_j - u_k$ shows that $u_j, u_{j|\mathbb{R}\times\partial\Omega}$ are Cauchy sequences in $e^{\gamma t}L^2(\mathbb{R}\times\Omega)$, $e^{\gamma t}L^2(\mathbb{R}\times\partial\Omega)$, and therefore that they tend respectively to limits v, w in these spaces. In particular, $Pu_j \to Pv$ in $\mathcal{D}'(\mathbb{R}\times\Omega)$, and therefore $Pv = f$. Since $u_j \to v$ and $Pu_j \to Pv$ in $e^{\gamma t}L^2(\mathbb{R}\times\Omega)$, Theorem 2.5 shows that $u_{j|\mathbb{R}\times\partial\Omega} \to v_{|\mathbb{R}\times\partial\Omega}$ in $H^{-\frac{1}{2}}_{loc}(\mathbb{R}\times\partial\Omega)$, and therefore that $v_{|\mathbb{R}\times\partial\Omega} = w$ and $Bv = g$. Thus v is actually the (unique) solution u of (6.10.1), and (6.10.3) is obtained by proceeding to the limit in the energy inequality for the $u_j \in e^{\gamma t} H^1(\mathbb{R}\times\bar{\Omega})$.

THEOREM 6.11 (Support) *Suppose* $t_o \in \mathbb{R}$; *then in the statement of Theorem* 6.10, *we can choose* γ_0 *in such a way that we also have the following property: if* f *and* g *are zero for all* $t < t_0$, *then the solution* u *is zero for* $t < t_0$.

PROOF: By translation, it is sufficient to take $t_0 = 0$. We use the following lemma:

LEMMA 6.12 : *Suppose* $\gamma > 0$ *and let* $f \in e^{\gamma t} L^2(\Sigma)$ *be zero for* $t < 0$*; then* $\|f\|_{0;\gamma'} \leq \|f\|_{0;\gamma}$ *for* $\gamma' \geq \gamma$. *Conversely, if* γ_j *is a sequence tending towards* $+\infty$ *and if* $f \in \bigcap_j e^{\gamma_j t} L^2(\Sigma)$ *is such that* $\sup_j \|f\|_{0;\gamma_j} < +\infty$, *then* $f = 0$ *for* $t < 0$.

The first part of this lemma is obvious. For the converse, we take $\varepsilon > 0$ and we consider $\varphi \in \mathcal{D}_0^\infty(]-\infty,-\varepsilon[\times\Omega)$; we note that

$$|(f,\varphi)| \leq \|f\|_{0;\gamma_j} \|\varphi\|_{0;-\gamma_j} \leq C\, e^{-\gamma_j \varepsilon} ,$$

so that we have $(f,\varphi) = 0$.

We now return to the proof of Theorem 6.11. Suppose $\tilde{\gamma}_0 \geq \gamma_0$, where γ_0 is the number appearing in Theorem 6.10; let $\gamma \geq \tilde{\gamma}_0$, $f \in e^{\gamma t} L^2(\Sigma)$ and $g \in e^{\gamma t} L^2(\partial\Sigma)$ be zero for $t < 0$. From Theorem 6.10, for any integer $j \in \mathbb{N}$ there exists a unique $u_j \in e^{\gamma_j t} L^2(\Sigma)$ such that $Pu_j = f$, $Bu_j = g$ with $\gamma_j = \gamma + j$ and by virtue of (6.10.3), we have

$$\sup_j \|u_j\|_{o;\gamma_j} < +\infty .$$

We shall now show that if $\tilde{\gamma}_o$ is chosen from the beginning to be sufficiently large, then $u_{j+1} = u_j$ for all j, and this will prove Theorem 6.11 since application of Lemma 6.12 shows that u_j is zero for $t < o$. To do this, we put $v_j = u_{j+1} - u_j$, and we take $\theta \in C^\infty(\mathbb{R})$ such that $\theta > o$, $\theta(t) = 1$ for $t \leq o$ and $\theta(t) = e^{-t}$ for $t \geq 1$. It is clear that $\theta v_j \in e^{\gamma_j t} L^2(\Sigma)$, and we have

$$o = \theta P v = \left(P + i \frac{\theta'(t)}{\theta(t)}\right)(\theta v) \quad \text{and} \quad o = \theta D v = D(\theta v) .$$

By applying the uniqueness result of Theorem 6.10 to the problem corresponding to $P + i \frac{\theta'}{\theta}$ and B, we obtain $v_j = o$ when $\tilde{\gamma}_o$ is chosen to be sufficiently large.

7. SOLUTION OF THE MIXED PROBLEM

Suppose Ω is a regular open subset of $\mathbb{R}^n$ and let T be a number > o. If $u \in L^2(]o,T[\times\Omega)$ satisfies $Pu \in L^2(]o,T[\times\Omega)$, Theorem 2.5 and (2.7.1) allow us to define the traces $u|_{]o\,T[\times\partial\Omega} \in H^{-\frac{1}{2}}_{loc}(]o,T[\times\partial\Omega)$ and $u|_{t=o} \in H^{-\frac{1}{2}}_{loc}(\Omega)$, and we have the jump formula (see (2.7.3))

$$P(u^o) = (Pu)^o - i\, u|_{t=o} \otimes \delta(t) ,$$

where u^o denotes the extension of u by o for $t < o$.

In the case of initial data which is equal to zero, we have

THEOREM 7.1 (Existence, uniqueness and regularity). *We consider problem* (1.1) *in the systems case, assuming the uniform Lopatinski condition to be satisfied. Let* $k \in \mathbb{N}$, $f \in H^k([o\ T] \times \bar{\Omega})$, $g \in H^k([o\ T] \times \partial\Omega)$ *be such that the traces* $D_t^j f|_{t=o}$ *and* $D_t^j g|_{t=o}$ *are zero for* $o \leq j \leq k-1$. *Then there exists a unique* $u \in H^k([o\ T] \times \bar{\Omega})$ *such that*

$$(7.1.1) \qquad Pu = f,\ Bu = g,\ u|_{t=o} = o$$

PROOF: To prove the existence of u, we consider extensions $\tilde{f}$, $\tilde{g}$ of f, g such that $\tilde{f} \in H^k(\Sigma)$, $\tilde{g} \in H^k(\partial\Sigma)$ are zero for $t < o$, this being possible by virtue of the assumptions made for the traces of f and g. We note that $\tilde{f}$ and $\tilde{g}$ are in the spaces $e^{\gamma t} H^k$ for all $\gamma > o$; by taking γ sufficiently large, Theorem 6.10 shows that there exists $\tilde{u} \in e^{\gamma t} H^k(\Sigma)$ such that $P\tilde{u} = \tilde{f}$, $B\tilde{u} = \tilde{f}$.

Theorem 6.11 shows that $\tilde{u}$ is zero for $t < o$; consequently $\tilde{u} = (\tilde{u})^o$ and the jump formula implies that $\tilde{u}|_{t=o} = o$. It only remains to put $u = \tilde{u}|_{]o\ T[\times \Omega}$ to obtain a solution of (7.1.1) in $H^k([o\ T] \times \bar{\Omega})$.

It is sufficient to prove the uniqueness when k = o. Let $u \in L^2(]o\ T[\times\Omega)$ be such that Pu = o, Bu = o, $u|_{t=o} = o$; the jump formula implies that u^o again satisfies $P(u^o) = o$ in $]-\infty, T[\times\Omega$ and $B(u^o) = o$ in $]-\infty, T[\times\partial\Omega$. Let t_o be such that $o < t_o < T$, and take a function $\theta \in C^\infty(\mathbb{R})$ such that $\theta(t) = 1$ for $t < t_o$ and $\operatorname{supp}\theta \subset]-\infty, T[$. Then, for $\gamma > o$, we have $\theta u^o \in e^{\gamma t}L^2(\Sigma)$, $P(\theta u^o) \in e^{\gamma t}L^2(\Sigma)$, $B(\theta u^o) \in e^{\gamma t}L^2(\partial\Sigma)$. Since $P(\theta u^o)$ and $B(\theta u^o)$ are zero for $t < t_o$, Theorem 6.11 implies that $\theta u^o = o$ for $t < t_o$, so that $u = o$ for $t < t_o$, and finally u = o.

The conditions imposed in (7.1) on the traces of the data f and g at the instant t = o are in fact too strong. It is sufficient that they satisfy compatibility conditions; we shall state what these are in the case $k = +\infty$. For k = o, a theorem with non-zero initial data is given in RAUCH [1].

7.2 Compatibility relations for the data

We still consider the case of the matrix problem (1.1) with

$$P = D_t - A(t,x,D_x) .$$

By successive differentiations with respect to t, we obtain

$$D_t = A + P \quad , D_t^2 = (A^{(1)} + A_o A) + (A + D_t) \circ P \;,\ldots,$$

$$(7.2.1.) \quad D_t^k = A_k(t,x,D_x) + C_k(t,x,D_t,D_x) \circ P$$

with

$$A_{k+1} = A_k \circ A + (A_k)^{(1)}, \quad C_{k+1} = A_k + D_t \circ C_k$$

where $(A_k)^{(1)}$ is obtained by applying the operator D_t to the coefficients of A_k. Let $u \in C^\infty([0\ T] \times \bar{\Omega})$, we put

$$Pu = f, \quad Bu = g, \quad u|_{t=0} = h.$$

We have $D_t^\ell g(o,x') = (D_t^\ell Bu)(o, x')$ for all $\ell \in \mathbb{N}$, $x' \in \partial\Omega$; using (7.2.1), we then deduce that the derivatives of f, g, h on the edge $\{t=0, x' \in \partial\Omega\}$, must satisfy the following conditions, which are known as *compatibility relations:*

$$(7.2.2) \quad (D_t^\ell g)(x',0) = \sum_{k=0}^{\ell} \binom{\ell}{k} (D_t^{\ell-k} B)(0,x') \left[A_k(0,x',D_x)h\big|_{\partial\Omega} + C_k(0,x,D_t,D_x)f\big|_{\{0\}\times\partial\Omega}\right. \quad \text{for all } \ell \in \mathbb{N}.$$

Conversely, we have

THEOREM 7.3 : *Consider problem* (1.1) *in the systems case, assuming the uniform Lopatinski condition to be satisfied and assuming* Ω *to be bounded. Suppose we have* $f \in C^\infty([0\ T] \times \bar{\Omega})$, $h \in C^\infty(\bar{\Omega})$ *satisfying the compatibility relations* (7.2.2). *Then, there exists a unique* $u \in C^\infty([0\ T] \times \bar{\Omega})$ *such that*

$$Pu = f, \quad Bu = g, \quad u|_{t=0} = h.$$

PROOF: Let $\tilde{f} \in C^\infty([o\ T] \times \mathbb{R}^n)$ and $\tilde{h} \in C^\infty(\mathbb{R}^n)$ be extensions of f, g; we know from Theorem 4.5 of Chapter VI that there exists $\tilde{v} \in C^\infty([o\ T] \times \mathbb{R}^n)$, a solution of the Cauchy problem $P\tilde{v} = f$, $\tilde{v}\big|_{t=o} = h$. Putting U = u-v, where $v = \tilde{v}\big|_{[o\ T] \times \Omega}$, the problem being studied is transformed into PU = F, BU = G, $U\big|_{t=o} = H$, where we have put

$$F = f - Pv = o, \quad G = g - Bv, \quad H = h - v\big|_{t=o} = o.$$

Since $v \in C^\infty([o\ T] \times \bar{\Omega})$, the functions Pv, Bv, $V\big|_{t=o}$ satisfy the compatibility relations, and by linearity, the same applies for the functions F, G, H. Thus, it is sufficient to establish the existence of u in the particular case where f = o, h = o.

The compatibility relations (7.2.2) then indicate that g is flat for t = o, and since Ω is bounded we can use Theorem 7.1 to obtain the existence of u in

$$\bigcap_{k \geq o} H^k([o\ T] \times \bar{\Omega}) = C^\infty([o\ T] \times \bar{\Omega}).$$

Finally, the uniqueness follows from Theorem 7.1 .

REMARK 7.4 : The property of finite propagation speed which we shall establish in Section 8 (see Theorem 8.5) shows that the assumption that Ω is bounded can be dropped from the statement of Theorem 7.3. In fact, it is sufficient to proceed as in the proof of Theorem 4.14 of Chapter VI to obtain the existence of a C^∞ solution.

REMARK 7.5 : In the scalar case, the same type of argument allows us to prove the analogue of Theorem 7.3. We also refer the reader to BALABAN [1], SAKAMOTO [1] and J. KATO [1].

REMARK 7.6 : For application to certain equations of Physics, the assumptions on (P, B, Ω) need to be generalised in several directions. For example, when the Lopatinski condition is no longer uniform, or when P is not strictly hyperbolic, or when the boundary $\partial\Omega$ is characteristic or contains corners. For a number of results in these directions we refer for example to MAJDA-OSHER [1], OHKUBO-SHIROTA [1], SARASON [1], and to their bibliographies. See also Exercise 9.4.

8. FINITE PROPAGATION SPEED

As for the Cauchy problem, we shall obtain a finite propagation speed property for the mixed problem by suitably deforming the hypersurface $\{ t = o \}$. In this context, it will be useful to present the uniform Lopatinski condition in a more intrinsic fashion.

8.1 Construction We take F, p, V, ν as follows: F is a real-valued vector space of dimension n+1; p is a polynomial which is homogeneous of degree 1 in F, with values in the space of $N \times N$ matrices with complex coefficients (in the following, F will be the fibre of $T^*\mathbb{R}^{n+1}$ at a point of $\mathbb{R} \times \partial\Omega$, and p will be the restriction to F of the principal symbol of the operator P); V and ν are two independent elements from the space F; we suppose that p is strictly hyperbolic with respect to V (see Proposition 3.7, Chapter VI) and that $p(\nu) \neq o$.

Then let $\eta_o \in \frac{F}{\mathbb{R}\nu}$ and $\gamma_o \geq o$ be such that $(\eta_o, \gamma_o) \neq o$. We define the subspace $E^+(\eta_o, \gamma_o)$ of $\mathbb{C}^N$ as follows: we choose $e_1, \ldots, e_{n-1}$ such that $(V, e, \ldots, e_{n-1}, \nu)$ is a basis of F, and we denote by $(\sigma, \xi_1, \ldots, \xi_{n-1}, \xi_n)$ the corresponding coordinates in F; we put $\xi' = (\xi_1, \ldots, \xi_{n-1})$ and we denote by $\pi = \pi(\sigma, \xi', \xi_n)$ the expression for the polynomial p, and by (σ_o, ξ'_o) the coordinates of η_o; let $E^+(\sigma_o, \xi'_o, \gamma_o)$ be the subspace corresponding to the operator $\pi(D_t, D_{x'}, D_{x_n})$. [See Section 3; in particular, for $\gamma_o > o$, $E^+(\sigma_o, \xi'_o, \gamma_o)$ is the space of the $v(o)$ when v describes the space of bounded solutions for $x_n \geq 0$ of the differential equation $\pi(\sigma_o - i\gamma_o, \xi'_o, D_{x_n}) v(x_n) = o$]; exactly as in the proof of Proposition 3.2, it can be shown that the subspace $E^+(\sigma_o, \xi'_o, \gamma_o)$ is independent of the choice of the vectors $e_1, \ldots, e_{n-1}$ and that it remains unchanged when ν is replaced by $\lambda\nu$ with $\lambda > 0$. We denote this subspace by $E^+(\eta_0, \gamma_0)$.

We now consider a linear mapping $b: \mathbb{C}^N \to \mathbb{C}^d$. We say that F, p, b, V, ν satisfy the uniform Lopatinski condition if, for all $\eta \in F/\mathbb{R}\nu$ and for all $\gamma \geq 0$ such that $(\eta, \gamma) \neq 0$, the restriction of b to $E^+(\eta, \gamma)$ is an isomorphism. By the homogeneity of $E^+(\eta, \gamma)$ in (η, γ), we see that this condition is not modified if V is replaced by λV with $\lambda > 0$.

ASSUMPTION 8.2

We now let M be a manifold of dimension $n+1$, and let P be a differential operator of degree 1 in M, with C^∞ $N \times N$ matrix coefficients. Suppose $y_0 \in M$ and let S, Y be two germs at y_0 of hypersurface of M. We suppose that P is strictly hyperbolic in the direction conormal to S at y_0, and that Y is non-characteristic for P at y_0.

We fix a connected component S^+ of $M \setminus S$ and a connected component Y^+ of $M \setminus Y$, i.e. orientations along the conormals at y_0 to S and Y. Let $B: Y \to \mathcal{L}(\mathbb{C}^N, \mathbb{C}^d)$ be a C^∞ mapping. We assume that the uniform Lopatinski condition is satisfied, i.e.: if V, ν are vectors conormal to S, Y at y_0, with positive orientation, then the uniform Lopatinski condition is satisfied by F, p, b, V, ν where $F = T^*_{y_0} M$, and where p is the restriction to F of the principal symbol of P, and where $b = B(y_0)$. As we have seen, this condition is independent of the choice of the conormal vectors V, ν. With these assumptions, we have

PROPOSITION 8.3 (Local uniqueness)

Let W *be a neighbourhood of* y_0 *in* M. *Then there exists a neighbourhood* W_0 *of* y_0 *in* M *such that: if* $u \in L^2_{loc}(W, \mathbb{C}^N)$ *satisfies*

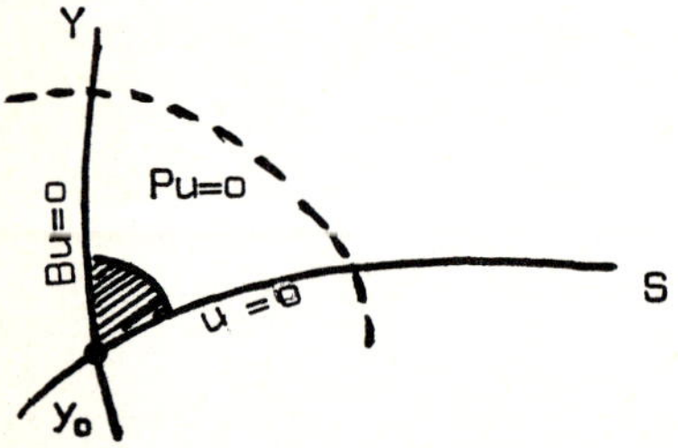

$$\begin{cases} Pu = 0 & \textit{in } W \cap S^+ \cap Y^+ \\ Bu = 0 & \textit{in } W \cap S^+ \cap Y \\ u = 0 & \textit{in } W \cap S \cap Y^+ \end{cases}$$

then u = o *in* $W_0 \cap S^+ \cap Y^+$

PROOF: By transport by a local chart of M transforming S and Y into hyperplanes, the problem reduces to $M = \mathbb{R}^{n+1}$,

$y_0 = 0$, $S = \{t = 0\}$, $Y = \{x_n = 0\}$, $S^+ = \{t > 0\}$, $Y^+ = \{x_n > 0\}$.

As in the proof of Holmgren's Theorem (see 1.10, Chapter VI), the diffeomorphism $x \mapsto x$, $t \mapsto t + |x|^2$ then allows the problem to be reduced to the case $M = \mathbb{R}^{n+1}$, $y_0 = 0$,

$S = \{t = |x|^2\}$, $Y = \{x_n = 0\}$, $S^+ = \{t > |x|^2\}$, $Y^+ = \{x_n > 0\}$.

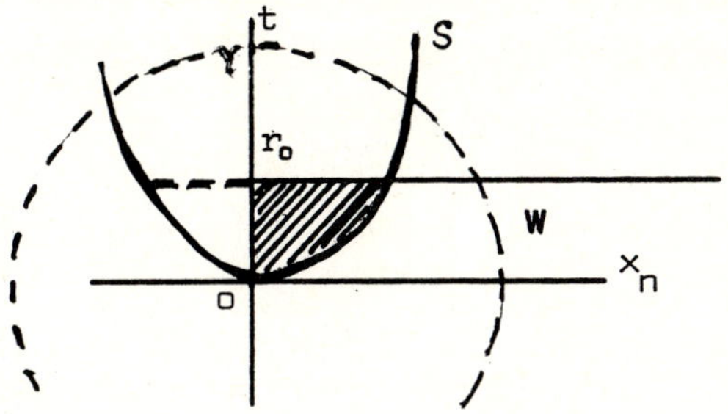

For $(t, x) = o$, the operator $P(t, x, D_t, D_x)$ is strictly hyperbolic with respect to t; the same therefore applies when (t, x) is fixed in a sufficiently small neighbourhood of o. For $(t, x') = o$, the restriction of $B(t, x')$ to $E^+(t,x';\sigma,\xi',\gamma)$ is an isomorphism for all $(\sigma, \xi') \in \mathbb{R}^n$, $\gamma \geq o$ such that $\sigma^2+|\xi'|^2+\gamma^2 = 1$. By continuity, the same applies when (t, x') is sufficiently near o. An argument based on extending B and the coefficients of P, analogous to that used during the proof of Lemma 4.12 of Chapter VI, thus allows the problem to be reduced to the case where: P, B are defined in $\mathbb{R}^{n+1}$, $\mathbb{R}^n$ the coefficients of P and B are constant outside a compact set, P is strictly hyperbolic with respect to t, and the uniform Lopatinski condition is satisfied. Then let $r_o > o$ be sufficiently small that $S^+ \cap \{t<r_o\} \Subset W$. We shall conclude by showing that $u = o$ in $S^+ \cap Y^+ \cap \{t<r_o\}$. Let $u^o \in L^2(\mathbb{R}_+^{n+1} \cap \{t<r_o\})$ be defined by $u^o = u$ in $S^+ \cap Y^+ \cap \{t<r_o\}$, and by $u^o = o$ otherwise. Since the trace of u on S is zero, the jump formula (see (2.7.2), (2.7.3)) shows

that $P(u^0) = 0$. We obviously have $B(u^0) = 0$ and $u^0 = 0$ for all $t < 0$. The uniqueness property in Theorem 7.1 then shows that $u^0 = 0$, and this concludes the proof.

PROPOSITION 8.4 : *We consider problem* (1.1) *in the systems case, assuming the uniform Lopatinski condition to be satisfied. Then there exists* $\rho > 0$ *such that: for all* $(t, x') \in \mathbb{R} \times \partial\Omega$, *the uniform Lopatinski condition is satisfied by* F, p, b, V, ν *with* $F = T^*_{t,x'}(\mathbb{R}^{n+1})$, p = *restriction to* F *of the principal symbol of* P, b = B(t, x'), ν = *inward conormal vector to* $\mathbb{R} \times \partial\Omega$ *at* (t, x'), V = *a vector from* F *belonging to the cone* $\{(\sigma, \xi) \in \mathbb{R} \times \mathbb{R}^n \mid |\xi| < \rho\sigma\}$.

PROOF : By using local charts of $\bar{\Omega}$ at the boundary, we see that it is sufficient to study the case in which $\Omega = \mathbb{R}^n_+$. We recall (see Theorem 3.13, Chapter VI) that there exists $v > 0$ such that for all $(t, x') \in \mathbb{R} \times \partial\Omega$, p is strictly hyperbolic with respect to any V from the cone $\Gamma = \{(\sigma, \xi) \mid |\xi| < \frac{1}{v}\sigma\}$. For $V \in \Gamma$ such that $|V| = 1$ and for $(\sigma, \xi', \gamma) \in \mathbb{R}^{n+1}$ such that $\gamma \geq 0$, $\sigma^2 + |\xi'|^2 + \gamma^2 = 1$, we put $X = (t, x'; \sigma, \xi', \gamma)$ and we denote by $E^+_V(X)$ the subspace $E^+(\sigma, \xi', \gamma)$ corresponding to F, p, V, ν. To study this subspace, we again use Construction 8.1; we see that, in this, we can choose for the vectors $e_1, \ldots, e_{n-1}$ the canonical basis vectors of $\mathbb{R}^{n-1}$, and that we then have

$\pi(\sigma,\xi',\xi_n) = p(\sigma V + (0,\xi',\xi_n))$. We can clearly suppose that the coefficient of ξ_n in p is the identity matrix so that

$\pi(\sigma - i\gamma, \xi', \xi u) = \xi_n I - a_V(X)$, where the matrix $a_V(X)$ is C^∞ with respect to (V, X). Once again we employ the same arguments for the matrix a_V as those used after the statement of Definition 3.3: we factorise the characteristic polynomial Δ_V of a_V into $\Delta_V = \Delta_V^+ \Delta_V^-$, the coefficients of the polynomials $\Delta_V^\pm$ being continuous with respect to (V,X); if we put

$Q_V^+(X) = \Delta_V^-(X, a_V(X))$, we have $E_V^+(X) = \text{Image } Q_V^+(X)$, and the uniform Lopatinski condition is expressed by the fact that the matrix $B(t, x')\, Q_V^+(X)$ is of maximal rank $d = \mu$, i.e. by the condition

$$(8.4.1) \qquad F(V,X) > 0 \qquad \text{for all } X$$

where we have put

$$F(V,X) = \operatorname*{Max}_{1 \le j_1 < \dots < j_d \le N} \left| \det\left(C_{j_1}(V,X), \dots, C_{j_d}(V,X)\right)\right|$$

and where we have denoted by $C_1, \dots, C_N$ the columns of the matrix BQ_V^+.

The function F is continuous, and the uniform Lopatinski condition for problem (1.1) indicates that (8.4.1) is satisfied for $V = V_0$, where we have put $V_0 = (1,0) \in \mathbb{R} \times \mathbb{R}^n$.

From the compactness in $(V, \sigma, \xi', \gamma)$ and from properties of the type (2.3.4) which are valid for $|t|$ large or $|x'|$ large, we see that $\operatorname{Inf}_X F(V_o, X) > o$ and that F is uniformly continuous. Thus (8.4.1) is satisfied when V is sufficiently near V_o, and this concludes the proof.

From Propositions 8.3 and 8.4 , and by using the same type of "sweeping" argument as that used for Theorem 4.13 in Chapter VI, we obtain

THEOREM 8.5 (Propagation Speed)

We consider problem (1.1) in the systems case, assuming the uniform Lopatinski condition to be satisfied. Suppose $\rho > o$ *as in Proposition* 8.4. *For* $(t_o, x_o) \in]o,T] \times \bar{\Omega}$, *we consider the backward cone*

$$C = \{(t,x) \in \mathbb{R}^{n+1} \mid |x - x_o| < \frac{1}{\rho}(t_o - t)\} .$$

Let $u \in L^2(C \cap (]o,T[\times \Omega))$ *be such that*

$$\begin{cases} Pu = o & \textit{in } C \cap (]o,T[\times \Omega) \\ Bu = o & \textit{in } C \cap (]o,T[\times \partial\Omega) \\ u = o & \textit{in } C \cap (\{o\} \times \Omega) \end{cases} .$$

Then $u = o$ *in* $C \cap (]o,T[\times \Omega)$.

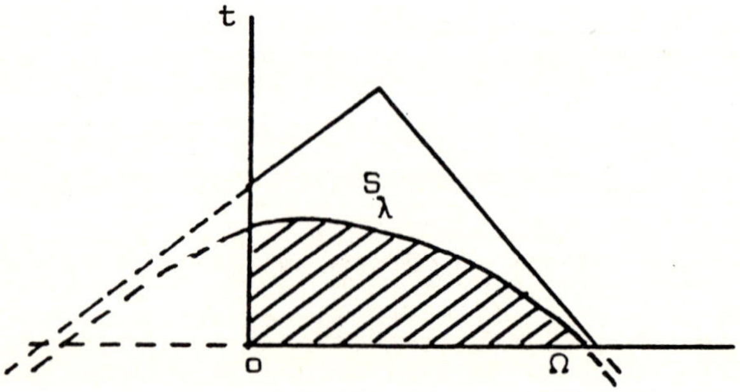

9. ADDITIONAL NOTES AND EXERCISES

9.1 SPACES $e^{\gamma t} H^s_\gamma$

a/ We put $\mathbb{R}^{n+1} = \mathbb{R} \times \mathbb{R}^n$, with generic point (t,x), and we denote the dual variable by (σ,ξ). For $\gamma \in \mathbb{R}\setminus 0$, $s \in \mathbb{R}$, we consider the Hilbert space $H^s_\gamma(\mathbb{R}^{n+1})$ defined in Section 11.2, Chapter IV.

We recall that the norm $|||u|||_{s;\gamma}$ in this space is defined by:

$$|||u|||^2_{s;\gamma} = \int (\gamma^2 + \sigma^2 + |\xi|^2)^s \, |\hat{u}(\sigma, \xi)|^2 \, d\!\!\!/\sigma \, d\!\!\!/\xi \quad .$$

We consider the space $e^{\gamma t} H^s_\gamma(\mathbb{R}^{n+1})$ of the $u \in \mathcal{D}'(R^{n+1})$ such that $e^{-\gamma t} u \in H^s_\gamma(R^{n+1})$, endowed with the norm

$$\|u\|_{s;\gamma} = |||e^{-\gamma t} u|||_{s;\gamma} \quad .$$

b/ Let $k \in \mathbb{N}$. Show that the following three quantities are equivalent:

$$(\gamma^2 + \sigma^2 + |\xi|^2)^k \;, \quad \sum_{\ell=0}^{k} (\gamma^2 + \sigma^2)^\ell \, |\xi|^{2(k-\ell)} \;,$$

$$\sum_{\ell+|\beta|=k} |(\sigma - i\gamma)^\ell \, \xi^\beta|^2 \quad .$$

By noting that $(\sigma - i\gamma)^\ell \, \widehat{e^{-\gamma t} u} = \widehat{e^{-\gamma t} D^\ell_t u}$ for $\ell \in \mathbb{N}$ and for $u \in \mathcal{D}'(R^{n+1})$ such that $e^{-\gamma t} u \in \mathcal{S}'(R^{n+1})$, show that:

(9.1.1.) $e^{\gamma t} H^k_\gamma(\mathbb{R}^{n+1})$ is the space of the $u \in e^{\gamma t} L^2(R^{n+1})$ such that $e^{-\gamma t} D^\alpha_{t,x} u \in L^2(R^{n+1})$ for $|\alpha| = k$, and we have the following equivalence, uniform with respect to γ:

$$\|u\|_{k;\gamma} \sim \sum_{|\alpha|=k} \|e^{-\gamma t} D^{\alpha} u\|_0 \quad .$$

c/ Consider a symbol $A \in PS^m(R^{n+1} \times R^{n+1} \times [\gamma_0, +\infty[)$ with $\gamma_0 > 0$. Deduce from Section 11.2, Chapter IV, that for $s \in \mathbb{R}$ there exists $C \geq 0$ such that:

(9.1.2.) $\|A_\gamma u\|_{s-m;\gamma} \leq C \|u\|_{s;\gamma}$ for all $u \in C_0^\infty(R^{n+1})$ and $\gamma \geq \gamma_0$.

d/ Let $k \in \mathbb{N}$. Deduce from (9.1.1), (9.1.2) a property of invariance of the spaces $e^{\gamma t} H^k_\gamma(R^{n+1}) \cap \mathcal{E}'$ under diffeomorphism with respect to x. Generalise this property to the case of spaces $e^{\gamma t} H^s_\gamma(R^{n+1}) \cap \mathcal{E}'$ when $s \in \mathbb{R}$ (use the symbolic calculus for the operators A_γ, which is deduced immediately from Section 11.2, Chapter IV).

e/ Suppose Ω is a regular open set in $\mathbb{R}^n$. As in the case of ordinary Sobolev spaces, we define the Hilbert space $e^{\gamma t} H^s_\gamma(\mathbb{R} \times \bar{\Omega})$ by restriction. Show that we have the analogue of (9.1.1) when $s = k \in \mathbb{N}$.

f/ When $\Omega = \mathbb{R}^n_+$ or when $\partial\Omega$ is compact, we define the Hilbert spaces $e^{\gamma t} H^s_\gamma(R \times \partial\Omega)$: when $\partial\Omega$ is compact, we proceed as in the case of ordinary Sobolev spaces by choosing once and for all a finite atlas of $\partial\Omega$ and a partition of unity (different choices lead to equivalent norms on $e^{\gamma t} H^s_\gamma(R \times \partial\Omega)$, the equivalence being uniform with respect to γ for $|\gamma| \geq \gamma_0$, $\gamma_0 > 0$). We denote by $|u|_{s;\gamma}$ the norm of u in $e^{\gamma t} H^s_\gamma(\mathbb{R} \times \partial\Omega)$.

g/ The general point of $\overline{\mathbb{R}}_+^{n+1}$ is denoted by $x = (x', x_n)$, with $x' \in \mathbb{R}^{n-1}$, $x_n \geq 0$. We again use the tangential symbol $\Lambda^s = (\gamma^2 + \sigma^2 + |\xi'|^2)^{s/2}$. By using (9.1.1), show that we have, for $k \in \mathbb{N}$, the following equivalence of norms, uniform with respect to γ:

$$(9.1.3.) \quad \|u\|^2_{k;\gamma} \sim \sum_{j=0}^{k} \|\Lambda_\gamma^{k-j} D^j_{x_n} u\|^2_{0;\gamma} \quad \text{for} \quad u \in C_0^\infty(\overline{R}_+^{n+1}).$$

9.2. ENERGY INEQUALITY IN THE SCALAR CASE.

The notation is that used in the statement of Theorem 4.4.

a/ We again use the reduction of the scalar operator P of order m to a system of first order, as in Example 4.8 of Chapter VI. We thus have, for $u \in C_0^\infty(R^{n+1})$:

$$\|U\|^2_{0;\gamma} = \sum_{j=1}^{m} \int \|e^{-\gamma t} D_t^{j-1} u\|^2_{m-j} \, dt .$$

Deduce from this, by using (9.1.1), that there exists $C' \geq 0$ such that $\|u\|^2_{m-1;\gamma} \leq C' \|U\|^2_{0;\gamma}$ for all $u \in C_0^\infty(R^{n+1})$. By using Theorem 4.1, infer from this that there exist $C \geq 0$ and $\gamma_0 > 0$ such that:

$$(9.2.1.) \quad \gamma \|u\|_{m-1;\gamma} \leq C \|P u\|_{0;\gamma} \quad \text{for} \quad \gamma \geq \gamma_0$$
$$\text{and} \quad u \in C_0^\infty(R^{n+1}) .$$

b/ Suppose that $\Omega = \mathbb{R}^n_+$ and apply the reduction 2.2 Using (9.1.3), show that there exists $c > 0$ such that:

$$|U|^2_{0;\gamma} \geq c\, |u|^2_{m-1;\gamma} \quad \text{for} \quad u \in C_0^\infty(\overline{R}_+^{n+1}) \text{ and all } \gamma .$$

Using Theorem 4.2, deduce from this that there exist

$C \geq 0$ and $\gamma_0 > 0$ such that:

$$(9.2.2.)\quad \gamma \|u\|^2_{m-1;\gamma} + \sum_{j=0}^{m-1} |\gamma_j u|^2_{m-1-j;\gamma} \leq C\Big(\frac{1}{\gamma}\|P u\|^2_{0;\gamma} + \sum_{j=1}^{d} |B_j u|^2_{m-1-r_j;\gamma}\Big)$$

for $\gamma \geq \gamma_0$ and $u \in C_0^\infty(\overline{\mathbb{R}}^{n+1}_+)$.

c/ By partition of unity, deduce Theorem 4.4 from (9.2.1), (9.2.2).

9.3 ALGEBRAIC PRELIMINARY TO EXERCISE 9.4

Suppose F is a vector subspace of $\mathbb{C}^N$ and M is an $N \times N$ Hermitian matrix. We denote by (,) the canonical hermitian structure on $\mathbb{C}^N$ and by $\perp$ the associated orthogonality relation.

We suppose that the subspace F is maximal amongst those for which M is positive, i.e. which satisfy $(Mu, u) \geq 0$ for all $u \in F$. We put $G = (M.F)^\perp$.

a/ Show that $-M$ is positive on G.

[take $v_0 \in G$, and study the sign of M on the subspace $(v_0) + F$] .

b/ Prove the inclusion $\operatorname{Ker} M \subset F \cap G$.

[the inclusion $\operatorname{Ker} M \subset G$ follows from the fact that M is hermitian; for the inclusion $\operatorname{Ker} M \subset F$, consider a vector $v_0 \in \operatorname{Ker} M$ and study the sign of M on the subspace $(v_0) + F$].

c/ Use the inclusion $F^{\perp} \subset (\text{Ker } M)^{\perp}$ to show that there exists a basis $(e_1, \ldots, e_N)$ of $\mathbb{C}^N$ in which M is written in the form

$$\left(\begin{array}{c|c} M_I & 0 \\ \hline 0 & 0 \end{array}\right)$$

where M_I is an invertible $p \times p$ matrix ($p = \text{rank } M$), and F is written $F = \{ u \in \mathbb{C}^N \mid u_1 = \ldots u_r = 0\}$ ($r = \text{codim } F$)). Also prove an analogous result relating to M and G.

9.4 POSITIVE SYMMETRIC SYSTEMS

Let Ω be a regular bounded open subset of $\mathbb{R}^n$. We consider the matrix differential operator $P = \partial_t + \sum_{j=1}^{n} A_j(x)\, \partial_j + C(x)$ where $C(x)$ is an $N \times N$ matrix and the $A_j(X)$ are $N \times N$ hermitian matrices with C^∞ dependence on $x \in \bar{\Omega}$. We suppose that the matrix $M(x) = \sum_{1}^{n} \nu_j(x)\, A_j(x)$ is of constant rank p, where $x \in \partial\Omega$ and $\nu(x) = (\nu_1(x), \ldots, \nu_n(x))$ denotes the outward unit normal to $\partial\Omega$. We take a complex vector bundle of rank r over $\partial\Omega$, and we denote by B_x its fibre over x. We suppose that for all $x \in \partial\Omega$ the hermitian matrix $M(x)$ is positive on the subspace B_x and that B_x is maximal amongst the subspaces for which $M(x)$ is positive. We now propose to investigate the mixed problem:

$$(9.4.1.) \quad \begin{cases} Pu = f & \text{in} \quad [0, +\infty[\times \Omega \\ u(t, x)_{|\partial\Omega} \in B_x & \text{in} \quad [0, +\infty[\times \partial\Omega \\ u(0, .x) = u_0 & \text{in} \quad \Omega \end{cases}$$

by showing that the operator $- Au = - \sum A_j(x)\, \partial_j u - C(x)\, u$

is a semigroup generator.

In the following, all functions and distributions have values in $\mathbb{C}^N$.

a/ Show that $C^\infty(\bar{\Omega})$ is dense in the space $V_A = \{ u \in L^2(\Omega) | Au \in L^2(\Omega) \}$ normed by $\|u\|^2 = \|u\|_0^2 + \|Au\|_0^2$. Show that the mapping $C^\infty(\bar{\Omega}) \ni u \longrightarrow M(x)u_{|\partial\Omega} \in C^\infty(\partial\Omega)$ extends continuously into a mapping of V_a into $H^{-\frac{1}{2}}(\partial\Omega)$. Deduce from this that the Green's formula

$$(9.4.2.) \qquad (Au, v)_{L^2} = (u, A^* v) + \int_{\partial\Omega} (M(x)u, v)_{\mathbb{C}^N} dx$$

remains valid for $u \in V_A$ and $v \in H^1(\bar{\Omega})$; we have put

$$A^* = -A + \Big(C(x) + C^*(x) - \sum_1^n \frac{\partial A_j}{\partial x_j}(x)\Big)$$

[proceed exactly as in Theorem 2.5].

b/ Let $x_0 \in \partial\Omega$; show that there exists a local chart of $\bar{\Omega}$ to the boundary and a trivialisation of B over this chart such that, after transport, the matrix M(x) and the bundle B_x are respectively written

$$\tilde{M}(x) = \left(\begin{array}{c|c} M_I(x) & 0 \\ \hline 0 & 0 \end{array} \right)$$

and $\tilde{B}_x = \{ u \in \mathbb{C}^N | u_1 = \dots = u_{N-r} = 0 \}$ where $M_I(x)$ is an invertible $p \times p$ matrix and $N - r \leq p$ [use Exercise 9.3].
Deduce from this that in the neighbourhood of x_0 there exists

an $N \times N$ matrix $\beta(x)$ such that $u \in B_x$ if and only if $\beta(x)\,M(x)u = 0$. Use this formulation of the boundary condition and the first part of a/ to prove that it is possible to define a closed subspace of V_A by

$$D = \{ u \in V_A \mid u(x)_{|\partial\Omega} \in B_x \quad \text{for all} \quad x \in \partial\Omega \}$$

c/ Use b/ to prove that the space $D_0 = D \cap H^1(\overline{\Omega})$ is dense in D. This also enables D to be considered as the closure of

$\{ u \in H^1(\overline{\Omega}) \mid u(x)_{|\partial\Omega} \in B_x \ , \ x \in \partial\Omega \}$ in V_A. [By partition of unity, reduce to the case of $\mathbb{R}^n_+$ as in b/, then for $\varepsilon > 0$ and $\eta > 0$ construct approximations $u_{\varepsilon,\eta} = \left(u^I_\varepsilon, \ u^{II}_{\varepsilon,\eta} \right)$ by putting $u^I_\varepsilon = u^I *' \rho_\varepsilon$, $u^{II}_\varepsilon = u^{II} * (\rho_\varepsilon(x_1,\dots, x_{n-1})\otimes\theta_\eta(x_n))$

where: u^I and u^{II} denote the components in $\mathbb{C}^P$ and $\mathbb{C}^{N-P}$ of $u \in D$, ρ_ε is a regularising sequence in the variables $(x_1,\dots, x_{n-1})$ and the θ_η are a regularising sequence in the variable x_n and with support in $x_n < 0$. Show that

$$u^I_\varepsilon \in L^2(\overline{\mathbb{R}}^+; H^1(\mathbb{R}^{n-1})), \ \frac{\partial}{\partial x_n} u^I_\varepsilon \in L^2(\mathbb{R}^n_+) \ , \ u^{II}_\varepsilon\big|_{\mathbb{R}^n_+} \in C^\infty(\overline{\mathbb{R}}^n_+)$$

and $u_{\varepsilon,\eta} \xrightarrow[\eta \to 0]{\varepsilon \to 0} u$ in D.

d/ Using (9.4.2), show that there exists c_0 such that

$\mathrm{Re}((\lambda + A)u, u)_{L^2} \geq (\lambda - c_0)\,\|u\|_0^2$ for $u \in D_0$ and $\lambda > c_0$; deduce from this that:

$$\|(\lambda + A)u\|_0 \geq (\lambda - c_0)\,\|u\|_0 \quad \text{for all} \quad u \in D\,.$$

Note that the maximality of the B_x is not used here.

e/ Show that the space $(\lambda + A)D$ is dense in L^2. [Let $v \in L^2$ be such that $((\lambda + A)u, v) = 0$ for all $u \in D_o$, and show that $(\lambda + A^*)v = 0$, $v \in V_{A^*} = \{ u \in L^2 \mid A^*u \in L^2 \}$ and

$$\int_{\partial\Omega} (u(x), M(x)\, v(x))_{\mathbb{C}^N}\, dx = 0 .]$$

Use Exercise 9.3 to show that the spaces $\tilde{B}_x = (M(x)\, B_x)^{\perp}$ define a vector bundle over which $-M$ is positive and show that

$$v \in \tilde{D} = \{ w \in V_{A^*} \mid w(x)_{|\partial\Omega} \in \tilde{B}_x \quad \text{for all} \quad x \in \partial\Omega \}$$

Finally, deduce from this that $v = 0$ by applying d/ to $(A^*, \tilde{D})$.

f/ Deduce that $(-A, D)$ is a generator of a strongly continuous semigroup in $L^2(\Omega)$ and apply this to the solution of the mixed problem (9.4.1).

g/ Show that the results of this exercise can be applied to the case of the Maxwell equations (Example 4.6 in Chapter VI) in an open set Ω, by adding as boundary condition the condition

$$\nu(x) \wedge E(x) = \alpha\, \nu(x) \wedge H(x) \qquad \text{for all} \quad x \in \partial\Omega$$

where α is a given real number and where $\wedge$ denotes the vector product in $\mathbb{R}^3$.

h/ Show that the results of this exercise can also be

applied to the case of the wave equation with Neumann boundary conditions. [If we put $U = (\frac{\partial u}{\partial t}, \frac{\partial u}{\partial x_1}, \dots, \frac{\partial u}{\partial x_n})$, then U is a solution of the symmetric system

$$\partial_t U = \begin{bmatrix} 0, & -\partial_{x_1}, & \dots, & -\partial_{x_n} \\ -\partial_{x_1}, & & & \\ \vdots & & & \\ -\partial_{x_n}, & & & \end{bmatrix} . U$$

with the boundary condition $(0, \nu_1(x), \dots, \nu_n(x)).U(x) = 0$].

9.5 EXAMPLE OF A BOUNDARY-VALUE PROBLEM FOR A FIRST-ORDER DIFFERENTIAL SYSTEM SATISFYING THE UNIFORM LOPATINSKI CONDITION.

In $\mathbb{R} \times \mathbb{R}^3$, we consider the (2 x 2 matrix) differential operator $P = p(D_t, D_x, D_y, D_z)$ defined by

$$P = D_t - \begin{pmatrix} 1 & 0 \\ 0 & -1 \end{pmatrix} D_x - \begin{pmatrix} 0 & 1 \\ 1 & 0 \end{pmatrix} D_y - \begin{pmatrix} 0 & i \\ -i & 0 \end{pmatrix} D_z$$

We denote by τ, ξ, η, ζ the dual variables of t,x,y,z. Show that $\det p(\tau, \xi, \eta, \zeta) = \tau^2 - \xi^2 - \eta^2 - \zeta^2$, and therefore that P is strictly hyperbolic with respect to t.

Let $\Omega = \{ (x, y, z) \in \mathbb{R}^3 \mid x > 0 \}$. Consider the boundary value problem

$$(9.5.1.)\quad \begin{cases} P\begin{pmatrix} u_1 \\ u_2 \end{pmatrix} = \begin{pmatrix} f_1 \\ f_2 \end{pmatrix} & \text{in} \quad \mathbb{R}\times\Omega \\ u_2|_{x=0} = g & \text{in} \quad \mathbb{R}\times\partial\Omega \end{cases} .$$

Show that for $\gamma > 0$ the subspace $E^+(\sigma, \eta, \zeta, \gamma)$ defined in Section 3 is $\{ (\alpha, \beta) \in \mathbb{C}^2 \mid \alpha = \frac{Z}{\sqrt[+]{\tau^2 - |Z|^2} - \tau} \beta \}$

where $Z = \eta + i\zeta$, $\tau = \sigma - i\gamma$, and where $\sqrt[+]{\quad}$ denotes the square root with positive imaginary part. Deduce that problem (9.5.1) satisfies the uniform Lopatinski condition.

CHAPTER 8

MICROLOCALISATION

1. GENERAL PROPERTIES OF THE SINGULAR SPECTRUM (WF)

Let X be an open set in $\mathbb{R}^n$ and suppose $u \in \mathcal{D}'(x)$. The definition of the singular spectrum of u (which we denote by WFu; see Definition 3.1, Chapter I) can be formulated as follows, remembering that a subset W of $\mathbb{R}^n \setminus 0$ is conical if $t\xi \in W$ when $\xi \in W$ and $t > 0$ (see Chapter I, Section 1):

DEFINITION 1.1 *Let* $(x_0, \xi_0) \in X \times (\mathbb{R}^n \setminus 0)$. *We say that* $(x_0, \xi_0) \notin WFu$ *if there exists a neighbourhood* V *of* x_0 *in* X *and a conic neighbourhood* W *of* ξ_0 *in* $\mathbb{R}^n \setminus 0$ *such that, for any function* $\alpha \in C_0^\infty(V)$, $\widehat{\alpha u}$ *has rapid decrease in* W *(i.e. such that for all* $N \in \mathbb{R}$, *there exists* $C \geq 0$ *with* $|\widehat{\alpha u}(\xi)| \leq C(1 + |\xi|)^{-N}$ *when* $\xi \in W$).

The singular spectrum of u is a conic closed subset of $X \times (\mathbb{R}^n \setminus 0)$, and we have $WF(u_Y) = (WFu)|_Y$ when Y is an open subset of X. We have also seen ((3.1.3), Chapter I):

PROPOSITION 1.2: *Let* $\pi_X : X \times (\mathbb{R}^n \setminus 0) \longrightarrow X$ *be the projection* $(x,\xi) \longmapsto x$. *For* $u \in \mathcal{D}'(X)$, *we have* $\pi_X(WFu) = \text{supp sing } u$.

We can also obtain partial regularity properties:

PROPOSITION 1.3: *Let* $u \in \mathcal{D}'(\mathbb{R}^n)$; *the generic point of* $\mathbb{R}^n$ *is denoted by* $x = (x', x_n)$ *with* $x' \in \mathbb{R}^{n-1}$; *we similarly put* $\xi = (\xi', \xi_n)$. *We assume that* WFu *does not intersect* $\{(0,0,0,\xi_n) \mid \xi_n \in \mathbb{R} \setminus 0\}$. *Then, for* $\alpha \in C_0^\infty(\mathbb{R}^n)$ *with support sufficiently near* (0, 0), αu *is* C^∞ *in* x_n *with values which are* x' *distributions (see Section* 1.5, *Chapter* 1).

PROOF: By hypothesis, there exist a neighbourhood V of (0, 0) in $\mathbb{R}^n$ and C > 0 such that $\widehat{\alpha u}$ is rapidly decreasing in $\{\xi \in \mathbb{R}^n \mid |\xi'| \leq C|\xi_n|\}$ for $\alpha \in C_0^\infty(V)$. We fix such a function α. Since $\widehat{\alpha u}$ is slowly growing (see (2.8.1), Chapter I), it can be shown that there exists $\mu \in \mathbb{R}$ such that, for all N, there exists C_N with $|\widehat{\alpha u}(\xi)| \leq C_N(1 + |\xi_n|)^{-N} (1 + |\xi'|)^{\mu+N}$ for $\xi = (\xi', \xi_n) \in \mathbb{R}^n$ (this follows immediately by examining in turn the two cases $|\xi'| \leq C|\xi_n|$, $|\xi'| > C|\xi_n|$) .

We deduce from this that, for $\varphi \in C_0^\infty(\mathbb{R}^{n-1})$ and $\psi \in C_0^\infty(\mathbb{R})$, we have:

$$< \alpha u, \varphi \otimes \psi > = \int \widehat{\alpha u}(\xi)\, \hat{\varphi}(-\xi')\, \hat{\psi}(-\xi_n)\, d\!\!/\xi'\, d\!\!/\xi_n$$

$$= \int \psi(x_n)dx_n \int \hat{\psi}(-\xi')d\!\!/\xi' \int e^{ix_n\xi_n}\, \widehat{\alpha u}(\xi',\xi_n) d\!\!/\xi_n .$$

Thus αu is identified with the function $U \in C_0^\infty(\mathbb{R}, \mathcal{E}'(\mathbb{R}^{n-1}))$ defined, for $x_n \in \mathbb{R}$, by

$$\widehat{U(x_n)}(\xi') = \int e^{ix_n\xi_n} \widehat{\alpha u}(\xi', \xi_n) d\!\!{}^-\xi_n .$$

We shall now give a variant of Definition 1.1. For this purpose, we start with the following proposition:

PROPOSITION 1.4: *Suppose we have* $v \in \mathcal{E}'(X)$ *and let* W *be an open cone in* $\mathbb{R}^n \setminus 0$ *such that* $\hat{v}$ *is rapidly decreasing in* W. *Then, for* $f \in C^\infty(X)$, $\widehat{fv}$ *is rapidly decreasing in any cone* W' *in* $\mathbb{R}^n \setminus 0$ *such that* $\overline{W'} \subset W$.

PROOF: Clearly, we can assume that $f \in C_0^\infty(X)$; we then have

$$\widehat{fv}(\xi) = \int \hat{f}(\xi - \eta)\, \hat{v}(\eta)\, d\!\!{}^-\eta = I_1(\xi) + I_2(\xi) + I_3(\xi)$$

with

$$I_1(\xi) = \int_{\eta \notin W} \hat{f}(\xi - \eta)\, \hat{v}(\eta)\, d\!\!{}^-\eta$$

$$I_2(\xi) = \int_{\eta \in W,\, |\eta| \leq |\xi|/2} \hat{f}(\xi - \eta)\hat{v}(\eta)\, d\!\!{}^-\eta$$

$$I_3(\xi) = \int_{\eta \in W,\, |\eta| \geq |\xi|/2} \hat{f}(\xi - \eta)\hat{v}(\eta)\, d\!\!{}^-\eta .$$

By hypothesis, there exists $c > 0$ such that $|\xi - \eta| \geq c(|\xi| + |\eta|)$ when $\xi \in W'$ and $\eta \notin W$. In view of the rapid decrease of $\hat{f}$ and the slow growth of $\hat{v}$, we see that there exists $\mu \in \mathbb{R}$ and, for all N, a constant C_N such that, for $\xi \in W'$:

$$|I_1(\xi)| \leq C_N \int (1 + |\xi| + |\eta|)^{-N}(1 + |\eta|)^{\mu}\, d\eta$$

$$\leq C_N(1 + |\xi|)^{-N/2} \int (1 + |\eta|)^{-N/2 + \mu} d\eta ,$$

and hence the rapid decrease of I_1 in W'.

When $|\eta| \leq |\xi|/2$, we have $|\xi - \eta| \geq |\xi|/2$, and therefore, similarly,

$$|I_2(\xi)| \leq C'_N(1 + |\xi|)^{-N} \int_W |\hat{v}(\eta)|\, d\eta .$$

Finally, since $\hat{f}$ is bounded, we have:

$$|I_3(\xi)| \leq C''_N \int_{|\eta| \geq |\xi|/2} (1 + |\eta|)^{-N} d\eta \leq C'''_N (1 + |\xi|)^{-N/2} \int (1 + |\eta|)^{-N/2} d\eta ,$$

and hence the rapid decrease of $\widehat{fv}$ in W'.

COROLLARY 1.5: *Let* $u \in \mathcal{D}'(X)$ *and* $(x_o, \xi_o) \in X \times (\mathbb{R}^n \setminus 0)$. *In order that* $(x_o, \xi_o) \notin WFu$, *it is necessary and sufficient that there exist* $\alpha \in C_o^\infty(X)$, *non-zero at* x_o, *and a conic neighbourhood* W *of* ξ_o *in* $\mathbb{R}^n \setminus 0$ *such that* $\widehat{\alpha u}$ *is rapidly decreasing in* W.

PROOF: The condition is clearly necessary; to see that it is sufficient, we consider a neighbourhood V of x_o in X such that

$\alpha(x) \neq 0$ for $x \in V$; if $\beta \in C_o^\infty(V)$, we have $\beta u = fv$ with $f = \beta/\alpha \in C_o^\infty(X)$, $v = \alpha u$. The conclusion thus follows from Proposition 1.4.

COROLLARY 1.6: *Let* $f \in C_o^\infty(X)$, $u \in \mathcal{D}'(X)$ *and let* W *be a closed cone in* $\mathbb{R}^n \setminus 0$ *such that* $(\operatorname{Supp} f \times W) \cap WFu = \emptyset$. *Then* $\widehat{fu}$ *is rapidly decreasing in* W.

PROOF: Let $\widetilde{W}$ be a closed conic neighbourhood of W in $\mathbb{R}^n \setminus 0$ such that $(\operatorname{Supp} f \times \widetilde{W}) \cap WFu = \emptyset$. For $x_o \in \operatorname{Supp} u$, there exists a neighbourhood V of x_o in X such that $\widehat{\alpha u}$ is rapidly decreasing in $\widetilde{W}$ when $\alpha \in C_o^\infty(V)$ (this follows from the compactness of $\widetilde{W} \cap S^{n-1}$). Let there be a finite covering of supp f by V_j's of this type, and let there be functions $\alpha_j \in C_o^\infty(V_j)$ such that $\Sigma\, \alpha_j = 1$ in supp f. We have $fu = \sum_j f(\alpha_j u)$, and hence the rapid decrease of $\widehat{fu}$ in W by applying Proposition 1.4.

DEFINITION 1.7: *Let* Γ *be a closed cone in* $X \times (\mathbb{R}^n \setminus 0)$. *We put* $\mathcal{D}'_\Gamma(X) = \{u \in \mathcal{D}'(X) \mid WFu \subset \Gamma\}$. *We say that a sequence* (u_j) *converges in* $\mathcal{D}'_\Gamma(X)$ *to* $u \in \mathcal{D}'_\Gamma(X)$ *if*

(1.7.1) (u_j) *converges in* $\mathcal{D}'(X)$ *to* u

(1.7.2) *For all* $(x_o, \xi_o) \in X \times (\mathbb{R}^n \setminus 0)$ *not belonging to* Γ, *there exists* $\alpha \in C_o^\infty(X)$ *with* $\alpha(x_o) \neq 0$, *and a conic neighbourhood* W *of* ξ_o *in* $\mathbb{R}^n \setminus 0$ *such that, for all* N,

$$\sup_{\xi \in W} (1 + |\xi|)^{-N} \, |\widehat{\alpha u}(\xi) - \widehat{\alpha u}_j(\xi)| \longrightarrow 0 \quad \text{if} \quad j \longrightarrow +\infty .$$

By arguing as in the proof of Corollary 1.6, it can be shown that condition (1.7.2) is equivalent to

(1.7.3) For any $\alpha \in C_0^\infty(X)$ and for any closed cone W in $\mathbb{R}^n \setminus 0$ such that $(\operatorname{supp} \alpha \times W) \cap \Gamma = \emptyset$, then, for all N,

$$\sup_{\xi \in W} (1 + |\xi|)^{-N} \, | \widehat{\alpha u}(\xi) - \widehat{\alpha u_j}(\xi)| \longrightarrow 0 \quad \textit{if} \quad j \longrightarrow +\infty .$$

By arguing as in the proof of Proposition 1.4, it can be shown that for $f \in C_0^\infty(X)$:

(1.7.4) The mapping $u \longmapsto fu$ is sequentially continuous from $\mathcal{D}'_\Gamma(X)$ into $\mathcal{D}'_\Gamma(X)$.

PROPOSITION 1.8: *Let* $u \in \mathcal{D}'_\Gamma(X)$; *then there exists a sequence* $u_j \in C_0^\infty(X)$ *such that* $u_j \to u$ *in* $\mathcal{D}'_\Gamma(X)$. *Furthermore, it is possible to choose the* u_j*'s with support in an arbitrary neighbourhood of* supp u .

PROOF: First we assume that $u \in \mathcal{D}'_\Gamma(X) \cap \mathcal{E}'(X)$. Let $\rho \in C_0^\infty(\mathbb{R}^n)$ with integral 1; we put $\rho_j(x) = j^n \rho(jx)$ and $u_j = \rho_j * u$. Let $f \in C_0^\infty(X)$ and let W be a closed cone in $\mathbb{R}^n \setminus 0$ such that $(\operatorname{supp} f \times W) \cap \Gamma = \emptyset$. Let V be a neighbourhood of supp f in X and $\widetilde{W}$ a closed conical neighbourhood of W in $\mathbb{R}^n \setminus 0$ such that $(V \times \widetilde{W}) \cap \Gamma = \emptyset$. Let $\alpha \in C_0^\infty(X)$ be equal to unity in the neighbourhood of supp f . For j sufficiently large, we have: $f u_j = f(\rho_j * u) = f(\rho_j * (\alpha u))$, and thus

$$\widehat{fu}(\xi) - \widehat{fu_j}(\xi) = \int \hat{f}(\xi - \eta) \, \widehat{\alpha u}(\eta)(1 - \hat{\rho}(\eta / j)) \, d\!\!\!/\eta \quad .$$

Since $1 - \hat{\rho}(\eta/j) \longrightarrow 0$, and remains bounded, when $j \to +\infty$, we proceed exactly as in the proof of Proposition 1.4 to establish that

$$\sup_{\xi \in W} (1 + |\xi|)^{-N} \, |\widehat{fu}(\xi) - \widehat{fu_j}(\xi)| \longrightarrow 0 \quad ,$$

and thus that $u_j \to u$, in $\mathcal{D}'_\Gamma(X)$. When u no longer has compact support, we consider a locally finite covering of X by relatively compact open sets X_k, and also α_k , $\beta_k \in C_0^\infty(X_k)$ with $\Sigma\, \alpha_k = 1$ in X and $\beta_k = 1$ in the neighbourhood of supp α_k. Suppose $\varphi_{k,j} \in C_0^\infty(X_k)$ such that $\varphi_{k,j} \longrightarrow \alpha_k u$ in $\mathcal{D}'_\Gamma(X)$ when $j \to +\infty$; then $\psi_j = \sum_k \beta_k \varphi_{kj}$ is in $C^\infty(X)$ and tends towards u in $\mathcal{D}'_\Gamma(X)$; let K_j be an exhaustive sequence of compact subsets of X and let $\chi_j \in C_0^\infty(X)$ be equal to unity in K_j; it then follows immediately that $\varphi_j = \chi_j \psi_j$ tends towards u in $\mathcal{D}'_\Gamma(X)$. Finally, the support property for the φ_j's is obtained by means of a new truncation, starting from (1.7.4).

DEFINITION 1.9: *Let* Γ *be a closed cone in* $X \times (\mathbb{R}^n \setminus 0)$ *and let* F *be a closed subset of* X. *We put*

$$\mathcal{D}'_{\Gamma,F}(X) = \{u \in \mathcal{D}'_\Gamma(X) \mid \text{supp } u \subset F\} \quad .$$

We shall usually assume that $\text{Proj}_X \Gamma \subset F$ *(this is not an actual restriction in view of Proposition* 1.2*), and in* $\mathcal{D}'_{\Gamma,F}(X)$ *we shall consider the notion of sequential convergence induced by* $\mathcal{D}'_\Gamma(X)$.

2. THE FUNDAMENTAL THEOREMS

We start by studying the singular spectrum of an integral distribution associated with a phase and an amplitude (see Chapter III, Section 7).

THEOREM 2.1: *Let* $A \in \mathcal{D}'(X)$ *be the distribution defined by*

$$< A , u > = \int e^{i\varphi(x,\theta)} a(x,\theta) u(x) dx\, d\theta \qquad (u \in C_0^\infty(X))$$

where φ *is a phase and* a *an amplitude. Then*

$$(2.1.1) \quad WF A \subset \{(x, \varphi'_x(x,\theta)) \mid \theta \neq 0 , \; (x,\theta) \in \text{cone supp } a , \; \varphi'_\theta(x,\theta) = 0\} .$$

PROOF: Let F denote the right-hand side of (2.1.1). Suppose we have $(x_0 , \xi_0) \in X \times (\mathbb{R}^n \setminus 0)$ which does not belong to F. It can easily be shown that there exists a neighbourhood V of x_0 in X, a closed conic neighbourhood W of ξ_0 in $\mathbb{R}^n \setminus 0$ and a conic neighbourhood $\tilde{C}$ of $C = \{(x,\theta) \in \text{cone supp } a \mid \theta \neq 0 , \varphi'_\theta(x,\theta) = 0\}$ such that $\varphi'_x(x,\theta) \notin W$ when $x \in V$ and $(x,\theta) \in \tilde{C}$. (Note that this proves that F is closed).

By truncation, we can assume that $a(x,\theta) = 0$ for $|\theta| \leq 1$. Let $\chi(x,\theta)$ be a C^∞ function in $X \times (\mathbb{R}^N \setminus 0)$, homogeneous of degree 0 in θ, such that $\chi = 1$ in the neighbourhood of C, and $\text{supp } \chi \subset \tilde{C}$. From Proposition 7.7 of Chapter III, we have $WF A = WF A_1$, where A_1 is defined by the phase φ and the amplitude χa. For $\alpha \in C_0^\infty(V)$, we have

$$\widehat{\alpha A_1}(\xi) = \int e^{i(\varphi(x,\theta) - x.\xi)}\, b(x,\theta)dx\, d\theta$$

with $b(x,\theta) = \alpha(x)\chi(x,\theta)a(x,\theta)$. By construction, we have $\varphi'_x(x,\theta) - \xi \neq 0$ for $\xi \in W$ and $(x,\theta) \in$ cone supp b; we then deduce, by continuity and homogeneity with respect to (θ,ξ), that there exists a constant $c > 0$ such that $|\varphi'_x(x,\theta) - \xi| \geq c(|\theta| + |\xi|)$ for $\xi \in W$ and $(x,\theta) \in$ cone supp b.

Consider the differential operator

$$L = \sum_{j=1}^{n} \frac{1}{|\varphi'_x(x,\theta)-\xi|^2} (\varphi'_{x_j}(x,\theta) - \xi_j)D_{x_j} \quad ;$$

For $\xi \in W$ and $k \in \mathbb{N}$, we obtain

$$\widehat{\alpha A_1}(\xi) = \int e^{i(\varphi(x,\theta)-x.\xi)}\, {}^tL^k b(x,\theta)dx\, d\theta \; .$$

It can easily be shown, denoting by m the degree of the amplitude a , that $|{}^tL^k b(x,\theta)| \leq C(1+|\theta|)^m(|\theta| + |\xi|)^{-k}$ for $\xi \in W$, so that we have

$$\widehat{\alpha A_1}(\xi) \leq C\, |\xi|^{-k/2} \int (1 + |\theta|^{m-k/2}\, d\theta$$

for $\xi \in W$, $|\xi| \geq 1$ and $m - k/2 < - N$, and this shows that $(x_0\, ,\, \xi_0) \notin WF\, A_1$.

REMARK 2.2: By truncation, it can easily be deduced from Theorem 2.1 that, with the same assumptions, we have more precisely

$$\mathrm{WF}\,A \subset \{(x, \varphi'_x(x,\theta)) \mid \theta \neq 0 ,$$
$$(x,\theta) \in \text{ess supp } a \ , \ \varphi'_\theta(x,\theta) = 0\} ,$$

where ess supp a denotes the smallest closed cone in $X \times \mathbb{R}^n$ outside of which a is $S^{-\infty}$.

REMARK 2.3: The proof of Theorem 2.1 also proves that, for fixed φ, the mapping $a \mapsto A$ is sequentially continuous from $S^m(X \times \mathbb{R}^n)$ into $\mathcal{D}'_\Lambda(X)$ where

$$\Lambda = \{(x,\varphi'_x(x,\theta)) \mid \theta \neq 0 , \ \varphi'_\theta(x,\theta) = 0\} .$$

Let us now consider the tensor product of two distributions.

THEOREM 2.4: *Let* X *be an open set in* $\mathbb{R}^n$, *and* Y *an open set in* $\mathbb{R}^p$, *and suppose* $u \in \mathcal{D}'(X)$, $v \in \mathcal{D}'(Y)$. *Then*

(2.4.1) $\mathrm{WF}(u \otimes v) \subset$

$$(\mathrm{WF}u \times \mathrm{WF}v) \cup (\mathrm{supp}_0 u \times \mathrm{WF}v) \cup (\mathrm{WF}u \times \mathrm{supp}_0 v)$$

where we have put for example $\mathrm{supp}_0 u = \mathrm{supp}\, u \times \{0\}$, *and identified* $(X \times Y) \times \mathbb{R}^{n+p}$ *with* $(X \times \mathbb{R}^n) \times (Y \times \mathbb{R}^p)$.

PROOF: Suppose we have $(x_0, \xi_0, y_0, \eta_0)$, not belonging to the right-hand side of (2.4.1). First we assume that $\xi_0 \neq 0$ and $(x_0, \xi_0) \notin \mathrm{WF}u$; we can then consider $\alpha \in C_0^\infty(X)$ with $\alpha(x_0) \neq 0$ and $\widehat{\alpha u}$ rapidly decreasing in a conic neighbourhood W of ξ_0 in $\mathbb{R}^n \setminus 0$; suppose $\beta \in C_0^\infty(Y)$ is such that $\beta(y_0) \neq 0$. We have $\widehat{\alpha\beta(u \otimes v)}(\xi,\eta) = \widehat{\alpha u}(\xi)\, \widehat{\beta v}(\eta)$. Since $\xi_0 \neq 0$, we can find a conic neighbourhood C of (ξ_0, η_0) in $\mathbb{R}^{n+p} \setminus 0$ such that $(\xi, \eta) \in C$ implies $\xi \in W$ and $|\xi| + |\eta| \leq c\,|\xi|$, where c is a positive constant. Thus $\widehat{\alpha u}(\xi)\ \widehat{\beta v}(\eta)$ is rapidly decreasing with respect to (ξ, η) in C, and this shows that $(x_0, y_0, \xi_0, \eta_0) \notin WF(u \otimes v)$. We argue similarly when $\eta_0 \neq 0$ with $(y_0, \eta_0) \notin WFv$. Finally, when, for example, we have $\xi_0 = 0$ and $(y_0, \eta_0) \in WFv$, then $x_0 \notin \operatorname{supp} u$ by hypothesis; hence $u \otimes v$ is zero in the neighbourhood of (x_0, y_0) so that $(x_0, y_0, \xi_0, \eta_0) \notin WF(u \otimes v)$.

REMARK 2.5: Suppose Γ_1, Γ_2 are closed cones in $X \times (\mathbb{R}^n \setminus 0)$, $Y \times (\mathbb{R}^p \setminus 0)$. Let F_1, F_2 be closed subsets of X, Y. The above proof also shows that the mapping $(u,v) \to u \otimes v$ is sequentially continuous from $\mathcal{D}'_{\Gamma_1,F_1}(X) \times \mathcal{D}'_{\Gamma_2,F_2}(Y)$ into $\mathcal{D}'_{\Gamma,F}(X \times Y)$ where $F = F_1 \times F_2$ and

$\Gamma = (\Gamma_1 \times \Gamma_2) \cup ((F_1 \times \{0\}) \times \Gamma_2) \cup (\Gamma_1 \times (F_2 \times \{0\}))$.

EXAMPLE 2.6: From Example 3.2 of Chapter I, putting $x = (x', x_n)$, $x' = (x_1, \ldots, x_{n-1})$, we have

$WF(\delta(x') \otimes Y(x_n)) \subset$

$$\{x = 0, \xi \neq 0\} \cup \{x' = 0, x_n > 0, \xi' \neq 0, \xi_n = 0\}$$

(it is shown in Exercise (5.1) that this inclusion is actually an equality).

We shall now show that, under certain assumptions, it is possible to define in a natural manner the inner product of two distributions. The argument employed is analogous to that which allows $H^{-s}(\mathbb{R}^n)$ to be identified with the dual of $H^s(\mathbb{R}^n)$. For

$E \subset X \times \mathbb{R}^n$, we put $\check{E} = \{(x, -\xi) \mid (x, \xi) \in E\}$.

THEOREM 2.7: *The mapping of* $C_0^\infty(X) \times C_0^\infty(X)$ *into* $\mathbb{C}$, *which associates* $\langle u_1, u_2 \rangle = \int u_1(x)\, u_2(x)\, dx$ *with* (u_1, u_2), *extends uniquely into a mapping (also denoted by* $\langle \cdot, \cdot \rangle$ *) of*

$$\{(u_1, u_2) \in \mathcal{D}'(X) \times \mathcal{D}'(X) \mid WF\, u_1 \cap \check{WF}\, u_2 = \emptyset,$$

$\text{supp}\, u_1 \cap \text{supp}\, u_2$ compact$\}$ *into* $\mathbb{C}$ *which, for all closed cones* Γ_1, Γ_2 *in* $X \times (\mathbb{R}^n \setminus 0)$ *such that* $\Gamma_1 \cap \check{\Gamma}_2 = \emptyset$, *for any compact subset* K *of* X, *is sequentially continuous with respect to* $(u_1, u_2) \in \mathcal{D}'_{\Gamma_1}(X) \times \mathcal{D}'_{\Gamma_2}(X)$ *when* $\text{supp}\, u_1 \cap \text{supp}\, u_2 \subset K$.

PROOF: a) *Uniqueness:* Let $u_1, u_2 \in \mathcal{D}'(X)$ be such that $WF\, u_1 \cap \check{WF}\, u_2 = \emptyset$, $\text{supp}\, u_1 \cap \text{supp}\, u_2$ compact. We can find a neighbourhood V_1 of supp u_1 in X and a neighbourhood V_2 of supp u_2 in X such that $V_1 \cap V_2$ is relatively compact in X. From Proposition 1.8, we can find two sequences $(u_{j,k})_{k \in \mathbb{N}}$ such that $u_{j,k} \longrightarrow u_j$ in $\mathcal{D}'_{\Gamma_j}(X)$ with $\text{supp}\, u_{j,k} \subset V_j$ (for $j = 1,2$, and $\Gamma_j = WF\, u_j$). We must therefore have

$$\langle u_1, u_2 \rangle = \lim_{k \to +\infty} \int u_{1k}(x) u_{2k}(x) dx .$$

b) *Existence:* Let Γ_1, Γ_2 and K be as in the statement of the theorem. For $x_0 \in K$ we put $\Gamma_{jo} = \{\xi \mid (x_0, \xi) \in \Gamma_j\}$ and we consider an open conic neighbourhood $\tilde{\Gamma}_{jo}$ of Γ_{jo} such that $\tilde{\Gamma}_{10} \cap \tilde{\Gamma}_{20} = \emptyset$, along with a neighbourhood V of x_o in X such that $V \times \gamma_j$ does not intersect Γ_j, where γ_j denotes the complement of $\tilde{\Gamma}_{jo}$ in $\mathbb{R}^n \setminus 0$. Suppose we have a finite covering of the compact set K by neighbourhoods $V = V_k$ of this type, and that we have functions $\alpha_k \in C_o^\infty(V_k)$ such that $\sum_k \alpha_k = 1$ in the neighbourhood of K, along with functions $\beta_k \in C_o^\infty(V_k)$ such that $\beta_k = 1$ in the neighbourhood of supp α_k. For $(u_1, u_2) \in \mathcal{D}'_{\Gamma_1}(X) \times \mathcal{D}'_{\Gamma_2}(X)$ with supp $u_1 \cap$ supp $u_2 \subset K$, we put

$$(2.7.1) \qquad < u_1 , u_2 > = \sum_k \int \widehat{\alpha_k u_1}(\xi)\, \widehat{\beta_k u_2}(-\xi)\, d\!\!{}^-\xi$$

observing that the integrals are absolutely convergent : in view of Corollary 1.6, $\widehat{\alpha_k u_1}$ is rapidly decreasing in γ_1, $\widehat{\beta_k u_2}$ is rapidly decreasing in γ_2, and thus $\widehat{\alpha_k u_1}(\xi)\, \widehat{\beta_k u_2}(-\xi)$ is rapidly decreasing in $\gamma_1 \cup \check{\gamma}_2 = \mathbb{R}^n \setminus 0$. Using (1.7.3), it follows immediately that the integrals depend (sequentially) continuously on $(u_1, u_2) \in \mathcal{D}'_{\Gamma_1}(X) \times \mathcal{D}'_{\Gamma_2}(X)$, and finally, that the right-hand side of (2.7.1) coincides with $\int u_1(x)\, u_2(x)\, dx$ when $u_1, u_2 \in C_o^\infty(X)$.

CONSTRUCTION 2.8: Let X, Y be open subsets of $\mathbb{R}^n$, $\mathbb{R}^p$; if $E \subset (X \times Y) \times \mathbb{R}^{n+p}$, we define $E' \subset (X \times \mathbb{R}^n) \times (Y \times \mathbb{R}^p)$ by $E' = \{(x, \xi, y, \eta) \mid (x, \xi, y, -\eta) \in E\}$. A subset E' of $(X \times \mathbb{R}^n) \times (Y \times \mathbb{R}^p)$ defines a relation of $Y \times \mathbb{R}^p$ with $X \times \mathbb{R}^n$:

if $G \subset Y \times \mathbb{R}^p$, we put $E' \circ G = \{(x,\xi) \in X \times \mathbb{R}^n \mid$ there exists $(y,\eta) \in G$ with $(x , \xi , y , \eta) \in E'\}$.

We consider a continuous linear operator $A : C_0(Y) \longrightarrow \mathcal{D}'(X)$. If $K_A \in \mathcal{D}'(X \times Y)$ denotes the kernel of A, we have by definition

(2.8.1) $$\langle Av , u \rangle = \langle K_A , u \otimes v \rangle$$

for $u \in C_0^\infty(X)$, $v \in C_0^\infty(Y)$.

We now seek to extend the definition of Av to certain distributions $v \in \mathcal{D}'(Y)$ by using the equality (2.8.1) and Theorem 2.7. For $v \in \mathcal{D}'(Y)$ and $u \in C_0^\infty(X)$, we recall that $WF(u\otimes v) \subset \operatorname{supp}_0 u \times WF\, v$ and $\operatorname{supp}(u\otimes v) = \operatorname{supp} u \times \operatorname{supp} v$. Theorem 2.7 thus allows the inner product $\langle K_A , u\otimes v \rangle$ to be defined when

$$\begin{cases} WF' K_A \cap (\operatorname{supp}_0 u \times WF\, v) = \emptyset \\ \operatorname{supp} K_A \cap (\operatorname{supp} u \times \operatorname{supp} v) \ \text{compact} . \end{cases}$$

These conditions are realised for all $u \in C_0^\infty(X)$ if and only if

(2.8.2) $$WF' K_A \circ WF\, v \subset X \times (\mathbb{R}^n \setminus 0)$$

(2.8.3) The projection $\pi_X : \operatorname{supp} K_A \cap (X \times \operatorname{supp} v) \longrightarrow X$ is proper.

Let us assume that these conditions are realised; if u_j is a sequence which tends to 0 in $C_o^\infty(X)$, Remark 2.5 shows that $u_j \otimes v$ tends to 0 in $\mathcal{D}'_C(X \times Y)$, where $C = (X \times \{0\}) \times WF\, v$, and Theorem 2.7 then shows that $< K_A , u_j \otimes v >$ tends to 0. Thus formula (2.8.1) defines Av as an element of $\mathcal{D}'(X)$. Suppose we have Γ, a closed cone in $Y \times (\mathbb{R}^p \setminus 0)$ and F, a closed subset of Y such that $WF' K_A \circ \Gamma \subset X \times (\mathbb{R}^n \setminus 0)$, and such that the projection $\text{supp}\, K_A \cap (X \times F) \longrightarrow X$ is proper. We see similarly that if $v_k \to v$ in $\mathcal{D}'_{\Gamma,F}(Y)$, then $Av_k \to Av$ in $\mathcal{D}'(X)$, and this proves the first two parts of:

THEOREM 2.9: *Suppose we have an operator* $A : C_o^\infty(Y) \longrightarrow \mathcal{D}'(X)$.

i) *We define an extension (itself denoted by* A*) of* A *into an operator from*

$\{v \in \mathcal{D}'(Y) \mid WF' K_A \circ WF\, v \subset X \times (\mathbb{R}^n \setminus 0)$,

$\text{supp}\, K_A \cap (X \times \text{supp}\, v) \xrightarrow{\pi_X} X$ *proper\}* *into* $\mathcal{D}'(X)$, *by putting* $< Av , u > = < K_A , u \otimes v >$ *for all* $u \in C_o^\infty(X)$.

ii) *Let* Γ *be a closed cone in* $Y \times (\mathbb{R}^p \setminus 0)$ *and let* F *be a closed subset of* Y *such that*

$$WF' K_A \circ \Gamma \subset X \times (\mathbb{R}^p \setminus 0)$$

$$\text{supp}\, K_A \cap (X \times F) \xrightarrow{\pi_X} X \text{ proper}.$$

Then the operator A *is sequentially continuous from* $\mathcal{D}'_{\Gamma,F}(Y)$ *into* $\mathcal{D}'(X)$.

iii) *In situation* (i), *we have*

(2.9.1) $$WF(Av) \subset (WF'K_A \circ WF\, v) \cup (WF'K_A \circ \mathrm{supp}_0 v) ,$$

where the right-hand side is a conic closed subset of $X \times (\mathbb{R}^n \setminus 0)$.

We shall actually establish a more general result than (iii):

THEOREM 2.10: *Let* Σ *be a closed cone in* $(X \times Y) \times (\mathbb{R}^{n+p} \setminus 0)$, Γ *a closed cone in* $Y \times (\mathbb{R}^p \setminus 0)$, S *a closed subset of* $X \times Y$, *and* F *a closed subset of* Y. *Suppose that*

(2.10.1) $$\Sigma' \circ \Gamma \subset X \times (\mathbb{R}^n \setminus 0) ,$$

(2.10.2) *The projection* $S \cap (X \times F) \xrightarrow{\pi_X} X$ *proper*,

$$\mathrm{Proj}_{X \times Y} \Sigma \subset S , \quad \mathrm{Proj}_Y \Gamma \subset F .$$

We put $\Lambda = (\Sigma' \circ \Gamma) \cup (\Sigma' \circ (F \times \{0\}))$. *Then* Λ *is a closed cone in* $X \times (\mathbb{R}^n \setminus 0)$, *and the mapping* $(K_A , v) \longmapsto Av$ *is sequentially continuous from* $\mathcal{D}'_{\Sigma,S}(X \times Y) \times \mathcal{D}'_{\Gamma,F}(Y)$ *into* $\mathcal{D}'_{\Lambda}(X)$.

PROOF: First, we prove that Λ is closed in $X \times (\mathbb{R}^n \setminus 0)$; suppose $(x_j , \xi_j) \in \Lambda$ is a sequence tending to (x,ξ) in $X \times (\mathbb{R}^n \setminus 0)$; we consider a sequence $(y_j , \eta_j) \in \Gamma$ such that $(x_j , \xi_j , y_j , \eta_j) \in \Sigma'$; we have $(x_j , y_j) \in S$, $y_j \in F$ and, since $x_j \to x$ in X, (2.10.2) shows that y_j remains in a compact subset of Y; by extracting a subsequence, we can therefore assume that $y_j \to y$ in Y. If the $|\eta_j|$ are not bounded, we can assume, after extracting a different subsequence, that $|\eta_j| \to \infty$, and therefore that $(x_j, \xi_j / |\eta_j|, y_j, \eta_j / |\eta_j|) \to (x, 0, y, \eta)$ with $|\eta| = 1$; we then obtain $(y, \eta) \in \Gamma$ and $(x, 0, y, \eta) \in \Sigma'$ which is impossible in view of (2.10.1). The $|\eta_j|$ are therefore bounded, and furthermore, after extracting a subsequence, we can suppose that $\eta_j \to \eta$ in $\mathbb{R}^p$, $(x_j, \xi, y_j, \eta_j) \longrightarrow (x, \xi, y, \eta) \in \Sigma'$. If $\eta \neq 0$ we have $(y, \eta) \in \Gamma$, and therefore $(x, \xi) \in \Sigma' \circ \Gamma \subset \Lambda$. If $\eta = 0$ we have $(y, \eta) \in F \times \{0\}$, and therefore $(x, \xi) \in \Sigma' \circ (F \times \{0\}) \subset \Lambda$. We shall now prove that if $K_A \in \mathcal{D}'_{\Sigma, S}(X \times Y)$ and $v \in \mathcal{D}'_{\Gamma, F}(Y)$, then $Av \in \mathcal{D}'_{\Lambda}(X)$. Let $(x_0, \xi'_0) \in X \times (\mathbb{R}^n \setminus 0)$, not belonging to Λ; we shall now show that $(x_0, \xi'_0) \notin WF(Av)$. Let K_1 be a compact neighbourhood of x_0 in X, and $K_2 = \mathrm{Proj}_Y \pi_X^{-1} K_1$; in view of (2.10.2), K_2 is a compact subset of F; for $y_0 \in K_2$, we denote by Σ_0 the fibre of Σ at (x_0, y_0) and by Γ_0 the fibre of Γ at y_0; by using an argument analogous to that employed to prove that Λ is closed, it is easy to show that there exists a conic open neighbourhood $\tilde{\Gamma}_0$ of Γ_0 in $\mathbb{R}^p \setminus 0$, a conic open neighbourhood $\tilde{\Sigma}_0$ of Σ_0 in $\mathbb{R}^{n+p} \setminus 0$, a conic neighbourhood W of ξ'_0 in $\mathbb{R}^n \setminus 0$, a number $\varepsilon > 0$, a neighbourhood V_1 of x_0 in X and a neighbourhood V_2 of y_0 in Y, such that

(2.10.3) $$W \cap (\tilde{\Sigma}'_0 \circ \tilde{\Gamma}_0) = \emptyset$$

(2.10.4) $$V_1 \times V_2 \times \{(\xi, \eta) \mid \xi \in W, \ |\eta| \leq \varepsilon|\xi|\}$$

does not intersect Σ,

and also, in view of (2.10.1), such that

(2.10.5) $$\tilde{\Sigma}_o \cap \{(\xi,\eta) \mid -\eta \in \tilde{\Gamma}_o, \quad |\xi| \leq \epsilon|\eta|\} = \emptyset .$$

By shrinking V_1 and V_2, we can also impose

$$V_1 \subset K_1$$

(2.10.6) $$(V_2 \times \gamma) \cap \Gamma = \emptyset$$

(2.10.7) $$(V_1 \times V_2 \times \sigma) \cap \Sigma = \emptyset$$

where γ denotes the complement of $\tilde{\Gamma}_o$ in $\mathbb{R}^p \setminus 0$ and σ the complement of $\tilde{\Sigma}_o$ in $\mathbb{R}^{n+p} \setminus 0$.

By partition of unity in the neighbourhood of the compact set K_2, we see that when $\alpha \in C_o^\infty(X)$ has support sufficiently near x_o, $< Av , \alpha(x) e^{-ix.\xi'} >$ is a finite sum of terms of the form

(2.10.8) $$\int < f K_A , \alpha(x)e^{-ix.\xi'} \beta(y)v >$$

where $\beta \in C_o^\infty(V_2)$, $f \in C_o^\infty(V_1 \times V_2)$, V_1 and V_2 being of the above type. In accordance with the proof of Theorem 2.7, the inner product (2.10.8) is equal to

(2.10.9) $$\int \widehat{fK_A}(\xi, \eta) \, [\hat{\alpha}(\xi' - \xi) \, \widehat{\beta v}(-\eta)] \, d\!\!{}^-\xi \, d\!\!{}^-\eta \quad .$$

We note that, in view of (2.10.7), $\widehat{fK_A}$ is rapidly decreasing outside $\tilde{\Sigma}_o$, and that, for ξ' fixed, $\hat{\alpha}(\xi' - \xi) \, \widehat{\beta v}(-\eta)$ is rapidly decreasing with respect to (ξ,η) outside $\{(\xi, \eta) \mid -\eta \in \tilde{\Gamma}_o , |\xi| \leq \epsilon|\eta|\}$ in view of (2.10.6), and therefore that the integral is absolutely convergent in view of (2.10.5). Let W' be a conic open neighbourhood of ξ'_o in $\mathbb{R}^n \setminus 0$ such that $\overline{W}' \subset W$; we shall assume henceforth that $\xi' \in W'$. Since we have, for all N, an upper bound of the form $|\hat{\alpha}(\xi' - \xi)| \leq C_N(1 + |\xi'| + |\xi|)^{-N}$ when $\xi \notin W$, it is easy to show that

$$\int_{\xi \notin W} \widehat{fK_A}(\xi, \eta)\ \hat{\alpha}(\xi' - \xi)\ \widehat{\beta v}(-\eta) d\eta$$

is rapidly decreasing in ξ'. We shall therefore replace (2.10.9) by the integral

$$(2.10.10) \qquad \int_{\xi \in W} [\widehat{fK_A}(\xi, \eta)\ \widehat{\beta v}(-\eta)]\ \hat{\alpha}(\xi' - \xi)\ d\xi\, d\eta\ .$$

(2.10.11) <u>*Lemma*</u>: $\widehat{fK_A}(\xi, \eta)\ \widehat{\beta v}(-\eta)$ *is rapidly decreasing in* (ξ, η) *for* $\xi \in W$, $\eta \in \mathbb{R}^n$.

This lemma implies that (2.10.10) is rapidly decreasing in ξ': in order to see this, we argue as in the proof of Proposition 1.4, by performing separate integrations for $|\xi| \leq |\xi'|/2$ and for $|\xi| \geq |\xi'|/2$. In order to prove Lemma (2.10.11), we first note that $\widehat{fK_A}(\xi,\eta)$ is rapidly decreasing in (ξ,η) for $\xi \in W$ and $|\eta| \leq \epsilon|\xi|$, in view of (2.10.4); when $|\eta| \geq \epsilon|\xi|$ and $-\eta \notin \tilde{\Gamma}_o$, $\widehat{\beta v}(-\eta)$ is rapidly decreasing in η (and therefore in (ξ,η) in view of (2.10.6)); when $\xi \in W$ and $-\eta \in \tilde{\Gamma}_o$, then $(\xi, \eta) \notin \tilde{\Sigma}_o$ from (2.10.3), and thus $\widehat{fK_A}(\xi,\eta)$ is rapidly decreasing in (ξ,η) in view of (2.10.7).

We have thus finally shown that $Av \in \mathcal{D}'_{\Lambda}(X)$; the sequential continuity with respect to (K_A , v) is proved in exactly the same way.

REMARK 2.11: We now consider the situation of Theorem 2.9. If $u \in C_o^{\infty}(X)$, we can construct a neighbourhood $\tilde{F}$ of F in Y such that $\operatorname{supp} K_A \cap (\operatorname{supp} u \times \tilde{F})$ is compact. For $v \in \mathcal{D}'_{\Gamma,F}(Y)$, we consider a sequence $v_k \in C_o^{\infty}(Y)$ such that $v_k \to v$ in $\mathcal{D}'_{\Gamma}(Y)$ with $\operatorname{supp} v_k \subset \tilde{F}$ (see Proposition 1.8). Then

$$< Av , u > = \lim_{k \to +\infty} < Av_k , u > .$$

REMARK 2.12: For many operators A, the set $WF_X K_A = WF' K_A \circ (Y \times \{0\})$ is empty; then formula (2.9.1) simplifies to

$$(2.12.1) \qquad WF(Av) \subset WF' K_A \circ WFv .$$

As a particular case of the two preceding theorems, we now study the inverse image of a distribution by a C^{∞} mapping; let $f : X \to Y$ be a C^{∞} mapping and let $f^* : C_o^{\infty}(Y) \longrightarrow C^{\infty}(X)$ be the operator defined by $f^* v = v \circ f$. From Example 8.5 in Chapter III and from Theorem 2.1, we have $WF' K_{f^*} \subset$ $\{(x , {}^t f'(x)\eta , f(x), \eta) \mid x \in X , \eta \in \mathbb{R}^p \setminus 0 \}$.

Since $\operatorname{supp} K_{f^*} = \{(x , f(x)) \mid x \in X\}$, the projection $\operatorname{supp} K_{f^*} \longrightarrow X$ is proper, and Theorems 2.9 and 2.10 therefore imply:

THEOREM 2.13 (Inverse Image) *Suppose* $f : X \to Y$ *is a* C^∞ *mapping and that* $f^* : C_0^\infty(Y) \longrightarrow C^\infty(X)$ *is the inverse image operator defined by* $f^* v = v \circ f$. *We put*

$$N_f = \{(f(x), \eta) \mid x \in X, \quad \eta \in \mathbb{R}^p \setminus 0, \quad {}^t f'(x)\eta = 0\} .$$

Then f^* *extends uniquely into an operator (itself denoted by* f^**) from* $\{v \in \mathcal{D}'(Y) \mid WF\, v \cap N_f = \emptyset\}$ *into* $\mathcal{D}'(X)$ *such that, for any closed cone* Γ *in* $Y \times (\mathbb{R}^p \setminus 0)$ *which does not intersect* N_f, f^* *is sequentially continuous from* $\mathcal{D}'_\Gamma(Y)$ *into* $\mathcal{D}'(X)$. *If* Γ *is as above, then*

$\Lambda = \{(x, {}^t f'(x)\eta) \mid x \in X, \quad (f(x), \eta) \in \Gamma\}$ *is a closed cone in* $X \times (\mathbb{R}^n \setminus 0)$, *and* f^* *is sequentially continuous from* $\mathcal{D}'_\Gamma(Y)$ *into* $\mathcal{D}'_\Lambda(X)$. In particular, we have

$$\text{(2.13.1)} \quad WF(f^* v) \subset \{(x, {}^t f'(x)\eta) \mid x \in X, \quad (f(x), \eta) \in WF\, v\}$$

REMARK 2.14: The assumption $WF\, v \cap N_f = \emptyset$ is satisfied for all $v \in \mathcal{D}'(Y)$ when f is a submersion, since we then have $N_f = \emptyset$. When f is a diffeomorphism, the sequential continuity of the operator $f^* : \mathcal{D}'(Y) \longrightarrow \mathcal{D}'(X)$ defined in Theorem 2.13 shows that this operator coincides with the operator defined in Section 7.1, Chapter I.

It can be seen similarly that if χ is a diffeomorphism of an open set Z in $\mathbb{R}^n$ onto X, then, under the assumptions of Theorem 2.13, we have

$$(2.14.1) \qquad \chi^*(f^*v) = (f \circ \chi)^*v \quad .$$

EXAMPLE 2.15 (Convolution Kernel). Let $f : R^n \times R^n \longrightarrow R^n$ be the submersion defined by $f(x,y) = x-y$. For $v \in \mathcal{D}'(R^n)$, we consider $u = f^*v \in \mathcal{D}'(R^n \times R^n)$, and by an abuse of notation we write $u(x,y) = v(x-y)$. u is the kernel of the operator $\varphi \longmapsto v * \varphi$ of convolution with v, and we have

$$WF'u \subset \{(x , \xi , y , \xi) \mid x , y \in R^n , (x-y , \xi) \in WF\, v\} .$$

(Exercise 5.2 shows that this inclusion is actually an equality). In the case where $v = \delta(x') \otimes Y(x_n)$, Example 2.6 therefore shows that

$$WF'u \subset \{(x , \xi , x , \xi) \mid \xi \neq 0\} \quad \cup$$
$$\{(x' , x_n , \xi' , 0 ; x' , y_n , \xi' , 0) \mid x_n > y_n , \xi' \neq 0\} .$$

EXAMPLE 2.16: Let $f \in C^\infty(X , R)$ be such that $f'(x) \neq 0$. If δ is the Dirac measure at 0 in $\mathbb{R}$, then Theorem 2.13 allows us to define $f^*\delta = \delta(f) \in \mathcal{D}'(X)$, and we have

$$(2.16.1) \qquad WF(\delta(f)) \subset \{(x , \eta f'(x)) \mid f(x) = 0 \quad , \quad \eta \in R\backslash 0\} .$$

Approximating δ by a sequence of C^∞ functions, and using the sequential continuity of the operator f^*, it can be shown that $\delta(f)$ can be defined in terms of oscillatory integrals as follows:

$$< \delta(f) , \varphi > = \int e^{if(x).\eta} \varphi(x)\, dx\, đ\eta \qquad \text{for } \varphi \in C_0^\infty(X) .$$

This expression shows that in the case $f(x) = x_n$, $\delta(f)$ coincides with the distribution $\delta(x_n = 0)$ defined in Section 7.2, Chapter I, and consequently (2.14.1) implies that $\delta(f)$ coincides in the general case with the distribution defined in Section 7.2, Chapter I.

3. THE CASE OF MANIFOLDS

LEMMA 3.1: *Let* X, Y *be open sets in* $\mathbf{R}^n$ *and let* $f : X \to Y$ *be a diffeomorphism. For* $v \in \mathcal{D}'(Y)$, *we have*

$$(3.1.1) \quad WF(f^*v) = \{(x, {}^t f'(x)\eta) \mid x \in X, (f(x), \eta) \in WF v\}.$$

PROOF: The inclusion of the left-hand side of (3.1.1) in the right-hand side follows from Theorem 2.13, and the equality (3.1.1) is then deduced by replacing f by f^{-1}. This lemma shows that, in the following definition, the choice of the local chart is immaterial:

DEFINITION 3.2: Suppose X is a C^∞ manifold and suppose $u \in \mathcal{D}'(X)$. We define $WF\,u \subset T^*X \setminus 0$ by the condition: $(x, \xi) \in WF\,u$ if and only if there exists a local chart $\chi : X \supset V \longrightarrow \tilde{V} \subset \mathbf{R}^n$ of X to the neighbourhood of x such that $\tilde{\chi}(x, \xi) \in WF[\chi^{-1*}(u_{|V})]$, where $\tilde{\chi}$ is the chart of T^*X induced by χ.

Exercise 5.4 gives an intrinsic characterisation of the singular spectrum.

It is an immediate result that WFu is a conic closed set in $T^* X \setminus 0$, and that its projection onto X is sing supp u. If Γ is a closed cone in $T^* X \setminus 0$, we again put
$\mathcal{D}'_\Gamma(X) = \{u \in \mathcal{D}'(X) \mid WFu \subset \Gamma\}$. We say that a sequence $u_j \in \mathcal{D}'_\Gamma(X)$ converges in $\mathcal{D}'_\Gamma(X)$ towards $u \in \mathcal{D}'_\Gamma(X)$ if, for any chart χ of X (denoted as above), $\tilde{u}_j$ converges towards $\tilde{u}$ in $\mathcal{D}'_{\tilde{\Gamma}}(\tilde{V})$, where $\tilde{u}_j$, $\tilde{u}$, $\tilde{\Gamma}$ are obtained by transport of $u_j|_V$, $u|_V$, $\Gamma|_V$.

The generalisation of Theorem 2.4 (tensor product) to the case in which X and Y are manifolds is immediate. By partition of unity and local charts, we can also generalise Proposition 1.8 (relating to the density of $C_0^\infty(X)$) and Theorem 2.7 (relating to the coupling of two distributions) to the case in which X is a manifold. For the generalisation of Theorem 2.7, it is necessary to suppose that X is endowed with a positive density μ, and to replace $\int u_1(x)\, u_2(x)dx$ by $\int_X u_1(x)\, u_2(x)\mu$, and $X \times (R^n \setminus 0)$ by $T^*X \setminus 0$.

Let X, Y be two manifolds endowed with positive densities μ, ν. If $A : C_0^\infty(Y) \longrightarrow \mathcal{D}'(X)$ is a continuous linear operator, we recall (see Section 8, Chapter I) that the kernel $K_A \in \mathcal{D}'(X \times Y)$ of A is characterised, for $\varphi \in C_0^\infty(X)$, $\psi \in C_0^\infty(Y)$, by $< K_A , (\varphi \otimes \psi)(\mu \otimes \nu) > = < A\psi , \varphi\mu >$.

In view of (8.2.1), Chapter I, the singular spectrum of K_A is independent of ν. Construction 2.8 and Theorems 2.9 and 2.10 relating to the extension of A generalise word for word if we replace $(X \times Y) \times R^{n+p}$ by $T^*(X \times Y)$, $X \times R^n$ by T^*X and $Y \times R^p$ by T^*Y. [To prove part (iii) of Theorem 2.9 and Theorem 2.10, consider partitions of unity $\Sigma\, \alpha_i = 1$ in X, $\Sigma\, \beta_j = 1$ in Y by C^∞ functions with compact support which are locally finite in open coordinate patches. From the analogue of assumption (2.8.3), we have $Av = \sum_j A(\beta_j v)$, the sum being locally finite; thus $Av = \sum_{i,j} \alpha_i A\, \beta_j v$, the sum being once again locally finite. It is therefore sufficient to study the singular spectrum of a term $\alpha_i A\, \beta_j v$. Now in this case we can use local charts to reduce to the known situation $X = R^n$, $Y = R^p$]. If $f : X \to Y$ is a C^∞ mapping, we then deduce the generalisation of Theorem 2.13 relating to the inverse image operator f^*, by putting $N_f = \{(y, \eta) \in T^*Y \setminus 0 \mid$ there exists $x \in X$ with $y = f(x)$, ${}^t f'(x)\eta = 0\}$. We shall now describe the particular case where X is a sub-manifold of Y and $f : X \to Y$ is the canonical injection.

THEOREM 3.3 (Restriction to a sub-manifold): *Suppose* Y *is a manifold and* X *is a sub-manifold of* Y. *Let* $N_X = \{(y, \eta) \in T^*Y \setminus 0 \mid y \in X , \eta_{|T_y X} = 0\}$ *be the bundle conormal to* X, *deprived of its null section. Then the operator* γ

of restriction to X *(defined by* $\gamma v = v_{|X}$ *for* $v \in C_0^\infty(Y)$*) extends uniquely into an operator (also denoted by* γ*) from* $\{v \in \mathcal{D}'(Y) \mid WFv \cap N_X = \emptyset\}$ *into* $\mathcal{D}'(X)$ *such that, for any closed cone* Γ *of* $T^*Y \setminus 0$ *not intersecting* N_X*,* γ *is sequentially continuous from* $\mathcal{D}'_\Gamma(Y)$ *into* $\mathcal{D}'(X)$*. If* Γ *is as above, the projection* Λ *of* $\Gamma_{|X}$ *into* T^*X *is a closed cone in* $T^*X \setminus 0$*, and* γ *is sequentially continuous from* $\mathcal{D}'_\Gamma(Y)$ *into* $\mathcal{D}'_\Lambda(X)$*. In particular, we have*

$$(3.3.1) \quad WF(\gamma v) \subset \{(y, \eta_{|T_yX}) \mid y \in X \ , \ (y, \eta) \in WFv\} .$$

The inverse image theorem now allows us, in certain cases, to define the product of two distributions. For $\Gamma_1 , \Gamma_2 \subset T^*X$, we put

$$\Gamma_1 + \Gamma_2 = \{(x , \xi_1 + \xi_2) \mid (x , \xi_1) \in \Gamma_1 , (x , \xi_2) \in \Gamma_2\} .$$

THEOREM 3.4 (Multiplication): *Let* X *be a manifold. The multiplication operator*

$$C_0^\infty(X) \times C_0^\infty(X) \ni (u_1 , u_2) \longmapsto u_1u_2 \in C_0^\infty(X)$$

extends uniquely into an operator from

$\{(u_1 , u_2) \in \mathcal{D}'(X) \times \mathcal{D}'(X) \mid WF u_1 + WF u_2 \subset T^*X \setminus 0\}$ *into* $\mathcal{D}'(X)$ *which, for all closed cones* Γ_1*,* Γ_2 *in* $T^*X \setminus 0$ *satisfying* $\Gamma_1 + \Gamma_2 \subset T^*X \setminus 0$ *, is sequentially continuous from* $\mathcal{D}'_{\Gamma_1}(X) \times \mathcal{D}'_{\Gamma_2}(X)$ *into* $\mathcal{D}'(X)$ *.*

If Γ_1*,* Γ_2 *are as above, then* $\Lambda = (\Gamma_1 + \Gamma_2) \cup \Gamma_1 \cup \Gamma_2$ *is a closed* *cone in* $T^*X \setminus 0$, *and the multiplication is sequentially continuous from* $\mathcal{D}'_{\Gamma_1}(X) \times \mathcal{D}'_{\Gamma_2}(X)$ *into* $\mathcal{D}'_{\Lambda}(X)$. *In particular, we have*

$$(3.4.1) \quad WF(u_1u_2) \subset (WF\,u_1 + WF\,u_2) \cup WF\,u_1 \cup WF\,u_2 .$$

PROOF: The uniqueness follows from the density property of $C_o^\infty(X)$. Let $f : X \longrightarrow X \times X$ be the mapping defined by $f(x) = (x,x)$; we have
$N_f = \{(x,x,\xi_1,\xi_2) \mid (\xi_1,\xi_2) \neq 0 \ , \ \xi_1 + \xi_2 = 0\}$. For $u_1 , u_2 \in C_o^\infty(X)$, we clearly have $u_1u_2 = f^*(u_1 \otimes u_2)$. For $u_j \in \mathcal{D}'_{\Gamma_j}(X)$ $(j = 1,2)$, we have $u_1 \otimes u_2 \in \mathcal{D}'_{\Gamma}(X)$ with
$\Gamma = (\Gamma_1 \times \Gamma_2) \times (X \times \{0\} \times \Gamma_2) \times (\Gamma_1 \times X \times \{0\})$ from Remark 2.5, and the condition $N_f \cap \Gamma = \emptyset$ signifies that
$\Gamma_1 + \Gamma_2 \subset T^*X \setminus 0$; if such is the case, we put
$u_1u_2 = f^*(u_1 \otimes u_2)$, and this concludes the proof in view of Remarks 2.5 and 2.12.

Suppose X,Y are two manifolds endowed with positive densities μ, ν. Let $f : X \to Y$ be a C^∞ mapping. By transposition of the inverse image operator $f^* : C_o^\infty(Y) \longrightarrow C^\infty(X) \subset \mathcal{D}'(X)$, we obtain the direct image operator (which depends on the densities μ, ν) $f_* : C_o^\infty(X) \longrightarrow \mathcal{D}'(X)$ defined by

$$(3.4.1) \qquad \langle f_* u \ , \ v\nu \rangle = \int_X u(x)\, v(f(x))\, \mu$$

for $u \in C_o^\infty(X)$, $v \in C_o^\infty(Y)$.

We have $\operatorname{supp} K_{f_*} = \{(f(x), x) \mid x \in X\}$

$WF' K_{f_*} =$

$$\{(y, \eta ; x, \xi) \mid (y,\eta) \in T^*Y \setminus 0 , \ y = f(x) , \ \xi = {}^t f'(x)\eta\} .$$

Theorems 2.9 and 2.10 (generalised to manifolds) thus imply:

THEOREM 3.5 (Direct image): *Let* X, Y *be two* C^∞ *manifolds endowed with positive densities* μ, ν ; *let* $f : X \to Y$ *be a* C^∞ *mapping and* $f_* : C_0^\infty(X) \longrightarrow \mathcal{D}'(X)$ *the direct image operator defined by* (3.4.1). *Then* f_* *extends (in accordance with Theorem* 2.9) *into an operator (also denoted by* f_* *) from* $\{u \in \mathcal{D}'(X) \mid f_{|\operatorname{supp} u} : \operatorname{supp} u \to Y$ *proper*$\}$ *into* $\mathcal{D}'(Y)$. *For any closed cone* Γ *in* $T^*X \setminus 0$ *and for any closed subset* F *of* X *such that* $f_{|F} : F \to Y$ *is proper and* $\operatorname{Proj}_X \Gamma \subset F$, *then* $\Lambda = \{(y, \eta) \in T^*Y \setminus 0 \mid \exists x \in X$ *with* $y = f(x)$, $(x, {}^t f'(x)\eta) \in \Gamma \cup (F \times \{0\})\}$ *is a closed cone in* $T^*Y \setminus 0$, *and* f_* *is sequentially continuous from* $\mathcal{D}'_{\Gamma,F}(X)$ *into* $\mathcal{D}'_\Lambda(Y)$. *In particular, we have*

$$(3.5.1) \quad WF(f_*u) \subset \{(y, \eta) \in T^*Y \setminus 0 \mid \exists x \in X \text{ with } y = f(x), \ (x, {}^t f'(x)\eta) \in WFu \cup \operatorname{supp}_0 u\} .$$

REMARK 3.6: In the case where f is a diffeomorphism, it can be shown that $f_* = (f^{-1})^*$ when the measures μ and ν correspond through f . When $X = Y \times Z$, $\mu = \nu \otimes \theta$, where ν, θ are positive densities on the manifolds Y, Z, and when

$p : Y \times Z \to Y$ is the projection, we obtain the following special case of Theorem 3.3:

COROLLARY 3.7 (Integration along fibres): *Suppose* Y, Z *are two manifolds endowed with positive densities* ν, θ. *Let* $p_* : C_0^\infty(Y \times Z) \longrightarrow C_0^\infty(Y)$ *be the operator defined by*

$$(p_* u)(y) = \int_Z u(y,z)\nu \qquad (u \in C_0^\infty(Y \times Z)) \quad .$$

Then p_* *extends (in accordance with Theorem 2.9) into an operator (also denoted by* p_**) from*

$\{u \in \mathcal{D}'(Y \times Z) \mid p_{|\mathrm{supp}\, u} : \mathrm{supp}\, u \longrightarrow Y$ *proper*$\}$ *into* $\mathcal{D}'(Y)$.

For any closed cone Γ *in* $T^*(Y \times Z) \setminus 0$ *and for any closed subset* F *of* $Y \times Z$ *such that* $p_{|F} : F \longrightarrow Y$ *is proper, and* $\mathrm{Proj}_{Y \times Z}\Gamma \subset \Gamma$, *then* $\Lambda = \{(y, \eta) \in T^*Y \setminus 0 \mid \exists z \in Z$ *with* $(y, z, \eta, 0) \in \Gamma\}$ *is a closed cone in* $T^*Y \setminus 0$ *and* p_* *is sequentially continuous from* $\mathcal{D}'_{\Gamma,F}(Y \times Z)$ *into* $\mathcal{D}'_\Lambda(Y)$. *In particular, we have*

$$(3.7.1) \quad WF(p_* u) \subset \{(y,\eta) \in T^*Y \setminus 0 \mid \exists z \in Z \text{ with } (y, z, \eta, 0) \in WF u\} \; .$$

We shall now investigate the composition of operators. Suppose X_2, X, X_1 are manifolds endowed with the positive densities μ_1, μ, μ_2. Suppose we have operators

$A_2 : C_0^\infty(X_2) \longrightarrow \mathcal{D}'(X)$, $A_1 : C_0^\infty(X) \longrightarrow \mathcal{D}'(X_1)$. We seek the conditions under which $A_1 \circ A_2$ is meaningful as an operator from $C_o^\infty(X_2)$ into $\mathcal{D}'(X_1)$. For $u_2 \in C_o^\infty(X_2)$, we have $u = A_2 u_2 \in \mathcal{D}'(X)$, with

$$WF\,u \subset WF'\,K_{A_2} \circ \operatorname{supp}_o u_2 \qquad \text{(cf. (2.9.1))}$$

and

$$\operatorname{supp} u \subset \operatorname{supp} K_{A_2} \circ \operatorname{supp} u_2 \qquad \text{(cf. (5.1.1),}$$

Chapter I).

Theorem 2.9 therefore allows $A_1 u$ to be defined when

$$\begin{cases} WF'\,K_{A_1} \circ (WF'\,K_{A_2} \circ \operatorname{supp}_o u_2) \subset T^*X_1 \setminus 0 . \\ \text{The projection} \\ \operatorname{supp} K_{A_1} \cap (X_1 \times (\operatorname{supp} K_{A_2} \circ \operatorname{supp} u_2)) \longrightarrow X_1 \text{ is proper.} \end{cases}$$

These conditions are satisfied for all $u_2 \in C_o^\infty(X_2)$ if and only if

$$\begin{cases} WF'_X\,K_{A_1} \cap WF_X\,K_{A_2} = \emptyset . \\ \text{The projection} \\ (\operatorname{supp} K_{A_1} \times \operatorname{supp} K_{A_2}) \cap (X_1 \times \operatorname{Diag}(X \times X) \times X_2) \rightarrow X_1 \times X_2 \\ \text{is proper.} \end{cases}$$

where we have put

$$(3.7.2)\quad \begin{cases} WF'_X K_{A_1} = \{(x,\xi) \in T^*X\setminus 0 \mid \text{ there exists } x_1 \in X_1 \\ \qquad \text{with } (x_1, 0, x, \xi) \in WF'K_{A_1}\} \\ WF_X K_{A_2} = \{(x,\xi) \in T^*X\setminus 0 \mid \text{ there exists } x_2 \in X_2 \\ \qquad \text{with } (x, \xi, x_2, 0) \in WF\, K_{A_2} \quad . \end{cases}$$

Under the above conditions, $A_1(A_2 u_2) \in \mathcal{D}'(X_1)$ is well defined and Theorem 2.10 allows us to see that if (u_{2k}) is a sequence which tends to 0 in $C_o^\infty(X_2)$, then $A_1(A_2u_2) \longrightarrow 0$ in $\mathcal{D}'(X_1)$; we therefore have

THEOREM 3.8 (Composition): *Suppose X_1, X, X_2 are manifolds, and $A_1 : C_o^\infty(X) \longrightarrow \mathcal{D}'(X_1)$, $A_2 : C_o^\infty(X_2) \longrightarrow \mathcal{D}'(X)$ are operators such that*

$$WF'_X K_{A_1} \cap WF_X K_{A_2} = \emptyset \qquad (\text{cf } 3.7.2)\ .$$

We assume that the projection

$$(\text{supp } K_{A_1} \times \text{supp } K_{A_2}) \cap (X_1 \times \text{Diag}(X \times X) \times X_2) \longrightarrow X_1 \times X_2$$

is proper.

Then $A_1 \circ A_2$ is well defined (in accordance with Theorem 2.9), as an operator from $C_o^\infty(X_2)$ into $\mathcal{D}'(X_1)$.

REMARK 3.9: Theorem 2.10 also allows the dependence of $A_1 \circ A_2$ on A_1, A_2 to be investigated: suppose that Σ_1 is a closed cone in $T^*(X_1 \times X)\setminus 0$, Σ_2 is a closed cone in

$T^*(X \times X_2) \setminus 0$, F_1 is a closed subset of $X_1 \times X$ and that F_2 is a closed subset of $X \times X_2$; we assume that

$\mathrm{Proj}_{X_1 \times X}\, \Sigma_1 \subset F_1$, $\mathrm{Proj}_{X \times X_2}\, \Sigma_2 \subset F_2$ and that the projection $(F_1 \times F_2) \cap (X_1 \times \mathrm{Diag}\,(X \times X) \times X_2) \longrightarrow X_1 \times X_2$ is proper, and that $\Sigma'_{1X} \cap \Sigma_{2X} = \emptyset$ (using notation analogous to that in (3.7.2)). Then for $u_2 \in C_o^\infty(X_2)$, the mapping $(K_{A_1}, K_{A_2}) \longmapsto A_1 \circ A_2 u_2$ is sequentially continuous from $\mathcal{D}'_{\Sigma_1,F_1}(X_1 \times X) \times \mathcal{D}'_{\Sigma_2,F_2}(X \times X_2)$ into $\mathcal{D}'(X_1)$.

We conclude this section by investigating $WF(K_{A_1 \circ A_2})$

THEOREM 3.10: *We adopt the conditions of Theorem 3.8 relating to operator composition; we put*

$$A = A_1 \circ A_2 \quad , \quad K_{A_j} = K_j \ (j = 1,2) \quad , \quad K_A = K \ .$$

Let $f : X_1 \times X \times X_2 \longrightarrow X_1 \times X \times X \times X_2$ *be the mapping defined by* $f(x_1 , x , x_2) = (x_1 , x , x , x_2)$, *and let* $p : X_1 \times X \times X_2 \longrightarrow X_1 \times X_2$ *be the projection. The manifolds* $X_1, X, X_2, X_1 \times X_2, X_1 \times X \times X_2$ *are endowed respectively with the positive densities* $\mu_1 , \mu , \mu_2 , \mu_1 \otimes \mu_2 , \mu_1 \otimes \mu \otimes \mu_2$. *We then have*

$$(3.10.1) \qquad K = p_* f^*(K_1 \otimes K_2)$$

i.e. formally

$$K(x_1\,,\,x_2) \;=\; \int_X K_1(x_1\,,\,x)\,K_2(x\,,\,x_2)\mu \quad .$$

Consequently, we obtain

$$WF'K \subset$$

$$(WF'K_1 \circ WF'K_2) \;\cup\; (WF'K_1 \circ \mathrm{supp}_0\, K_2) \;\cup\; (\mathrm{supp}_0\, K_1 \circ WF'K_2)\;.$$

PROOF: First, we note that if K_1 and K_2 are in C_0^∞, then A is defined by $(Au_2)(x_2) = \int_{X\times X_2} K_1(x_1, x)K_2(x, x_2)u_2(x_2)\mu \otimes \mu_2$ for $u_2 \in C_0^\infty(X_2)$, $x_2 \in X_2$. Thus

$$K(x_1\,,\,x_2) \;=\; \int_X K_1(x_1\,,\,x)\,K_2(x\,,\,x_2)\mu \quad ,$$

and this establishes (3.10.1) in this particular case. We now proceed to the general case; let $u_j \in C_0^\infty(X_j)$ $(j = 1\,,\,2)$.

We have

$$(3.10.2) \quad < K\,,\,u_1\mu_1 \otimes u_2\mu_2 > \;=\; < A_1\Lambda_2u_2\,,\,u_1\mu_1 >$$

$$= \;< K_1, u_1\mu_1 \otimes A_2u_2\mu_2 > \;.$$

It is easy to show that neighbourhoods V_1, V_2 of supp K_1, supp K_2 can be constructed such that

(3.10.3) $\quad V_1 \cap (\mathrm{supp}\; u_1 \times (V_2 \circ \mathrm{supp}\; u_2))$ is compact (see Remark 2.11).

In view of Proposition 1.8, there exist sequences

$K_{1k} \in C_0^\infty(X_1 \times X)$, $\quad K_{2k} \in C_0^\infty(X \times X_2)$ such that

$\operatorname{supp} K_{jk} \subset V_j$, $K_{jk} \longrightarrow K_j$ in $\mathcal{D}'_{\Gamma_j}$ where j = 1, 2 and $\Gamma_j = WF\ K_j$; let A_{jk} be the operator with kernel K_{jk}, and let K_k be the kernel of $A_{1k} \circ A_{2k}$. From (3.10.2), (3.10.3) and Theorem 2.7, we have

$$< K , u_1\mu_1 \otimes u_2\mu_2 > = \lim_{k \to \infty} < K_k , u_1\mu_1 \otimes u_2\mu_2 > .$$

Since $K_k = p_* f^*(K_{1k} \otimes K_{2k})$, (3.10.3) and Remark 2.11 imply that

$$\lim_{k \to \infty} < p_* f^*(K_{1k} \otimes K_{2k}) , u_1\mu_1 \otimes u_2\mu_2 >$$

$$= < p_* f^*(K_1 \otimes K_2) , u_1\mu_1 \otimes u_2\mu_2 >$$

which thus gives (3.10.1). The inclusion given for WF'K is obtained from (3.10.1) by applying in turn Theorems 2.4 and 2.13, and Corollary 3.7.

4. THE CASE OF PSEUDO-DIFFERENTIAL OPERATORS

Let X be a manifold and P a pseudo-differential operator in X. Since $\operatorname{Proj}_{X \times X} WF\ K_P = \operatorname{sing\ supp} K_P$ is contained in the diagonal of X × X, it is possible, by using local charts and taking account of (8.2.1) of Chapter I, to reduce the problem to the case where X is an open set in $\mathbb{R}^n$ when we investigate WF K_p. In particular, Theorem 2.1 shows that

$WF'K_P \subset \operatorname{Diag}(T^*X \setminus 0) \times (T^*X \setminus 0)$. We denote by WF P the projection of WF'K_p onto one or other of the two factors $T^*X \setminus 0$.

It can be seen that WF P is a closed cone in $T^*X \setminus 0$, and that $WF'K_P = \mathrm{Diag}\,(WFP \times WFP)$. The results of the preceding section then imply

PROPOSITION 4.1: *Let* P, Q *be two pseudo-differential operators in a manifold* X, *and let* $u \in \mathcal{D}'(X)$. *If* P *is proper or if* u *has compact support, we have*

$$(4.1.1) \qquad WF(Pu) \subset WF\,P \cap WF\,u \subset WF\,u \; .$$

If P *or* Q *is proper, we have*

$$(4.1.2) \qquad WF(PQ) \subset WF\,P \cap WF\,Q \; .$$

We shall now generalise the property of existence of a parametrix, encountered in Section 7.2, Chapter IV.

THEOREM 4.2 (Microlocal parametrix): *Suppose* X *is a manifold and suppose we have* $P \in L^m(X)$ *elliptic of degree* m *in a closed cone* Γ *in* $T^*X \setminus 0$ *(this means that the principal symbol* $\sigma_m(P)$ *of* P *is elliptic in a conic neighbourhood of* Γ*). Then there exists a proper* $Q \in L^{-m}(X)$ *such that*

$$\begin{cases} WF(PQ - I) \cap \Gamma = \emptyset & (4.2.1) \\ WF(QP - I) \cap \Gamma = \emptyset \; . & (4.2.2) \end{cases}$$

PROOF: Clearly, we can suppose that P is proper. Let

$p \in S^m(T^*X \setminus 0)$ be a principal symbol of P, and let $\tilde{\Gamma}$ be an open conic neighbourhood of Γ in $T^*X \setminus 0$ such that $\sigma_m(P)$ is elliptic in $\tilde{\Gamma}$. We know that there exists $q \in S^{-m}(T^*X \setminus 0)$ such that $pq - 1 \in S^{-1}(\tilde{\Gamma})$. Let $Q_o \in L^{-m}(X)$ be proper, with principal symbol q. We have $PQ_o = I + R_o$, where $R_o \in L^o(X)$ is proper and with principal symbol of degree 0 equal to zero in $\tilde{\Gamma}$; we can therefore decompose R_o into $R_o = R_1 - R_2$, with $R_1 \in L^o(X)$ proper, $WF\,R_1 \cap \Gamma = \emptyset$, $R_2 \in L^{-1}(X)$ proper (this decomposition is trivial when X is an open subset of $\mathbb{R}^n$, and in the general case is deduced from this by partition of unity). Let $B \in L^o(X)$ be proper such that $B \sim \sum_{k \geq 0} R_2^k$. Putting $Q = Q_oB$, we have $PQ = R_1B + (I - R_2)B = R_1B + (I + R)$, with $R \in L^{-\infty}(X)$. Since $WF(R_1B) \subset WF\,R_1$ does not intersect Γ, we obtain (4.2.1). Similarly, we construct $\tilde{Q} \in L^{-m}(X)$, proper, such that $WF(\tilde{Q}P - I) \cap \Gamma = \emptyset$; by calculating $\tilde{Q}PQ = \tilde{Q}(PQ) = (\tilde{Q}P)Q$, we see that $[WF(Q - \tilde{Q})] \cap \Gamma = \emptyset$, which thus gives (4.2.2).

COROLLARY 4.3: *Let* X *be a manifold, let* $P \in L^m(X)$ *and* $u \in \mathcal{D}'(X)$. *Then* $WF\,u \subset WF(Pu) \cup \mathrm{Char}\,P$.

PROOF: Suppose we have $(x_o, \xi_o) \in T^*X \setminus 0$ not belonging to $WF(Pu) \cup \mathrm{Char}\,P$. As P is elliptic at (x_o, ξ_o), the above theorem shows that there exists $Q \in L^{-m}(X)$ proper, such that $QP = I + R$, $(x_o, \xi_o) \notin WF\,R$. As $u = QPu - Ru$, Theorem 4.2 implies that $WF\,u \subset WF(Pu) \cup WF\,R$, and therefore that

$(x_o, \xi_o) \notin$ WFu. We shall now give a characterisation of the WF of a distribution on a manifold by means of p.d.o.'s; this characterisation could be taken as the definition of WF u (see HORMANDER [5]).

PROPOSITION 4.4: *Let* X *be a manifold and let* $u \in \mathcal{D}'(X)$. *Then*

$$\text{WF } u = \bigcap_{\begin{cases} P \in L^0(X) \text{ proper} \\ Pu \in C^\infty(X) \end{cases}} \text{Char } P \tag{4.4.1}$$

PROOF: The inclusion of the left-hand side of (4.4.1) in the right-hand side follows from Corollary 4.3. Conversely, suppose $(x_o, \xi_o) \in T^*X \setminus 0$, not belonging to WF u. Let Γ be a conic neighbourhood of (x_o, ξ_o) in $T^*X \setminus 0$ which does not intersect WF u, and let $P \in L^o(X)$ be elliptic at (x_o, ξ_o) such that WF $P \subset \Gamma$. From (4.1.1), we have

$\text{WF}(Pu) \subset \text{WF}P \cap \text{WF}u = \emptyset$, and thus $P u \in C^\infty(X)$.

When X is an open subset of $\mathbb{R}^n$, we can define the essential support of a pseudo-differential operator, precisely, by using its complete symbol:

THEOREM 4.5: *Let* X *be an open set in* $\mathbb{R}^n$ *and let* P *be a pseudo-differential operator in* X. *Then* WF P *is the essential support of the complete symbol* $\sigma(P)$ *of* P *(i.e. the smallest conic closed subset of* $X \times (\mathbb{R}^n \setminus 0)$ *outside which* $\sigma(P)$ *is* $S^{-\infty}$).

PROOF: We can write

$$K_P(x, y) \equiv \int e^{i(x-y)\xi} a(x, \xi) đ\xi \qquad (\text{mod } C^\infty(X \times X))$$

where a is a symbol representing $\sigma(P)$. From Remark 2.2, $WF'K_P \subset \{(x, \xi, x, \xi) \mid \xi \neq 0, (x, \xi) \in \text{ess supp } a\}$; thus $WF\, P \subset \text{ess supp } \sigma(P)$. In order to establish the inclusion in the other direction, we consider $(x_0, \xi_0) \in X \times (\mathbb{R}^n \setminus 0)$ such that $(x_0, \xi_0) \notin WF\, P$, i.e. such that $(x_0, \xi_0, x_0, -\xi_0) \notin WF\, K_P$. There exists a neighbourhood V of x_0 in X and a conic neighbourhood W of ξ_0 in $\mathbb{R}^n \setminus 0$ such that $\widehat{f K_p}$ is rapidly decreasing in $W \times W$ when $f \in C_0^\infty(X \times X)$ has support in $V \times V$; we take in particular $f(x,y) = \alpha(x) \cdot \beta(y)$ where $\alpha, \beta \in C_0^\infty(V)$ are equal to unity in the neighbourhood of x_0. We have

$$\begin{aligned}\widehat{\alpha \otimes \beta K_P}(\xi, \eta) &= \langle K_P, \alpha(x) e^{-ix\xi} \beta(y) e^{iy\eta} \rangle \\ &= \langle P(\beta(y) e^{iy\eta}), \alpha(x) e^{-ix\xi} \rangle \\ &= \langle e^{-ix\eta} \alpha(x) P(\beta(y) e^{iy\eta}), e^{-ix(\xi-\eta)} \rangle \\ &= \sigma(Q)\widehat{(\xi - \eta}, \eta)\end{aligned}$$

where $\sigma(Q)$ is the complete symbol of $Q = \alpha P \beta$. We therefore have

$$\begin{aligned}\sigma(Q)(x, \eta) &= \int e^{ix\zeta} \sigma(Q)(\hat{\zeta}, \eta)\, đ\zeta \\ &= \int e^{ix(\xi-\eta)} \sigma(Q)\widehat{(\xi - \eta}, \eta)\, đ\xi\end{aligned}$$

so that

$$|\sigma(Q)(x\,,\,\eta)| \le$$

$$\int_W |\sigma(Q)(\widehat{\xi - \eta}\,,\,\eta)|\, đ\xi + \int_{R^n \setminus W} |\sigma(Q)(\widehat{\xi - \eta}\,,\,\eta)|\, đ\xi\,.$$

Let W' be a conic neighbourhood of ξ_o in $\mathbb{R}^n \setminus 0$ such that $\overline{W}' \subset W$. We have seen that $\sigma(Q)(\xi - \eta\,,\,\eta)$ is rapidly decreasing for $\xi, \eta \in W$; thus in W the integral is rapidly decreasing for $\eta \in W$. By combining the immediate upper bound $|\sigma(Q)(\hat{\zeta}\,,\,\eta)| \le C_N(1 + |\zeta|)^{-N}(1 + |\eta|)^m$ where m is the degree of Q and N is arbitrary, with the inequality $|\xi - \eta| \ge c(|\xi| + |\eta|)$ for $\xi \notin W$ and $\eta \in W'$ (with $c > 0$), we see that in $\mathbb{R}^n \setminus W$ the integral is rapidly decreasing for $\eta \in W'$. It can be shown in similar fashion that the derivatives of $\sigma(Q)$ are rapidly decreasing in W'. We thus have finally $(x_o, \xi_o) \notin$ ess supp $\sigma(P)$ since $\sigma(P) \equiv \sigma(Q) \pmod{S^{-\infty}}$ for x sufficiently near x_o.

5. ADDITIONAL NOTES AND EXERCISES

5.1 CONCLUSION OF THE CALCULATION OF $WF(\delta(x') \otimes Y(x_n))$ (see Example 2.6)

a) We put $u = \delta(x') \otimes Y(x_n)$. Let $\alpha \in C_o^\infty(R^{n-1})$ and $\beta \in C_o^\infty(]0, +\infty[)$. Show that $\widehat{(\alpha \otimes \beta)u} = \alpha(0)\,\hat{\beta}(\xi_n)$.

Deduce from this that

$$\{(0, x_n\,;\,\xi', 0) | x_n > 0\,,\,\xi' \in R^{n-1} \setminus 0\} \subset WF\,u\,.$$

b) Let $\alpha \in C_0^\infty(\mathbb{R}^{n-1})$ and $\beta \in C_0^\infty(\mathbb{R})$ with $\beta(0) \neq 0$. Show that $\widehat{(\alpha \otimes \beta)u} = \alpha(0)\, \widehat{\beta Y}(\xi_n)$ and that $\widehat{\beta Y}(\xi_n) \sim \beta(0) \frac{1}{i\xi_n}$ when $|\xi_n| \to +\infty$. Deduce from this that

$$\{ (0, \xi) \mid \xi \in \mathbb{R}^n \setminus 0 \} \subset \mathrm{WF}\, u .$$

c) Conclude, taking account of Example 2.6, that

$\mathrm{WF}\, u =$

$$\{ x = 0, \xi \neq 0 \} \cup \{ x' = 0, x_n > 0, \xi' \neq 0, \xi_n = 0 \} .$$

5.2 CONCLUSION OF THE CALCULATION OF WF'(v(x-y)) FOR $v \in \mathcal{D}'(\mathbb{R}^n)$ (see Example 2.15)

Suppose $v \in \mathcal{D}'(\mathbb{R}^n)$ and $\varphi, \psi \in \mathcal{D}(\mathbb{R}^n)$.

a) Show that $\langle v(x-y), \varphi(x)\, \psi(x-y) \rangle = \langle 1, \varphi \rangle \langle v, \psi \rangle$.

b) Deduce from this that $\hat{T}(\xi, -\xi) = \langle 1, \varphi \rangle \widehat{\psi v}(\xi)$, where we have put $T(x, y) = \varphi(x)\, \psi(x-y)\, v(x-y)$.

c) Conclude, taking account of Example 2.15, that

$$\mathrm{WF}'(v(x-y)) = \{ (x, \xi, y, \xi) \mid x, y \in \mathbb{R}^n, (x-y, \xi) \in \mathrm{WF}\, v \} .$$

5.3 CONSTRUCTION OF A DISTRIBUTION WHOSE SINGULAR SPECTRUM IS A DIRECTION IN $T^*(X) \setminus 0$.

We put $\mathbb{R}^n = \mathbb{R} \times \mathbb{R}^{n-1}$, $x = (t, y)$, $\xi = (\tau, \eta)$. Let $f \in \mathcal{S}(\mathbb{R}^{n-1})$ with $f(0) \neq 0$, and let $g \in C^\infty(\mathbb{R})$ be such that $g(\tau) = 1$ for $\tau \geq 1$, $g(\tau) = 0$ for $\tau \leq 1/2$. We put:

$$h(\xi) = g(\tau)\, f(\tau^{-1/2} \eta) \quad \text{and} \quad u = \mathcal{F}^{-1} h .$$

We now propose to show that:

$$\text{WF } u = \{ (0, 0 ; \tau, 0) \mid \tau > 0 \} . \tag{5.3.1.}$$

a) Show that $h(\xi)$ is rapidly decreasing (as $|\xi| \to +\infty$) in a conic neighbourhood of any point (τ_o, η_o) in $\mathbb{R}^n$ such that $\eta_o \neq 0$, or such that $\eta_o = 0$ and $\tau_o < 0$.

b) Show that for any multi-index α, there exists $C_\alpha \geq 0$ such that

$$|\partial^\alpha h(\xi)| \leq C_\alpha (1 + |\xi|)^{-|\alpha|/2} \quad \text{for } \xi \in \mathbb{R}^n .$$

Deduce from this that for any multi-index β, the distribution $\widehat{x^\alpha D^\beta u}$ is in $L^1(\mathbb{R}^n)$ when $|\alpha|$ is sufficiently large, and thus that $u \in C^\infty(\mathbb{R}^n \setminus 0)$ with $\sup_{x \in \mathbb{R}^n} |x^\alpha D^\beta u(x)| < +\infty$.

c) Deduce from this that $u \notin C^\infty(\mathbb{R}^n)$ and that, using in particular Proposition 1.4, we have the equality (5.3.1).

5.4 INTRINSIC CHARACTERISATION OF THE SINGULAR SPECTRUM

Let X be a manifold of dimension n , and let $u \in \mathcal{D}'(X)$. Let $(x_o, \xi_o) \in T^*(X) \setminus 0$. We now propose to show that the following conditions (i), (ii), are equivalent:

(i) $(x_o, \xi_o) \notin \text{WF } u$

(ii) For all $p \geq 1$, for all $\lambda_o \in \mathbb{R}^p$, for any function $\psi \in C^\infty(X \times \mathbb{R}^p, \mathbb{R})$ such that $d_x \psi(x_o, \lambda_o) = \xi_o$, there exists a neighbourhood V of x_o in X and a neighbourhood Λ of λ_o in $\mathbb{R}^p$ such that: for all $\alpha \in C_o^\infty(V, \Omega_1(X))$, the quantity $< u(x), \alpha(x)\, e^{-i\tau\psi(x, \lambda)} >$ is rapidly decreasing with respect to τ when $\tau \to +\infty$, uniformly for $\lambda \in \Lambda$.

For the proof, it is clearly sufficient to consider the case $X = \mathbb{R}^n$.

a) Show that (ii) implies (i) by taking $p = n$, $\lambda_o = \xi_o$, $\psi(x, \lambda) = x.\lambda$.

b) Assume that (i) is satisfied, and consider ψ as in (ii). We know that there exists a neighbourhood V_o of x_o and a neighbourhood W_o of ξ_o such that, for $\alpha \in C_o^\infty(V)$: $\widehat{\alpha u}(\tau \xi)$ is rapidly decreasing with respect to τ when $\tau \to +\infty$, uniformly for $\xi \in W_o$.

Consider $\beta \in C_o^\infty(V)$ such that $\beta = 1$ in supp α. Show that:

$$(5.4.1.) \quad < u, \alpha\, e^{-i\tau\psi} > \; = \tau^n \int \widehat{\alpha u}(\tau \xi)\, I(\tau, \xi, \lambda)\, d\xi$$

where $$I(\tau, \xi, \lambda) = \int e^{i\tau(x.\xi - \psi(x,\lambda))}\, \beta(x)\, dx \; .$$

Show that there exists a neighbourhood $V \subset V_o$ of x_o, a neighbourhood $W \subset W_o$ of ξ_o, a neighbourhood Λ of λ_o and a number $c > 0$ such that:

(5.4.2.) $$|\xi - \psi'_x(x, \lambda)| \geq c(1 + |\xi|) \text{ for } x \in V, \quad \xi \in \mathbb{R}^n \setminus W, \quad \lambda \in \Lambda .$$

By integrating by parts, using the differential operator $|\xi - \psi'_x(x, \lambda)|^{-2} < \xi - \psi'_x(x, \lambda), \frac{\partial}{\partial x} >$, deduce from this that for any integer N there exists $C_N \geq 0$ with

$$|I(\tau, \xi, \lambda)| \leq C_N \tau^{-N}(1 + |\xi|)^{-N} \text{ for } \tau \geq 1, \quad \xi \in \mathbb{R}^n \setminus W, \quad \lambda \in \Lambda .$$

By decomposing the integral which appears in the left-hand side of (5.4.1) into the sum of an integral in $\mathbb{R}^n \setminus W$ and an integral in W, conclude that the condition (ii) is satisfied.

5.5 TRACE CALCULATIONS

Let X be a compact manifold endowed with a density dx. We take $f \in C^\omega(X; X)$ and we denote by $\mathrm{Fix}(f)$ the set of its fixed points. We suppose that each fixed point x is simple, i.e. that it satisfies $\det(I - f'(x)) \neq 0$. We denote f^* the operator of inverse image under f.

a) Let P be a p.d.o. of degree m on X. We associate with it the operator $f^* \circ P$, the kernel of which we denote by K_{f^*P}. Show, by using Example 8.5 of Chapter III and the theorems relating to the singular spectrum, that the distribution K_{f^*P} admits a restriction to the diagonal $\Delta = \{(x, x) | x \in X\}$. Furthermore, if Σ denotes the conic closed set

$\{ (x, \xi; x, \xi) | (x, \xi) \in T^*X \setminus 0\}$, then the mapping

$$P \in \bigcup_m L^m(X) \longrightarrow \mathrm{tr}(f^* P) = \langle K_{f^*P}|_\Delta , 1\,dx \rangle \in \mathbf{C}$$

is continuous when we endow $\cup L^m(X)$ with the notion of convergence induced by $\mathcal{D}'_\Sigma(X \times X)$. (We have identified $K_{f^*P}|_\Delta$ with a distribution on X, by the projection of Δ onto X).

In the particular case where P = Identity, show that we have

$$K_{f^*}|_\Delta = \sum_{x \in \mathrm{Fix}(f)} \frac{\delta_x}{|\det(I - f'(x))|} .$$

[Note that $\mathrm{supp}(K_{f^*}|_\Delta) \subset \mathrm{Fix}(f)$ and calculate the action of this distribution on a test function with support in a small coordinate patch domain near a fixed point of f, by using the local expression (8.5.1) of Chapter III].

b) More generally, let E be a vector bundle on X and let θ be a morphism of bundles: $f^*E \to E$. With any p.d.o. $P \in L^m(X; E, E)$ we associate the operator $T_P = \theta \circ f^* \circ P$ from $C^\infty(X ; E)$ which acts on a section u via the formula

$$(T_P\, u)(x) = \theta_x \cdot Pu(f(x)) , \quad x \in X .$$

The kernel distribution of T_P is a distribution $K_{T_P} \in \mathcal{D}'(X \times X; E \boxtimes E^*)$ (Exercise 11.5, Chapter IV). Show as in a) that it is possible to define the restriction $K_{T_P}|_\Delta \in \mathcal{D}'(X; E \otimes E^*)$ and that the trace mapping of $E \otimes E^*$ into $\mathbf{C}$ allows us to define the distribution $\mathrm{trace}(K_{T_P}|_\Delta) \in \mathcal{D}'(X)$.

Show that the mapping $\operatorname{trace}(T_P)$ defined by

$$P \in UL^m(X\ ;\ E,\ E) \longrightarrow \langle \operatorname{trace}(K_{T_P|\Delta}),\ 1\,dx\rangle \in \mathbb{C}$$

is continuous (in the same sense as in a)) and that for P = Identity we have:

$$\text{(5.4.1.)} \qquad \operatorname{trace}(T_P) = \sum_{x \in \operatorname{Fix}(f)} \frac{\operatorname{trace}(\theta_x)}{|\det(I - T_x f)|} .$$

5.6 THE LEFCHETZ-ATIYAH-BOTT FORMULA: Let (E_k, P_k) be an elliptic complex over a compact manifold X endowed with a density dx .

a) Prove that the identity endomorphism $I = (I_k)$ is the limit of endomorphisms (S_k^ε) of the complex (in the sense of Exercise 11.5, Chapter IV) with C^∞ kernels, the limit being understood in the sense of the convergence, as $\varepsilon \to 0$, of the kernels $\mathcal{S}_k^\varepsilon(x, y)$ in $\mathcal{D}'_\Sigma(X \times X,\ E_k \otimes E_k^*)$, where $\Sigma = \{(x, \xi; x, \xi) \mid (x, \xi) \in T^*X \setminus 0\}$.

[Let (Q_k) be a parametrix of the complex; show that the operators $Q_k^\varepsilon = (I - \rho_k^\varepsilon)\, Q_k$ again constitute a parametrix of the complex when ρ_k^ε denotes operators with C^∞ kernels which converge, as $\varepsilon \to 0$, towards I_k in the sense of $\mathcal{D}'_\Sigma(X \times X\ ;\ E_k \otimes E_k^*)$. Finally, define S_k^ε by the equality

$$P_{k-1}\, Q_{k-1}^\varepsilon + Q_k^\varepsilon\, P_k = I_k - S_k^\varepsilon \ ] .$$

b) Take a mapping $f \in C^\infty(X, X)$, of which all the fixed points are simple (see Exercise 5.4), and for any k take a morphism of bundles $\theta_k : f^* E_k \to E_k$. Associate with these the endomorphism T_k of $C^\infty(X ; E_k)$ defined by $T_k = \theta_k \circ f^*$. Assume that the (T_k)'s define an endomorphism of the complex. Then, show that we have the equality:

$$(5.5.1.) \quad \sum_k (-1)^k \text{ Trace } (H(T_k)) = \sum_k (-1)^k \text{ Trace } (T_k)$$

where we have put

$$(5.5.2.) \quad \text{Trace } (T_k) = \sum_{x \in \text{Fix}(f)} |\det(I - T_x f)|^{-1} \text{ Trace } (\theta_k(x)).$$

[Use Exercise 11.5 of Chapter IV to prove the equality (5.5.1) when T_k is replaced by the operator $T_k . S_k^\varepsilon$ with C^∞ kernel, and then deduce (5.5.1) by proceeding to the limit when $\varepsilon \to 0$. Use Exercise 5.4 to prove (5.5.2)].

c) In the case of the De Rham complex (Example 9.4, Chapter IV), take $\theta_k = \Lambda^k T^* f$, and obtain the classical Lefchetz formula

$$(5.5.3.) \quad \sum (-1)^k \text{ Trace } (H^k f) = \sum_{x \in \text{Fix}(f)} \text{sign } [\det(I - T_x f)]$$

where $H^k f$ denotes the endomorphism induced by f^* on the kth cohomology group $H^k(X, \mathbf{C})$.

[Let E be the vector-valued space of dimension m on $\mathbf{C}$ and let $\lambda \in \mathbf{C}$; by putting u in triangular form, show that we have the identity:

$$\det(\lambda I - u) = \sum_{k=0}^{m} \lambda^{m-k} (-1)^k \operatorname{Trace}(\Lambda^k u) .$$

Use this identity with $\lambda = 1$ to make the right-hand side of (5.5.2) explicit)].

BIBLIOGRAPHY

AGMON, S., DOUGLIS, A., NIRENBERG, L. [1] Estimates near the boundary for solutions of elliptic partial differential equations satisfying general boundary conditions. I, Comm. Pure Appl. Math. 12 (1959), 623-727; II, Comm. Pure Appl. Math. 17 (1964), 35-92.

AGRANOVITCH, M.S. [1] Boundary value problems for systems with a parameter. Mat. Sbornik (1971), 27-65.

ATIYAH, M.F., BOTT, R. [1] A Lefchetz fixed-point formula for elliptic complexes. I, Ann. of Math. 86 (1967), 374-407.

ATIYAH, M.F., BOTT, R., PATODI, V.K. [1] On the heat equation and the index theorem. Inv. Math. 19 (1973), 279-330.

ATIYAH, M.F., SINGER, I.M. [1] The index of an elliptic operator. I, Ann. of Math. 87 (1968), 484-530.

BALABAN, T. [1] On the mixed problem for a hyperbolic equation. Memoirs AMS No. 112, 1971.

BEALS, R. [1] A general calculus of pseudo-differential operators. Duke Math. J. 42 (1975), 1-42.

BEALS, R., FEFFERMAN, C. [1] Spatially inhomogeneous pseudo-differential operators. Comm. Pure Appl. Math. 27 (1974), pages 1-24.

BONY, M., SHAPIRA, P. [1] Solutions hyperfonctions du problème de Cauchy. Lecture Notes in Math., Springer 287 (1973), 82-98.

BOUTET DE MONVEL, L. [1] Comportement d'un opérateur pseudo-différentiel sur une variété à bord. J. Anal. Math. 17 (1966), 241-304.

[2] Boundary problems for pseudo-differential operators. Acta Math. 126 (1971), 11-51.

[3] Hypoelliptic operators with double characteristics and related pseudo-differential operators. Comm. Pure Appl. Math. 28 (1974), 585-639.

CALDERON, A.P., ZYGMUND, A. [1] Singular integral operators and differential equations. Amer. J. Math. 79 (1957), 901-921.

CALDERON, A.P., VAILLANCOURT, R. [1] On the boundedness of pseudo-differential operators. J. Math. Soc. Japan 23 (1971), 374-378.

CHAZARAIN, J. [1] Opérateurs hyperboliques à caractéristiques de multiplicité constante. Ann. Inst. Fourier 24 (1974), pages 173-202.

CHAZARAIN, J., PIRIOU, A. [1] Caractérisation des problèmes mixtes hyperboliques bien posés. Ann. Inst. Fourier 22, (1972), 193-237.

DIEUDONNE, J. [1] Eléments d'analyse, Vol. 4, page 333, Gauthiers-Villars, 1971.

DUISTERMAAT, J.J. [1] Fourier integral operators. Courant Institute of Mathematical Sciences, 1973.

FLASCHKA, H., STRANG, G. [1] The correctness of the Cauchy problem. Adv. in Math. 6 (1971), 347-379.

GELFAND, I.M., CHILOV, G.E. [1] Les distributions. Dunod 1962.

GILKEY, P. [1] The index theorem and the heat equation. Publish or Perish, 1974.

GUILLEMIN, V., STERNBERG, S. [1] Geometric asymptotics. Amer. Math. Soc. Math. Survey No. 14, 1977.

HERSH, R. [1] Mixed problems in general variables. J. Math. & Mech. 12 (1963), 317-334.

HORMANDER, L. [1] Linear partial differential operators. Springer Verlag, 1964.

[2] Pseudo-differential operators and non-elliptic boundary problems. Ann. Math. 83 (1966), 129-209.

[3] Pseudo-differential operators and hypoelliptic equations. Amer. Math. Soc. Symp. on Singular Integral Operators, 1966, pages 138-183.

[4] On the characteristic Cauchy problem. Ann. of Math. 88 (1968), 341-370.

[5] Fourier integral operators, I. Acta Math., 127 (1971), 79-183.

[6] On the existence and the regularity of solutions of linear pseudo-differential equations. Enseign. Math. 27 (1971), 99-163.

[7] The Cauchy problem for differential equations with double characteristics. J. Anal. Math. 32 (1977), pages 11-195.

INCE, E.L. [1] Ordinary differential equations. Dover Publications, Inc.

IVRII, PETKOV, V. [1] Necessary conditions for the correctness of the Cauchy problem for non-strictly hyperbolic equations. Uspehi Nat. Nauk. 29 (1974), 3-70.

KATO, J. [1] Mixed problems of hyperbolic equations in a general domain. Proc. Japan Acad. 47 (1971), 67-70.

KOHN, J.J., NIRENBERG, L. [1] An algebra of pseudo-differential operators. Comm. on Pure and Applied Math., Vol XVIII, (1965), 269-305.

KREISS, H.O. [1] Initial boundary value problems for hyperbolic systems. Comm. Pure Appl. Math. 13 (1970), 277-298.

KUMANO-GO, H. [1] On the uniqueness for the solution of the Cauchy problem. Osaka Math. J., 15 (1963), 151-172.

LEFCHETZ, S. [1] Algebraic geometry. Princeton University Press, 1953.

LIONS, J.L. [1] Equations différentielles opérationnelles et problèmes aux limites. Springer Verlag, 1961.

LIONS, J.L., MAGENES, E. [1] Problèmes aux limites non homogènes et applications. Vol. I. Dunod, 1968.

MAJDA, A., OSHER, S. [1] Initial/boundary-value problems for hyperbolic equations with uniformly characteristic boundary. Comm. Pure Appl. Math. 28 (1975), 607-675.

MELIN, A., SJOSTRAND, J. [1] Fourier integral operators with complex phase functions and parametrix for an interior boundary-value problem. Comm. in Partial Diff. Equations, 1 (1976), 313-400.

MIZOHATA, S. [1] Quelques problèmes au bord du type mixte pour des équations hyperboliques. Séminaire Leray, Collège de France, 1966-1967.

OHKUBO, T., SHIROTA, T. [1] On structures of certain L^2-well-posed mixed problems for hyperbolic systems of first order. Hokkaïdo Math. J. 4 (1975), 82-158.

RALSTON, J. [1] Note on a paper of Kreiss. Comm. Pure Appl. Math. 24 (1971), 759-762.

RAUCH, J. [1] $\mathcal{L}_2$ is a continuable initial condition for Kreiss' mixed problems. Comm. Pure Appl. Math. 25 (1972), 265-285.

SAKAMOTO, R. [1] Mixed problems for hyperbolic equations, I & II. J. Math. Kyoto Univ. 10 (1970), 375-401 & 403-417.

SATO, M., KAWAI, T., KASHIWARA, M. [1] Hyperfunction and pseudo-differential equations, Part II. Proc. Conf. Katata, 1971, Lecture Notes in Math. No. 287, Springer Verlag.

SARASON, L. [1] Hyperbolic and other symmetrisable systems in regions with corners and edges. Indiana Univ. Math. J. 26 (1977), 1-39.

SCHWARTZ, L. [1] Théorie des distributions. Hermann, 1966.

SVENSSON, S.L. [1] Necessary and sufficient conditions for the hyperbolicity of polynomials with hyperbolic principal part. Ark. Mat. 17 (1969), 145-162.

TARTAKOFF, D. [1] Regularity of solutions to boundary-value problems for first-order systems. Indiana Math. J. 21 (1972), 1113-1029.

TAYLOR, M. [1] Pseudo-differential operators. Lecture Notes in Mathematics No. 416, Springer Verlag.

TREVES, F. [1] Topological vector spaces, distributions and kernels. Academic Press, 1967.

[2] Linear partial differential equations with constant coefficients. Existence, approximation and regularity of solutions. Gordon and Breach, 1966.

UNTERBERGER, A., BOKOBZA, J. [1] Les operateurs de Calderón - Zygmund précisés. C.R. Acad. Sc. Paris. 259 (1964) p.1612.

INDEX